Requirements Engineering for Software and Systems

Requirements Engineering for Software and Systems

Fourth Edition

Phillip A. Laplante

Mohamad H. Kassab

CRC Press
Taylor & Francis Group

AN AUERBACH BOOK

Fourth Edition published 2022
by CRC Press
6000 Broken Sound Parkway NW, Suite 300, Boca Raton, FL 33487-2742

and by CRC Press
4 Park Square, Milton Park, Abingdon, Oxon, OX14 4RN

First edition published by CRC Press 2009

CRC Press is an imprint of Taylor & Francis Group, LLC

© 2022 Phillip A. Laplante and Mohamad H. Kassab

Library of Congress Cataloging-in-Publication Data
Names: Laplante, Phillip A., author. | Kassab, Mohamad, author.
Title: Requirements engineering for software and systems / Phillip A.
Laplante and Mohamad H. Kassab.
Description: Boca Raton, FL : CRC Press, 2022. | Series: Applied software
engineering series | Includes bibliographical references and index.
Identifiers: LCCN 2022000663 | ISBN 9780367654528 (hbk) |
ISBN 9781032275994 (pbk) | ISBN 9781003129509 (ebk)
Subjects: LCSH: Computer programs—Specifications. |
Computer software—Development. | Software engineering. | Requirements engineering.
Classification: LCC QA76.76.S73 L37 2022 | DDC 005.1068—dc23/eng/20220322
LC record available at https://lccn.loc.gov/2022000663

ISBN: 978-0-367-65452-8 (hbk)
ISBN: 978-1-032-27599-4 (pbk)
ISBN: 978-1-003-12950-9 (ebk)

DOI: 10.1201/9781003129509

Typeset in Garamond
by codeMantra

Contents

Preface

Solid requirements engineering has increasingly been recognized as the key to improved, on-time, and on-budget delivery of software and systems projects. Nevertheless, a few undergraduate engineering programs stress the importance of this discipline. Recently, however, some software and systems engineering programs have introduced requirements engineering as mandatory in their curricula. In addition, new software tools are emerging that are empowering practicing engineers to improve their requirements engineering habits. However, these tools are not usually easy to use without significant training, and many working engineers are returning for additional courses that will help them understand the requirements engineering process.

This book is intended to provide a comprehensive treatment of the theoretical and practical aspects of discovering, analyzing, modeling, validating, testing, and writing requirements for systems of all kinds, with an intentional focus on software-intensive systems. This book brings into play a variety of formal methods, social models, and modern requirements writing techniques to be useful to the practicing engineer.

Audience

This book is intended for professional software engineers, systems engineers, and senior and graduate students of software or systems engineering. Much of the material is derived from the graduate-level "Requirements Engineering" course taught at Penn State's Great Valley School of Graduate and Professional Studies and online through its World Campus, where the authors work. The typical student in that course has 5 years of work experience as a software or systems engineering professional and an undergraduate degree in engineering, science, or business. Typical readers of this book will have one of the following or similar job titles:

Software engineer
Systems engineer
Sales engineer
Systems analyst
[XYZ] engineer (where "XYZ" is an adjective for most engineering disciplines, such as "electrical" or "mechanical")
Project manager
Business analyst
Technical architect
Lead architect
Product Owner

Many others can benefit from this text including the users of complex systems and other stakeholders.

Exemplar Systems

Although various exemplar systems will be used to illustrate points throughout the book, four particular systems will be used repeatedly. These systems were selected because they involve application domains with which most readers are likely to be familiar and because they cover a wide range of applications from embedded to organic in both industrial and consumer implementations. Consider a domain, however, with which you (and the authors) are likely unfamiliar, say, mining of anthracite coal. Imagining the obvious difficulties in trying to communicate about such a system highlights the importance of domain understanding in requirements engineering. This topic will be discussed at length in Chapter 1.

The first of these running examples is an airline baggage handling system, probably similar to one found in every major airport. Check-in clerks and baggage handlers tag your bags at check-in with a barcode ID tag. Then the baggage is placed on a conveyor belt where it moves to a central exchange point and is redirected to the appropriate auxiliary conveyor for loading on an airplane-bound cart or a baggage carousel. Along the way, the system may conduct some kind of image scan and processing to detect the presence of unauthorized or dangerous contents (such as weapons or explosives). A baggage handling system is an embedded, real-time system, that is, the software is closely tied to the hardware, and deadline satisfaction is a key goal of the system.

The Denver International Airport tried to build a very sophisticated version of such a system several years ago. The system used PCs, thousands of remote-controlled carts, and a 21-mile-long track. Carts moved along the track, carrying luggage from check-in counters to sorting areas and then straight to the flights waiting at airport gates. After spending $230 million over 10 years, the project was canceled (de Neufville 1994). Much of the failure can be attributed to requirements engineering mistakes.

The second exemplar system is a point of sale (POS) system for a large pet store chain. This type of system would provide such capabilities as cashier functions and inventory tracking, tax reporting, and end-of-year closeout. It might handle self-checkout, coupon scanning, product returns, and more. This business domain is transaction-oriented and although there are many available off-the-shelf systems let's assume there is a need for a custom system. This type of system is organic, that is, it is not tied to any specialized hardware. PCs or PC-based cash registers, storage devices, and network support comprise the main hardware.

The third system that will be explored is for a "Smart Home," that is, a home in which one or more PCs control various aspects of the home's climate control, security, ambiance, entertainment, and so forth. A Smart Home is a consumer application and a semi-detached system—it uses some specialized but off-the-shelf hardware and software. We will imagine that this Smart Home is being built for someone else (not you) so that you can remain objective about its features.

The fourth exemplar is a wet-well pumping system for a wastewater treatment station. Well-pumping systems, which are found in industrial and residential applications, are commonly used to illustrate various concepts in embedded, real-time systems control. The third and fourth systems are described in Appendices A and B, which include comprehensive (though imperfect) requirements specifications.

For the purposes of experimentation and practices, or for use in a course project, you are encouraged to select another appropriate system. Possible choices include:

Passenger safety restraint system for an automobile of your choice.

Family tree maker and genealogical database system.

Game of your choosing (keep it simple—3-D chess or checkers or popular board games are best).

Simulation of some familiar enterprise (e.g., a traffic intersection, elevator control, manufacturing environment).

High-level requirements for a safety-critical system such as a power plant.

Any aspect of your work that seems suitable to automation.

You can use your imagination, consult the many resources that are available on the Web, and have fun as you learn to "scope out" one or another of these systems.

Course Adoption

This book was written to support both undergraduate and graduate requirements engineering courses. The authors and their colleagues have used previous editions of this book in more than 50-course offerings at Penn State. Previous editions of the book have also been adopted by instructors at numerous universities worldwide.

Each chapter includes simple, intermediate, and advanced exercises. Advanced exercises are suitable as a research assignment or Independent study and are denoted by an asterisk (*). Vignettes at different locations of each chapter provide mini-case studies showing how the learnings in the chapter can be employed in real systems.

Sample requirements specifications are found in Appendices A and B. The first is for a smart home system. The second specification is for a well-pumping system. Those working in real-time systems or control engineering will be familiar with the widespread use of a well-pumping system in papers and texts. Neither of these requirements specifications is perfect nor complete. But they are used throughout the text for discussion, examples, and exercises. A possible course assignment could involve improving one of the specifications based on lessons from the text.

For those interested in adopting this text for an undergraduate or graduate course in requirements engineering, the following describes how the authors use the text in their courses (Kilcay-Ergin and Laplante 2012). For instructors adopting this text for courses in either a university or industrial setting, the authors can provide lecture notes upon request. Please contact them at (Philip Laplante, plaplante@psu.edu) or (Mohamad Kassab, muk36@psu.edu) for further information.

New for the Fourth Edition

Since the first edition, there have been made many changes and improvements. Feedback from instructors, students, and corporate users of the text was used to correct, expand, and improve the materials. Notable additions in the fourth edition include two newly added chapters: "On Non-Functional Requirements" (Chapter 5) and "Requirements Engineering: Road Map to the Future" (Chapter 12). The latter provides a discussion on the relationship between requirements engineering and emerging disruptive technologies (Internet of Things, Cloud Computing, Blockchain, AI, and Affective Computing).

All chapters of the book were significantly expanded with new materials that keep the book relevant to the current industrial practices. Particularly, the reader will find expanded discussions on new elicitation techniques, agile approaches (e.g., Kanpan, SAFe, DEVOps), requirements tools, requirements representation, risk management approaches, and functional size measurement methods. The provided statistics on RE state of practices were also updated throughout the book with results derived from a 2020 survey conducted by the authors. Besides, the fourth edition also has significant additions of vignettes, exercises, references throughout the book. Of course, this edition also implemented corrections and the removal of outdated material from the previous edition.

Throughout the book, the reader will also find newly added scannable QR codes that will direct the reader to one of the following:

- **Stay Updated**: a link to where the reader may find more up-to-date content when they become available.
- **Share Your Opinion**: the reader is invited to share an opinion (or experience) on selected topics.
- **Quick Access**: a link to a quick summary on a topic or quick access to a tool.
- **Supplemental Materials**: a link to supplemental materials by the authors on the selected topic.
- **Watch:** a link to a short video on a selected topic.

Requirements engineering is a dynamic field and the authors intend that this text keep pace with these changes. If you have any ideas for new material to be included in future additions of the book, please send these to the authors at (Phillip Laplante, plaplante@psu.edu) or (Mohamad Kassab, muk36@psu.edu).

Notes on Referencing and Errors

The authors have tried to uphold the highest standards for giving credit where credit is due. Each chapter contains a list of related readings, which have been extensively updated for the fourth edition, and they should be considered the primary references for that chapter. Where direct quotes or nonobvious facts are used, an appropriate note or in-line citation is provided. In particular, it is noted where the authors published portions in preliminary form in other scholarly publications.

Despite these best efforts and those of the reviewers and publisher, there are still likely errors to be found. Therefore, if you believe that you have found an error—whether it is a referencing issue, factual error, or typographical error, please contact the authors at the emails addressed referenced previously. You can access the list of known errata at https://phil.laplante.io/requirements/errata.php.

Disclaimers

When discussing certain proprietary systems, every effort has been taken to disguise the identities of any organizations as well as individuals that are involved. In these cases, the names and even elements of the situation have been changed to protect the innocent (and guilty). Therefore, any similarity between individuals or companies mentioned herein is purely coincidental.

REFERENCES

de Neufville, R. (1994, December) The baggage system at Denver: Prospects and lessons. *Journal of Air Transport Management*, 1(4): 229–236.

Kilicay-Ergin, N., & Laplante, P. A. (2012) An online graduate requirements engineering course. *IEEE Transactions on Education*. Available http://ieeexplore.ieee.org/stamp/stamp.jsp?tp=&arnumber=625 3277&isnumber=4358717

Acknowledgments

Many individuals assisted during the preparation of the first, second, and third editions of this text and the minor revised versions in between. Several people are especially noteworthy.

Dr. George Hacken of the New York Metropolitan Transportation Authority contributed many ideas on formal methods used in Chapter 7. The late Professor Larry Bernstein of Stevens Institute of Technology reviewed the first edition and provided encouragement and many suggestions for improvement. Dr. Colin Neill and Dr. Raghu Sangwan of Penn State read portions of the first edition and provided valuable feedback in many discussions subsequently. Dr. Nil Kilcay-Ergin provided meaningful feedback that influenced parts of the second edition.

Several former students of the authors made significant contributions. For example, Brad Bonkowski contributed to the case study found in Appendix A and Christopher Garrell contributed to the case study in Appendix B. Nicholas Natale contributed to the vignette at the end of Chapter 3. Many other students have read drafts of portions of the text and provided ideas, exercises, critical feedback, and examples. They are too numerous to name individually but the authors wish to thank them all.

Many of the appendices were contributed by professional colleagues of the authors, either directly to this edition, or through the *Encyclopedia of Software Engineering* (Taylor & Francis Publishing, 2010), of which Dr. Laplante was editor-in-chief. In particular, Appendix C on the Universal Modeling Language (UML) was contributed through the Encyclopedia by Jim Rumbaugh. Similarly, Appendix D on User Stories was written for the Encyclopedia by Don Shafer. Appendix E on Use Cases by Emil Vassev also first appeared in the Encyclopedia. Finally, Gavin Arthurs of IBM provided the excellent overview of DOORS NG found in Appendix F.

Various users of previous versions of this text, including students, professors, and professionals, have pointed out errors and suggestions for improvement—these individuals are also too numerous to mention, but the authors are grateful to them for helping to improve this book.

Of course, any errors of commission or omission that remain are due to the authors alone. Please contact them at:

Philip Laplante: plaplante@psu.edu, or

Mohamad Kassab: muk36@psu.edu

The authors would also like to thank the staff at Taylor & Francis, especially our editor John Wyzalek.

On personal notes, Dr. Laplante expresses his deepest gratitude to Nancy, Chris, Charlotte, Henry, and Miles for providing love, companionship, and encouragement during the writing of this text, while Dr. Kassab acknowledges with gratitude the support and love provided by his family members—Hassan, Mariam, Essa, Eman, Sana, and Samar—during the journey of writing this fourth edition.

Phil Laplante and Mohamad Kassab, October 2021

Authors

Phillip A. Laplante is Professor of Software and Systems Engineering and a member of the graduate faculty at The Pennsylvania State University. His research, teaching, and consulting focus on software quality particularly with respect to requirements, testing, and project management. Before joining Penn State, he was a professor and senior academic administrator at other colleges and universities.

From 2010 to 2016, he was the founding chair of the Software Professional Engineer Licensure Committee for the National Council of Examiners of Engineers and Surveyors. This volunteer committee created maintains and scores the exam used throughout the United States to license Professional Software Engineers.

Dr. Laplante has consulted to Fortune 500 companies, startup ventures, the U.S. Department of Defense, NASA, and the National Institute for Standards and Technology (NIST). He is on the Board of Directors for a $100 million heavy infrastructure construction company and serves on various corporate technology advisory boards.

Prior to his academic career, Dr. Laplante spent nearly a decade as a software engineer and project manager working on avionics (including the Space Shuttle), CAD, and software test systems. He was also director of business development for a software consulting firm. He has authored or edited 37 books and more than 300 papers, articles, reviews, and editorials.

Dr. Laplante received his BS, MEng, and PhD in computer science, electrical engineering, and computer science, respectively, from Stevens Institute of Technology and an MBA from the University of Colorado at Colorado Springs. He is a licensed professional engineer in Pennsylvania and is a Certified Software Development Professional. He is a fellow of the IEEE and SPIE and a member of numerous professional societies and program committees. He has won international awards for his teaching, research, and service.

Watch: A short Introductory video from Dr. Laplante
https://phil.laplante.io/requirements/phil.intro.php

Mohamad H. Kassab is an associate research professor and a member of the graduate faculty at The Pennsylvania State University. He earned his PhD and MS degrees in computer science from Concordia University in Montreal, Canada. Dr. Kassab was an affiliate assistant professor in the Department of Computer Science and Software Engineering at Concordia University between 2010 and 2012, a postdoctoral researcher in software engineering at Ecole de Technologie Supérieure (ETS) in Montreal between 2011 and 2012, and a visiting scholar at Carnegie Mellon University (CMU) between 2014 and 2015.

Dr. Kassab has been conducting research projects jointly with the industry to develop formal and quantitative models to support the integration of quality requirements within software and systems development life cycles. The models are being further leveraged with the support of developed architectural frameworks and tools. His research interests also include bridging the gap between software engineering practices and disruptive technologies (e.g., IoT, blockchain). He has published extensively in software engineering books, journals, and conference proceedings. He is also a member of numerous professional societies and program committees, and the organizer of many software engineering workshops and conference sessions.

With over 20 years of global industry experience, Dr. Kassab has developed a broad spectrum of skills and responsibilities in many software engineering areas. Notable experiences include business unit manager at Soramitsu, senior quality engineer at SAP, senior quality engineer at McKesson, senior associate at Morgan Stanley, senior quality assurance specialist at NOKIA, and senior software developer at Positron Safety Systems. He is an Oracle Certified Application Developer, Sun Certified Java Programmer, and Microsoft Certified Professional.

Dr. Kassab has taught a variety of graduate and undergraduate software engineering and computer science courses at Penn State and Concordia University. He has won many awards for his excellence in teaching.

Watch: A short Introductory video from Dr. Kassab
https://phil.laplante.io/requirements/mohamad.intro.php

Chapter 1

Introduction to Requirements Engineering

Motivation

Very early in the drive to industrialize software development, Royce (1975) pointed out the following truths:

> There are four kinds of problems that arise when one fails to do adequate requirements analysis: top-down design is impossible; testing is impossible; the user is frozen out; management is not in control. Although these problems are lumped under various headings to simplify the discussion, they are all variations of one theme—poor management. Good project management of software procurement is impossible without some form of explicit (validated) and governing requirements.

Even though the top-down design approach has been largely replaced by object-oriented design, these truths still apply today.

A great deal of research has verified that devoting systematic effort to requirements engineering can greatly reduce the amount of rework needed later in the life of the software or software-intensive product and can cost-effectively improve various qualities of the system. Too often systems engineers forego sufficient requirements engineering activity either because they do not understand how to do requirements engineering properly, or because there is a rush to start coding (in the case of a software product). Clearly, these eventualities are undesirable, and it is a goal of this book to help engineers understand the correct principles and practices of requirements engineering.

What Is Requirements Engineering?

There are many ways to portray the discipline of requirements engineering depending on the viewpoint of the definer. For example, a bridge is a complex system but has a relatively small number

DOI: 10.1201/9781003129509-1

of patterns of design that can be used (e.g., suspension, trussed, cable-stayed). Bridges also have specific conventions and applicable regulations in terms of load requirements, materials that are used, and the construction techniques employed. So, when speaking with a customer (e.g., the state department of transportation) about the requirements for a bridge, much of its functionality can be captured succinctly:

> The bridge shall replace the existing span across the Brandywine River at Creek Road in Chadds Ford, Pennsylvania, and shall be a cantilever bridge of steel construction. It shall support two lanes of traffic in each direction and handle a minimum capacity of 100 vehicles per hour in each direction.

Of course, there is a lot of information missing from this "specification" (e.g., weight restrictions), but it substantially describes what this bridge is to do and the design choices available to achieve these requirements are relatively simple.

Other kinds of systems, such as biomechanical or nanotechnology systems with highly specialized domain language, have seemingly exotic requirements and constraints. Still, other complex systems have so many kinds of behaviors that need to be embodied (even word processing software can support thousands of functions) that the specification of said systems becomes very challenging indeed.

Since the authors are primarily software engineers, we refer to that discipline for a convenient, more-or-less universal definition for requirements engineering that is due to Pamela Zave:

> Requirements engineering is the branch of software engineering concerned with the *real-world goals* for, functions of, and constraints on software systems. It is also concerned with the relationship of these factors to *precise specifications* of software behavior, and to their *evolution over time and across software families*. (Zave 1997)

But we wish to generalize the notion of requirements engineering to include any system, whether it be software only, hardware only, or hardware and software (and many complex systems are a combination of hardware and software), so we rewrite Zave's definition as follows:

> Requirements engineering is the branch of **engineering** concerned with the *real-world goals* for, functions of, and constraints on **systems**. It is also concerned with the relationship of these factors to *precise* specifications of **system** behavior and to their evolution over time and across families of **related systems**.

The changes we have made to Zave's definition are in bold. We refer to this modified definition when we speak of "requirements engineering" throughout this text. Though some may argue that the term "Requirements Engineering" is a misnomer, the term "engineering" serves as a reminder that RE is an important part of software/system development that is concerned with anchoring development to solve a real-world problem. We will explore all the ramifications of this definition and the activities involved in great detail as we move forward.

You Probably Don't Do Enough Requirements Engineering

Research suggests that requirements engineering is not done well by the industry. For example, in a global survey of more than 3,000 practicing engineers (over 250 respondents), 37% responded that requirements engineering practices in their companies were less than satisfactory (Kassab et al. 2014). In a more recent survey (Kassab and Laplante 2022), the number has slightly decreased to 30%.

Stay Updated: For the up-to-date data from the RE state of practice survey.
https://phil.laplante.io/requirements/updates/survey.php

Another study of corporate IT projects performed by IAG Consulting found that of the companies surveyed, fully two-thirds of the respondents classified their projects as likely failures. The most frequently cited reason for failure by respondents was poor requirements gathering techniques (Kanaracus 2008). In a third study of seven embedded systems engineering companies in Europe, Sikora et al. (2012) found similar results; in particular, they concluded that "existing requirements engineering methods are insufficient for handling requirements for complex embedded systems."

One potential reason for inadequate requirements engineering is suggested by Wnuk et al. (2011) who observed that companies seem to have difficulty in scaling up requirements engineering practices in terms of activity scope and the structure of requirements artifacts. In their survey, Sikora et al. (2012) found another possible reason for the apparent inadequacy in industrial requirements engineering. They found that practitioners desire "requirements specifications that are properly integrated into the overall systems architecture but are solution-free with regard to the specified function or component itself." Survey respondents noted that existing method support is inadequate in supporting this goal. It seems, then, that requirements engineering has a long way to go in meeting the needs of practitioners.

The lack of proper requirements engineering practices is even more widespread in small businesses (Kassab 2021)—even though proper requirements engineering practices can mean the difference between survival or failure. We will return back to the discussion on requirements engineering in small businesses in Chapter 12.

What Are Requirements?

Part of the challenge in requirements engineering has to do with an understanding of what a "requirement" really is. Requirements can range from high-level, abstract statements, and back-of-the-napkin sketches to formal (mathematically rigorous) specifications. These varying representation forms occur because stakeholders have needs at different levels and hence depend on different abstraction representations. Stakeholders also have varying abilities to make and read these representations (e.g., a business customer vs. a design engineer), leading to diverse quality in the requirements. We will discuss the nature of stakeholders and their needs and capabilities in the next chapter.

Requirements vs. Features vs. Goals

A fundamental challenge for the requirements engineer is recognizing that customers often confuse requirements, features, and goals (and engineers sometimes do too).

While *Goals* are high-level objectives of a business, organization, or system, a requirement specifies how a goal should be accomplished by a proposed system. On the other hand, a feature is a set of logically related requirements that allows the user to satisfy the goal.

So, to the users of a "Smart home system," one goal is to automate that which does not need human interaction, to free the occupants to enjoy themselves with other activities. This goal can be satisfied by implementing some features. One of these features could be by providing automated water purification to the system. Then, there will be a set of specific requirements to describe the water purification mechanism (see Appendix A). So, the requirements tend to be more granular than a feature and tend to be written with the implementation in mind.

To treat a goal as a requirement is to invite trouble because the achievement of the goal will be difficult to prove. In addition, goals evolve as stakeholders change their minds and refine and operationalize goals into behavioral requirements.

Requirements Classifications

There are many schemes available to classify requirements. We will discuss two popular schemes: *Requirements Level Classification* and *Requirements Specifications Types*.

Requirements Level Classification

To deal with the diversity in requirements types, Sommerville (2005) suggests organizing them into three levels of abstraction:

- User requirements
- System requirements
- Design specifications

User requirements are abstract statements written in natural language with accompanying informal diagrams. They specify what services (user functionality) the system is expected to provide and any constraints. Collected user requirements often appear as a "concept of operations" (Conops) document. In many situations, user stories can play the role of user requirements.

System requirements are detailed descriptions of the services and constraints. System requirements are sometimes referred to as *functional specifications* or *technical annexes* (a term that is rarely used). These requirements are derived from analysis of the user requirements, and they should be structured and precise. The requirements are collected in a systems requirements specification (SRS) document. Use cases can play the role of system requirement in many situations.

Finally, design specifications emerge from the analysis and design documentation used as the basis for implementation by developers. The system design specification is essentially derived directly from the analysis of the system requirements specification.

To illustrate the differences in these specification levels, consider the following from the airline baggage handling system:

A user requirement

- The system shall be able to process 20 bags per minute.

Some related system requirements

- Each bag processed shall trigger a baggage event.
- The system shall be able to handle 20 baggage events per minute.

Finally, the associated system specifications

1.2 The system shall be able to process 20 baggage events per minute in operational mode.
1.2.1 If more than 20 baggage events occur in a 1-minute interval, then the system shall …
1.2.2 [more exception handling] …

For a pet store POS system, consider the following:
A user requirement

- The system shall accurately compute sale totals including discounts, taxes, refunds, and rebates; print an accurate receipt; and update inventory counts accordingly.

Some related system requirements

- Each sale shall be assigned a sales ID.
- Each sale may have one or more sales items.
- Each sale may have one or more rebates.
- Each sale may have only one receipt printed.

Finally, the associated software specifications

1.2 The system shall assign a unique sales ID number to each sale transaction.
1.2.1 Each sales ID may have zero or more sales items associated with it, but each sales item must be assigned to exactly one sales ID

The systems specification in the appendices also contains numerous specifications organized by level for your exploration.

The different specification levels guide progressive testing throughout the project lifecycle (Figure 1.1).

Thus, user requirements, which are discovered first, are used as the basis for final acceptance testing. The systems requirements, which are generally developed after the user requirements, are used as the basis for the integration testing that precedes acceptance testing. And, finally, the design specifications that are derived from the system requirements are used for unit testing as each code unit is implemented.

Requirements Specifications Types

Another taxonomy for requirements specifications focuses on the type of requirement from the following list of possibilities:

- Functional requirements (FRs)
- Nonfunctional requirements (NFRs)
- Domain requirements

Let's look at these more closely.

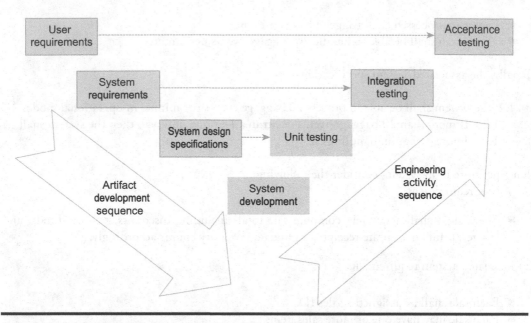

Figure 1.1 Relationship between requirements specification levels and testing.

Functional Requirements

Functional requirements (FRs) describe the services the system should provide and how the system will react to its inputs. In addition, FRs need to explicitly state certain behaviors that the system should not do (more on this later). FRs can be high level and general (in which case they are user requirements in the sense that was explained previously) or they can be detailed, expressing inputs, outputs, exceptions, and so on (in which case they are the system requirements described before).

There are many forms of representation for FRs, from natural language (in our case, the English language), visual models, and the more rigorous formal methods. We will spend much more time discussing requirements representation in Chapter 4.

To illustrate some FRs, consider the following sampling of ones for the baggage handling system.

2.1 The system shall handle up to 20 bags per minute. …
2.4 When the system is in idle mode, the conveyor belt shall not move. …
2.8 If the main power fails, the system shall shut down in an orderly fashion within 5 seconds. …
2.41 If the conveyor belt motor fails, the system shall shut down the input feed mechanism within 3 seconds.

For the pet store POS system, the following might be some FRs:

4.1 When the operator presses the "total" button, the current sale enters the closed-out state.
    4.1.1 When a sale enters the closed-out state, a total for each non-sale item is computed as the number of items times the list price of the item.
    4.1.2 When a sale enters the closed-out state, a total for each sale item is computed.

More FRs can be found in Section 3 of Appendix A and Section 2 of Appendix B.

Nonfunctional Requirements

While systems are characterized by functional behavior (what the system does). They are also characterized by their nonfunctional behavior (how the system behaves concerning some observable attributes like reliability, reusability, maintainability). In the software/system marketplace, in which functionally equivalent products compete for the same customer, nonfunctional requirements (NFRs) become more important in distinguishing between the competing products. They basically deal with the following issues:

- Security
- Reliability
- Maintainability
- Performance
- Usability
- Testability
- Interoperability
- Constraints (e.g., design/implementation constraints, financial constraints, resources constraints)

We will discuss NFRs in more detail in Chapter 5.

Domain Requirements

Domain requirements are derived from the application domain. These types of requirements may consist of new FRs or constraints on existing FRs, or they may specify how particular computations must be performed.

In the baggage handling system, for example, various domain realities create requirements. There are industry standards (we wouldn't want the new system to underperform vs. other airlines' systems). There are constraints imposed by existing hardware available (e.g., conveyor systems). And there may be constraints on performance mandated by collective bargaining agreements with the baggage handlers union.

For the pet store POS system, domain requirements are imposed by conventional store practices. For example:

- Handling of cash, credit cards, and coupons
- Display interface and receipt format
- Conventions in the pet store industry (e.g., frequent-buyer incentives, buy one get one free)
- Sale of items by the pound (e.g., horse feed) vs. by item count (e.g., dog leashes)

Appendices A and B also contain other examples of NFRs.

Domain Vocabulary Understanding

The requirements engineer must be sure to fully understand the application domain vocabulary (or have someone fluent in that vocabulary at the ready) as there can be subtle and profound differences in the use of terms in different domains. The following true incident illustrates the point. The first author was once asked to provide consulting services to a very large, international package delivery company. After a few hours of communicating with the engineers of the package delivery company, it became clear that the author was using the term "truck" incorrectly. While the author believed that the "truck" referred to the familiar vehicle that would deliver packages directly to customers' homes, the company used the term "package car" to refer to those vehicles. The term "truck" was reserved to mean any long-haul vehicle (usually an 18-wheeled truck) that carried large amounts of packages from one distribution hub to another. So there was a huge difference in the volume of packages carried in a "truck" and a "package car." Imagine, then, if a requirement was written involving the processing of "packages from 1,000 trucks" (when it was really meant "1,000 package cars"). Clearly, this difference in domain terminology understanding would have been significant and potentially costly.

VIGNETTE 1.1   Requirements for a Brick

The requirements engineer is sometimes challenged with the need for, or value of, requirements engineering activities, even the purpose of a requirements specification. The following is a real incident. A consulting requirements engineer was asked to provide guidance to the developers of a very large, complex, multimillion-dollar automated materials handling system. The system was already partially built but there were serious breakdowns in communications between the hardware and software engineers, jeopardizing the project. The consulting requirements engineer asked to see the requirements specification—but there was none. The chief engineer did not see the need for a specification document beyond a set of evolving Kanban cards (see chapter 8). To demonstrate the irrelevance of a requirements specification he challenged: "what are the requirements for a brick," hoping to trap the consulting requirements engineer into admitting that no such specifications exist. But there are standards-based requirements specifications for bricks. The ASTM International (formerly American Society for Testing and Materials) publishes technical standards for a wide range of materials, products, systems, and services, including bricks. ASTM C62-17 Standard Specification for Building Brick (Solid Masonry Units Made From Clay or Shale) specifies size, appearance, durability, strength, and more for bricks of all kinds (https://www.astm.org/Standards/C62.htm). This standard and the requirements derived from it are very important. For example, if nonstandard bricks lacked the expected compressive strength or weathering properties, a brick home might not be able to withstand severe weather conditions. Even unwanted and unexpected variations in size and coloration could be a problem for customers. So, yes, requirements are needed for even a system as simple as a brick, and well-developed requirements specifications are certainly needed in complex, multimillion-dollar systems. The chief engineer was embarrassed and unhappy with the consulting requirements engineer's answer because it exposed a serious project weakness. The consulting requirements engineer was never called back, and at this writing, the complex material handling system has not been deployed.

Requirements Engineering Activities

The requirements engineer is responsible for a number of activities. These include:

- Requirements elicitation/discovery
- Requirements analysis and reconciliation
- Requirements representation/modeling
- Requirements verification and validation
- Requirements management

We explore each of these activities briefly in the following sections and substantially more so in subsequent chapters.

Requirements Elicitation/Discovery

Requirements elicitation/discovery involves uncovering what the customer needs and wants. But elicitation is not like harvesting low-hanging fruit from a tree. While some requirements will be obvious (e.g., the POS system will have to compute sales tax), many requirements will need to be extricated from the customer through well-defined approaches. This aspect of requirements engineering also involves dis covering who the stakeholders are; for example, are there any hidden stakeholders? Elicitation also involves determining the NFRs, which are often overlooked.

Requirements Analysis and Agreement

Requirements analysis and requirements agreement involve techniques to deal with a number of problems with requirements in their "raw" form, that is, after they have been collected from the customers. Problems with raw requirements include the following:

- They don't always make sense.
- They often contradict one another (and not always obviously so).
- They may be inconsistent.
- They may be incomplete.
- They may be vague or just wrong.
- They may interact and be dependent on each other.

Many of the elicitation techniques that we will discuss subsequently are intended to avoid or alleviate these problems. Formal methods are also useful in this regard. Requirements analysis and agreement are discussed in Chapter 4.

Requirements Representation

Requirements representation (or modeling) involves converting the requirements processed raw requirements into some model (usually natural language, mathematics, and visualizations). Proper representations facilitate communication of requirements and conversion into a system architecture and design. Various techniques are used for requirements representation including informal (e.g., natural language, sketches, and diagrams), formal (mathematically sound representations), and semiformal (convertible to a sound representation or can be made fully formal by the addition of a semantic framework). Usually, some combination of these is employed in requirements representation, and we will discuss these further in Chapters 4 and 7.

Requirements Validation

Requirements validation is the process of determining if the specification is a correct representation of the customers' needs. Validation answers the question "Am I building the right product?" Requirements validation involves various semiformal and formal methods, text-based tools, visualizations, inspections, and so on, and is discussed in Chapter 6.

Requirements Management

One of the most overlooked aspects of requirements engineering, requirements management involves managing the realities of changing requirements over time. It also involves fostering traceability through appropriate aggregation and subordination of requirements and communicating changes in requirements to those who need to know.

Managers also need to learn the skills to intelligently push back when scope creep ensues. Using tools to track changes and maintain traceability can significantly ease the burden of requirements management. We will discuss software tools to aid requirements engineering in Chapter 9 and requirements management overall in Chapter 10.

Bodies of Knowledge

Three important structured taxonomies or bodies of knowledge exist for requirements engineering of software systems: the Software Engineering Body of Knowledge Version 3.0 (SWEBOK 2014), the Graduate Software Engineering Reference Curriculum (GSwE 2009), and the Principles and Practices (P&P) of Software Engineering Exam Specification (Kilcay-Ergin and Laplante 2013). These bodies of knowledge focus on the discipline of software engineering, but they can be applied to the engineering of any kind of system, whether software, electrical, mechanical, or hybrid.

SWEBOK was established to promote a consistent view of software engineering and to provide a foundation for curriculum development and for certification and licensing examinations. SWEBOK identifies the fundamental skills and knowledge that all graduates of a master's program in software engineering must possess.

The GSwE report highlights the strong link between software engineering and systems engineering and suggests that system engineering knowledge areas should be integrated into the software engineering curriculum (Kilcay-Ergin and Laplante 2013).

The Software Engineering P&P exam is used by states and jurisdictions of the United States as one component of licensure for those individuals that are working on software systems that affect the health, safety, and welfare of the public (Laplante et al. 2013).

Since the intent of the taxonomies is different, the scope of requisite knowledge also differs in some ways; for example, there is no clear coverage of GSwE's initiation and scope definition in SWEBOK. A comparison of the scope of knowledge for these bodies of knowledge is shown in Table 1.1.

This text is intended to provide reasonable coverage of most of the topics from the SWEBOK; however, it substantially covers the other two bodies of knowledge.

Table 1.1   Comparison of GSwE, SWEBOK, and Software Engineering P&P Knowledge Areas

GSwE 2009	SWEBOK Version 3.0	Software Engineering P&P
Fundamentals of RE • Relationship between systems engineering and software engineering • Definition of requirements • System design constraints • System design and requirements allocation • Product and process requirments • Functional and nonfunctional requirements • Emergent properties • Quantifiable requirements	**Software Requirements Fundamentals** • Definition of a software requirement • Product and process requirements • Functional and nonfunctional requirements • Emergent properties • Quantifiable requirements • System requirements and software requirements	**Software Requirements Fundamentals** • Concept of operations • Types of requirements • Product and process requirements • Functional and nonfunctional requirements • Quantifiable requirements • System requirements • Software requirements • Derived requirements • Constraints, service level
RE Process • Process models • Process actors • Process support and management • Process quality and improvement	**Requirements Process** • Process models • Process actors • Process support and management • Process quality and Improvement	
Initiation and Scope Definition • Determination and negotiation of requirements • Feasibility analysis • Process for requirements review/revision	No equivalent	

(Continued)

Table 1.1 (*Continued*)   Comparison of GSwE, SWEBOK, and Software Engineering P&P Knowledge Areas

GSwE 2009	SWEBOK Version 3.0	Software Engineering P&P
Requirements Elicitation • Requirements sources • Elicitation techniques	**Requirements Elicitation** • Requirements sources • Elicitation techniques	**Requirements Elicitation** • Requirements sources • Elicitation techniques • Requirements representation
Requirements Analysis • Requirements classification • Conceptual modeling • Heuristic methods • Formal methods • Requirements negotiation	**Requirements Analysis** • Requirements Classification • Conceptual modeling • Architectural design and requirements allocation • Requirements negotiation • Formal analysis	
Requirements Specification • Requirements specification techniques	**Requirements Specification** • The System definition document • System requirements specification • Software requirements specification	**Requirements Specification** • System definition document • System/subsystems specification • Software requirements specification • Interface requirements specification
Requirements Validation • Requirements reviews • Prototyping • Model validation • Acceptance tests	**Requirements Validation** • Requirements reviews • Prototyping • Model validation • Acceptance tests	**Requirements Verification and Validation** • Requirements reviews • Prototyping • Model validation • Simulation
Practical Considerations • Iterative nature of requirements process • Change management • Requirements attributes • Requirements tracing • Measuring requirements	**Practical Considerations** • Iterative nature of the requirements process • Change management • Requirements attributes • Requirements tracing • Measuring Requirements	**Requirements Management** • Iterative nature of the requirements process • Change management • Requirement attributes • Requirements traceability • Measuring requirements • Software requirements tools

Source: Adapted from Kilcay-Ergin, N., & Laplante, P.A., *IEEE Trans. Educ.*, 56: 199–207, 2013.

Also worth mentioning are other proposals, reference curricula, and bodies of knowledge that also focus significantly on requirements engineering processes and activities. These include:

■ The *Requirements Engineering Body of Knowledge (REBoK* 2013),
■ The *Project Management Body of Knowledge (PMBOK)*,

■ The *Business Analysis Body of Knowledge (PABOK)*,
■ The *Guide to the Systems Engineering Body of Knowledge* (SEBoK 2012),
■ The *Graduate Reference Curriculum for Systems Engineering* (GRCSE 2012),
■ The Association for Computing Machinery (ACM)/Institute for Electrical and Electronics Engineers (IEEE) Computer Society's *Curriculum Guidelines for Undergraduate Degree Programs in Software Engineering* (ACM/IEEE 2014). These also focus significantly on requirements engineering processes and activities.

The Requirements Engineer

What skills should requirements engineer have?

Since the 1990s, labor market observers (e.g., CNN Money) have consistently reported the Business Analysts' job as one of the highest paid—not only in the IT sector but also in general. Recently, the US Bureau of Labor Statistics announced that the employment of analysts is "projected to grow 21% until 2024, much faster than the average for all occupations." The growth in cloud computing, Artificial Intelligence (AI), the Internet of Things (IoT), cybersecurity, and mobile networks will only increase the demand for these professionals and is expected to outstrip the supply of talent. For example, the 2015–2019 report of Canada's ICT Council, a government-funded labor market intelligence and industry skills standard body, warns that "As a result of employment growth—combined with replacement demand due to skills mismatch, retirements, and other exits, demand-supply imbalances will affect some occupations more than others" and names "information systems analysts" as the first among the Top-7 so-called "high-demand occupations." Despite these market developments, relatively little has been published on the "requirements engineers" occupation in terms of in-demand competencies, skills, and professional experience as perceived by companies employing RE specialists.

There are studies that investigated the occupational aspects of the requirements engineer role in Canada (Wang et al. 2018), Germany (Herrmann 2013), in the Netherlands (Daneva et al. 2017), and in Mexico and Brazil (Calazans et al. 2017).

While well-established practitioners' communities framed RE as a profession from the body-of-knowledge perspective (e.g., PMBOK, SWEBOK, GSwE, REBOK, and BABOK). Each of these reflects the collective knowledge of the respective community, presents their most widely accepted practices, and positions RE in a unique way, as per the interfaces of RE with the critical activities of the respective professional community. Al-Ani and Sim (2006) proposed a model of the key RE areas of expertise and matched them against RE activities.

Christensen and Chang suggest that the requirements engineer should be organized, have experience throughout the (software) engineering lifecycle, have the maturity to know when to be general and when to be specific, and be able to stand up to the customer when necessary (Christensen and Chang 1996). Christensen and Chang further suggest that the requirements engineer should be a good manager (to manage the process), a good listener, fair, a good negotiator, and multidisciplinary (e.g., have a background in traditional hard sciences and engineering augmented with communications and management skills). Finally, the requirements engineer should understand the problem domain.

Gorla and Lam (2004) argued that engineers should be thinking, sensing, and judging in the Myers-Briggs sense. We could interpret this observation to mean that requirements engineers are structured and logical (thinking), focus on information gathered and do not try to interpret it (sensing), and seek closure rather than leaving things open (judging).

Klendauer et al. (2012) derived a RE competence model integrating contextual and situational factors tied to RE results variables. Penzenstadler et al. (2013) categorized the soft skills required for the execution of specific RE activities.

Finally, Ebert (2010) suggests that requirement engineers should have competency in the following areas:

- Requirements engineering
- Systems engineering
- Management
- Communication
- Cognition
- Social interaction

Ebert further notes that academic programs are insufficient to develop these competencies and that these must be acquired through years of practice and study. Academic courses can, however, start new requirements engineers off in the right direction.

Requirements Engineer Roles

The collective conclusion from both scholars' and practitioners' sources is that there is a significant incongruity regarding the perceptions of the RE role, tasks, and responsibilities in the marketplace. A major source of confusion is that a requirements engineer can be acting in different roles. We have identified the following role models:

- Requirements engineer as a software systems engineer
- Requirements engineer as a subject matter expert (SME)
- Requirements engineer as an architect
- Requirements engineer as a business process expert
- The ignorant requirements engineer

There are hybrid roles for the requirements engineer from the above as well.

Requirements Engineer as Software or Systems Engineer

It is likely that many requirements engineers are probably software engineers, electrical engineers, or systems engineers. When this is the case, the requirements engineer can positively influence the downstream development of models (e.g., the software design). The danger, in this case, is that the requirements engineer may begin to create a design when he should be developing requirements specifications.

Requirements Engineer as Subject Matter Expert

In many cases, the customer is looking to the requirements engineer to be an SME for expertise either in helping to understand the problem domain or in understanding the customers' own wants and desires. Sometimes the requirements engineer isn't an SME—they are an expert in requirements engineering. In those cases where the requirements engineer is not an SME, consider joining forces with an SME.

Requirements Engineer as Architect

Building construction is often used as a metaphor for software construction. In the authors' experience, architects and landscape architects play similar roles to requirements engineers (and this similarity is often used as an argument that software engineers need to be licensed). Daniel Berry has written about this topic extensively (Berry 1999, 2003). Berry noted that the analogous activities reduce scope creep and better involve customers in the requirements engineering process. In addition, Zachman (1987) introduced an architectural metaphor for information systems, though his model is substantially different from the one about to be presented.

The similarities between architecture (in the form of home specification) and software/systems specification are summarized in Table 1.2. If you have been through the process of constructing or renovating a home, you will appreciate the similarities.

Requirements Engineer as Business Process Expert

The activities of requirements engineering comprise a form of problem solving—the customer has a problem and the system must solve it. Often, solving the problem at hand involves the requirements engineer advising changes in business processes to simplify the expression of system behavior. While it is not the requirements engineer's role to conduct business process improvement, this side benefit is frequently realized.

Ignorance as Virtue

Berry (1995) suggested having both novices and experts in the problem domain involved in the requirements engineering process. His rationale is as follows. The "ignorant" people ask the "dumb" questions, and the experts answer these questions. Having the requirements engineer as

Table 1.2   Architectural Model for Systems Engineering

Homebuilding	Software/System Building
Architect meets with and interviews clients. Tours property. Takes notes and pictures.	Requirements engineer meets with customers and uses interviews and other elicitation techniques.
Architect makes rough sketches (shows to clients, receives feedback).	Requirements engineer makes models of requirements to show to customers (e.g., prototypes, draft SRS).
Architect makes more sketches (e.g., elevations) and perhaps more sophisticated models (e.g., cardboard, 3D-virtual models, fly-through animations).	Requirements engineer refines requirements and adds formal and semiformal elements (e.g., UML). More prototyping is used.
Architect prepares models with additional detail (floor plans).	Requirements engineer uses information determined above to develop complete SRS.
Future models (e.g., construction drawings) are for contractors' use.	Future models (e.g., software design documents) are for developers' use.

the most ignorant person in the group is not necessarily a bad thing, at least with respect to the problem domain because it forces him to ask the hard questions and to challenge conventional beliefs. Of course, the ignorant requirements engineer is completely in opposition to the role of SME.

Berry also noted that using formal methods in requirements engineering is a form of ignorance because a mathematician is generally ignorant about an application domain before he or she starts modeling it.

Role of the Customer

What role should the requirements engineer expect the customer to play? The customers' roles are varied and include:

- Helping the requirements engineer understand what they need and want (elicitation and validation)
- Helping the requirements engineer understand what they don't want (elicitation and validation)
- Providing domain knowledge when necessary and possible
- Alerting the requirements engineer quickly and clearly when they discover that they or others have made mistakes
- Alerting the requirements engineer quickly when they determine that changes are necessary (really necessary)
- Controlling their urges to have "aha moments" that cause major scope creep
- Sticking to all agreements

In particular, the customer is responsible for answering the following four questions, with the requirements engineer's help, of course:

1. Is the system that I want feasible?
2. If so, how much will it cost?
3. How long will it take to build?
4. What is the plan for building and delivering the system?

The requirements engineer must manage customers' expectations with respect to these issues. We will explore the nature and role of customers and stakeholders in the next chapter.

Problems with Traditional Requirements Engineering

Traditional requirements engineering approaches suffer from a number of problems, many of which we have already addressed (and others, which will be addressed) and many of these are not easily resolved. These problems include:

- Natural language problems (e.g., ambiguity, imprecision)
- Domain understanding

- Dealing with complexity (especially temporal behavior)
- Difficulties in enveloping system behavior
- Incompleteness (missing functionality)
- Over-completeness (gold-plating)
- Overextension (dangerous "all")
- Inconsistency
- Incorrectness
- and more

We will explore the resolution of these problems throughout the book and particularly in Chapter 6.

Natural language problems result from the ambiguities and context sensitivity of natural (human) languages. We know these language problems exist for everyone, not just requirements engineers, and lawyers and legislators make their living finding, exploiting, or closing the loopholes found in any laws and contracts written in natural language.

We have already covered the issue of domain understanding and have observed, as have others, that the requirements engineer may be an expert in the domain in which the system to be built will operate. System complexity is a pervasive problem that faces all systems engineers, and this will be discussed shortly. The problems of fully specifying system behavior and of missing behavior are also very challenging, but some techniques can be helpful in missing, at least, the most obvious functionality in a system.

Complexity

One of the greatest difficulties in dealing with the requirements engineering activities, particularly elicitation and representation, for most systems is that they are complex. Without trying to invent a definition for complexity, we contend that the challenges and complexity of capturing nontrivial behavior of any kind illustrate the notion of complex. Such difficulties are found in even the simplest of repeatable human endeavors.

Imagine, for example, someone asked you to describe the first 5 minutes of your morning from the moment you wake up. Could you do it with precision (try it)? No, you could not. Why? There are too many possible different paths your activities can take and too much uncertainty. Even with an atomic clock, you can't claim to wake up at the same time every day (because even atomic clocks are imperfect). But of course, you wake up differently depending on the day of the week, whether you are on vacation from work, if it is a holiday, and so on. You have to account for that in your description. But what if you wake up sick? How does that change the sequence of events? What if you accidentally knock over the glass of water on your nightstand as you get up? Does that change the specification of the activity? Or if you trip on your dog as you head to the bathroom? We could go on with this example and repeat this exercise with other simple tasks such as mowing the lawn or shopping for food. In fact, until you constrain the problem to the point of ridiculousness, you will find it challenging or even impossible to precisely capture any nontrivial human activities.

Now consider complex information or embedded processing system. Such a system will likely have to have interactions with humans. Even in those systems that do not directly depend on human interaction, it is the intricacy of temporal behavior as well as the problems of unanticipated events that make requirements elicitation and specification so difficult (and all the other requirements activities hard too).

Rittel and Webber (1973) defined a class of complex problems that they called "wicked." Wicked problems are complex problems that have ten characteristics:

- There is no definitive formulation of a wicked problem.
- Wicked problems have no stopping rule.
- Solutions to wicked problems are not true or false, but good or bad.
- There is no immediate and no ultimate test of a solution to a wicked problem.
- Every solution to a wicked problem is a "one-shot operation"; because there is no opportunity to learn by trial and error, every attempt counts significantly.
- Wicked problems do not have an enumerable (or an exhaustively describable) set of potential solutions, nor is there a well-described set of permissible operations that may be incorporated into the plan.
- Every wicked problem is essentially unique.
- Every wicked problem can be considered a symptom of another problem.
- The existence of a discrepancy representing a wicked problem can be explained in numerous ways. The choice of explanation determines the nature of the problem's resolution.
- The planner (designer) has no right to be wrong.

Rittel and Webber meant wicked problems to be of an economic, political, and societal nature (e.g., hunger, drug abuse, and conflict in the Middle East); therefore, they offer no appropriate solution strategy for requirements engineering. Nevertheless, it is helpful to view requirements engineering in the context of a wicked problem because it helps explain why the task is so difficult—because in many cases, real systems embody many of the characteristics of a wicked problem.

Even though "wicked" problems are complex problems, not every complex problem is a "wicked" problem. A requirements specification for a 3-D printing system which can print artificial bone components quickly during replacement surgery or an autonomous vehicle, are complex but these kinds of systems are already in development or deployed so their creation cannot be a wicked problem.

Gold-Plating and Ridiculous Requirements

In his outstanding book on systems design, Brooks (2010) cautions against "ridiculous requirements," that is requirements that would be difficult to deliver given the state of technology at the time they are proposed. Another kind of ridiculous requirement would be one that, though possible to realize, is unlikely to be used.

Brooks gives the example of the *Comanche, a* "self-ferrying helicopter," which was supposed to be able to fly itself over the Atlantic. This feature was not intended to be used frequently, yet it substantially complicated the design. Specifying features that are unlikely to be used, even those that are not difficult to implement, is sometimes derisively called gold-plating.

Obsolete Requirements

An *obsolete requirement* is one that has changed significantly (during system development) or no longer applies to the new system under consideration. For second, third, and so on generation systems and for systems derived from related products or in product lines (see Exercise 1.14), it is common practice to reuse portions of requirements specifications. When requirements specifications are reused, it is often the case that obsolete requirements are propagated. Wnuk et al. (2013)

reported survey data from 219 respondents concerning obsolete requirements. They found that obsolete requirements are a significant problem for software-intensive products and in large projects. They also found that most of the companies surveyed did not have processes or automated tools for handling obsolete software requirements—only 10% of those surveyed reported using tools to support the identification of obsolete requirements.

Dubious requirements are related to obsolete requirements. Dubious requirements are those whose origin is unknown and/or which serve an unknown purpose. Thus, a dubious requirement may not be applicable in the current system, and thus, obsolete.

Obsolete and dubious requirements must be aggressively identified and culled through all phases of the requirements engineering lifecycle. Maintaining traceability links for requirements to their source and to related requirements is one significant way to prevent obsolete and dubious requirements. Date stamping requirements (upon their creation and for any changes) can help prevent obsolete requirements. Another, more labor-intensive way to manage obsolete and dubious requirements is by holding regular requirements inspections/reviews (discussed in Chapter 6).

Requirements churn refers to the changes in requirements that occur after the requirements are agreed upon. Requirements churn can be measured as the ratio of requirements changed per unit of time or as the ratio of requirements changed per total number of requirements at a given time. Requirements will change throughout the lifecycle of a system, but these changes have to be managed carefully. Maintaining historical statistics for the churn rate for projects at various stages and monitoring these rates can help to identify problems, such as too rapidly changing requirements. A high churn rate can be a leading indicator of obsolete requirements.

Four Dark Corners

Many of the problems with traditional requirements engineering arise from "four dark corners" (Zave and Jackson 1997). We repeat the salient points here, verbatim, with commentary in italics.

1. All the terminology used in requirements engineering should be grounded in the reality of the environment for which a machine is to be built.
2. It is not necessary or desirable to describe (however abstractly) the machine to be built.
   Rather, the environment is described in two ways as it would be without, or in spite of, the machine and as we hope it will become because of the machine.
   *Specifications are the **what** to be achieved by the system, not the **how**.*
3. Assuming that formal descriptions focus on actions, it is essential to identify which actions are controlled by the environment, which actions are controlled by the machine, and which actions of the environment are shared with the machine.
   All types of actions are relevant to requirements engineering, and they might need to be described or constrained formally.
   If formal descriptions focus on states, then the same basic principles apply in a slightly different form.
   The method of formal representation should follow the underlying organization of the system. For example, a state-based system is best represented by a state-based formalization.
4. The primary role of domain knowledge in requirements engineering is in supporting the refinement of requirements to implementable specifications.
   Correct specifications, in conjunction with appropriate domain knowledge, imply the satisfaction of the requirements.

Failure to recognize the role of domain knowledge can lead to unfilled requirements and forbidden behavior.

Managing these dark corners is an important duty for requirements engineers and project managers.

Difficulties in Enveloping System Behavior

Imagine an arbitrary system with n inputs and m outputs.

Imagine that this system operates in an environment where the set of inputs, I, derives from human operators, sensors, storage devices, other systems, and so on. The outputs, O, pertain to display devices, actuators, storage devices, other systems, and so on. The only constraint we will place on the system is that the inputs and outputs are digitally represented within the system (if they are from/ to analog devices or systems, an appropriate conversion is needed). We define a behavior of the system as an input/output pair. Since the inputs and outputs are discrete, this system can be thought of as having an infinite but countable number of behaviors, $B \subseteq I \times O$.

Imagine the behavior space, B, is represented by the Venn diagram of Figure 1.2.

The leftmost circle in the diagram represents the desired behavior set, as it is understood by the customer. The area outside this circle is unwanted behavior and desirable behavior that, for whatever reason, the customer has not discovered.

The requirements engineer goes about his business and produces a specification that is intended to be representative of the behavior desired by the customer (the rightmost circle of Figure 1.2). Being imperfect, this specification captures some (but not all) of the desired behavior, and it also captures some undesirable behavior. The desired behavior not captured in the specified behavior is the missing functionality. The undesirable behavior that is captured is forbidden functionality.

The goal of the requirements engineer, in summary, is to make the left and right circles overlap as much as possible and to discover those missing desirable behaviors that are not initially encompassed by the specification (leftmost circle).

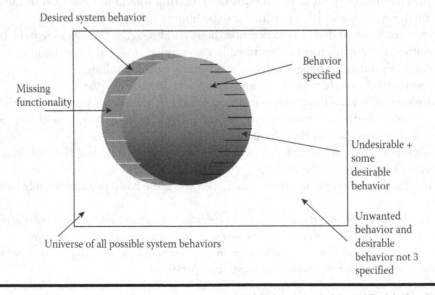

Figure 1.2   Universe of all possible behaviors, desired behavior, and specified behavior.

Recognizing that there is an infinite number of behaviors that could be excluded from the desired behaviors of the system, we need to, nevertheless, focus on the very worst of these. Therefore, some of the requirements that we will be discovering and communicating will take the form

The "system shall not ..."

A requirement that takes this form is sometimes called "unwanted functionality," "shall not behavior," or "forbidden behavior," and it can be used to address a "hazard" situation. The term "hazard" is used to address something (an incident) that can potentially occur while the system is running (at run time), which can lead to failure and then consequently poses a danger or risk. For spaceborne communications satellites, for example, "Loss of communications with the Earth station" is an example of a hazard. "Power Failure" is another example of a hazard too.

On the other hand, "Project cost overruns" is not usually considered as a hazard because it is not being realized at run time, but at development time (while the system is being developed). In addition, by the same definition, requirements change, or standards changes are not considered as hazards.

A hazard by itself is not a requirement. For example, "power failure" is a hazard but not a requirement. But when we write a requirement to address hazards, we often write these in the form of "The system shall not..." statements. It is also possible to address hazards in the affirmative requirement form. For example, "The system shall not lose power" is equivalent in form to "The system shall provide continuous power." Whether we chose to address the hazard with a requirement in the "shall not" or "shall" form is a matter of style and clarity.

On a snowy January 13 in 1982, an Air Florida Boeing 737 jet departed from Washington DC's National Airport and crashed into the 14th Street Bridge less than a mile from the airport. Of the 79 people on board the plane, only 5 survived. The plane also struck seven vehicles on the bridge, killing four motorists and injuring another four.

The unfortunate passengers on the plane should have known that there was a risk of the plane crashing, no matter how small that risk may have been. In some cases, the passengers may have even taken out special crash insurance. But what were the expectations of the motorists in the vehicles on the bridge who were killed or injured? Surely, the thought of an airplane crashing into the bridge never crossed their mind. For the motorists, therefore, this crash represented an unknown but manifested uncertainty.

As an example of forbidden behavior, consider the following for the baggage handling system:

When the "conveyor jam" signal is set to the high state, the feed mechanism shall not permit additional items to enter the conveyor system.

We will discuss how to identify and characterize these unwanted or "shall not" behaviors in Chapter 3.

The Danger of "All" in Specifications

Requirement specification sentences often involve some universal quantification (e.g., "all users shall be able to access ..."). But the use of "all" specifications is dangerous because they are usually not true (Berry and Kamsties 2000). For example, if a requirement for the baggage handling system states: *All baggage shall be assigned a unique identifier.*

It is worthwhile to challenge this requirement. That is, is there some form of "baggage" that we wish to not give a unique identifier? If a suspicious package is given to an airport check-in agent, should they print out a barcode tag and put it on the package? Or, should they put it aside for further processing, such as inspection by security personnel? Whether the answer is "yes" or "no" is irrelevant—we want to consider these kinds of situations and deal with them as we are writing requirements in draft form, rather than postpone the issue until requirements analysis occurs. Therefore, requirements involving the word "all" should be challenged and relaxed when possible.

Related to "all" specifications are "never," "always," "none," and "each" (because they can be formally equated to universal quantification), these words and their mathematical equivalents should be avoided as well.

VIGNETTE 1.2   Government Procurement

In the United States, the requirements engineering process required of a contractor providing some system to a government entity is quite rigorous. The government entity, such as the Department of Defense or Department of Energy, will often solicit project proposals based on an already well-defined set of requirements. For example, many of the requirements for a new nuclear power generation plant will be based on previous systems, which, in turn, are largely based on standards, regulations, and laws. These requirements are established by the contracting agency (and other stakeholders) before the work contract is awarded.

Contractors can participate in the contract proposal generation through industry days. During an industry day, the government entity will convene potential contractors to discuss the current requirements of the system. This process is iterative, and as the prebid requirements become more mature, the system becomes more well-defined.

After the contract is awarded, the requirements engineering process continues. Throughout, the winning contractor will use many of the techniques described in the book to refine system requirements. It is essential that during this process contractor customer communications are well structured. Therefore, some of the more informal approaches to requirements engineering will be prohibited (and often the contract states which methodologies can or cannot be used). For example, holding a number of JAD sessions (discussed in Chapter 3) is often a contractual requirement.

The process described is similar to government contracting in many countries.

Exercises

1.1   What are some of the major objections and deterrents to proper requirements engineering activities?

1.2   How is requirements engineering different for "small" systems?

1.3   How is requirements engineering different for "startup" businesses?

1.4   Emerging technologies including AI (artificial intelligence), 5G, IoT (Internet of Things), serverless computing, biometrics, AR (augmented reality)/VR (virtual reality), edge computing, blockchain, robotics, NLP (natural language processing), and quantum computing, whose development, practical applications, or both are still largely unrealized. How is requirements engineering different when developing "Emerging Technologies" based systems?

1.5   What are some factors that may cause customers to alter requirements?

1.6   What issues might arise when a requirements engineer, who is not a subject matter expert, enlists a subject matter expert to assist in defining requirements?

1.7   List some representative user requirements, system requirements, and software specifications for

    1.7.1   The pet store POS system

    1.7.2   The baggage handling system

1.8   List five typical functional requirements for

    1.8.1   The pet store POS system

    1.8.2   The baggage handling system

1.9   List five forbidden functional requirements for

    1.9.1   The pet store POS system

    1.9.2   The baggage handling system

1.10   Identify the nonfunctional requirements in

    1.10.1   The specification for the smart home system in Appendix A

    1.10.2   The specification for the wet-well pumping system in Appendix B

1.11   List five possible nonfunctional requirements for

    1.11.1   The pet store POS system

    1.11.2   The baggage handling system

1.12   Conduct some Web research to discover prevailing regulations or standards (NFR) for smart home systems.

1.13   For the smart home system make a list of some of the hazards (what this system shall not do) based on the regulations you discovered in Exercise 1.12 and any other information you might have.

*1.14   A product line is a set of related products that share certain features. Product lines may be planned or they may emerge. What are some of the requirements engineering challenges for product lines? How does the "one-shot solution" aspect of Ritter and Webber's wicked problem apply? You may wish to conduct some research to answer this question.

References

ACM/IEEE. (2014). Computer Society Curriculum Guidelines for Undergraduate Degree Programs in Software Engineering. https://www.acm.org/education/se2014.pdf (accessed January 2017).

Al-Ani, B., & Sim, S. E. (2006). So, you think you are a requirements engineer? In *14th IEEE International Requirements Engineering Conference (RE'06)*, Minneapolis/St. Paul, MN, USA (pp. 337–338). IEEE.

Berry, D. M. (1995). The importance of ignorance in requirements engineering. *Journal of Systems and Software*, 28(1): 179–184.

Berry, D. M. (1999). Software and house requirements engineering: Lessons learned in combating requirements creep. *Requirements Engineering Journal*, 3(3&4): 242–244.

Berry, D. M. (2003). More requirements engineering adventures with building contractors. *Requirements Engineering Journal*, 8(2): 142–146.

Berry, D. M., & Kamsties, E. (2000). The dangerous 'All' in specifications. In *Proceedings of the Tenth International Workshop on Software Specification and Design (IWSSD'00)*, San Diego, CA, 5–7 November.

Brooks, F. P., Jr. (2010). *The Design of Design: Essays from a Computer Scientist*. Pearson Education, Addison-Wesley Professional, Boston, MA.

Calazans, A. T. S., Paldês, R. Á., Masson, E. T. S., Brito, I. S., Rezende, K. F., Braosi, E., & Pereira, N. (2017). Software requirements analyst profile: A descriptive study of Brazil and Mexico. In *2017 IEEE 25th International Requirements Engineering Conference (RE)*, Lisbon, Portugal (pp. 204–212). IEEE.

Christensen, M., and Chang, C. (1996). Blueprint for the ideal requirements engineer. *IEEE Software*, 13(2): 12.

Daneva, M., Wang, C., & Hoener, P. (2017). What the job market wants from requirements engineers? An empirical analysis of online job ads from the Netherlands. In *2017 ACM/IEEE International Symposium on Empirical Software Engineering and Measurement (ESEM)* (pp. 448–453). IEEE.

Ebert, C. (2010). Requirements engineering: Management. In P. Laplante (Ed.), *Encyclopedia of Software Engineering* (pp. 932–948). Taylor & Francis, Boca Raton, FL.

Gorla, N., & Lam, Y. W. (2004). Who should work with whom? Building effective software project teams. *Communications of the ACM*, 47(6): 79–82.

Graduate Reference Curriculum for Systems Engineering (GRCSE). (2012). http://www.bkcase.org/grcse/ (accessed January 2017).

Graduate Software Engineering Reference Curriculum (GSwE). (2009). https://www.acm.org/binaries/content/assets/education/gsew2009.pf (accessed January 2017).

Guide to the Systems Engineering Body of Knowledge (SEBoK). (2012). http://www.sebokwiki.org/ (accessed January 2017).

Herrmann, A. (2013). Requirements engineering in practice: There is no requirements engineer position. In *International Working Conference on Requirements Engineering: Foundation for Software Quality* (pp. 347–361). Springer, Berlin, Heidelberg.

The Information and Communications Technology Council. The Smartest Economy Reshaping Canada"s Workforce: Labor Market Outlook 2015–2019.

Institute for Business Analysis. (2010). A Guide to the Business Analysis Body of Knowledge (BABOK Guide), Version 3.

Kanaracus, C. (2008). Study: Bad Requirements-Gathering Hurts IT Projects. NetworkWorld.com.

Kassab, M. (2021, September). How requirements engineering is performed in small businesses? In *2021 IEEE 29th International Requirements Engineering Conference Workshops (REW)* (pp. 220–223). IEEE.

Kassab, M., & Laplante, P. (2022). The current and evolving landscape of requirements engineering in practice. *IEEE Software*. DOI: 10.1109/MS.2022.3147692.

Kassab, M., Neill, C., & Laplante, P. (2014). State of practice in requirements engineering: Contemporary data. *Innovations in Systems and Software Engineering*, 10(4): 235–241.

Kilcay-Ergin, N., & Laplante, P. A. (2013). An online graduate requirements engineering course. *IEEE Transactions on Education*, 56(2): 199–207.

Klendauer, R., Berkovich, M., Gelvin, R., Leimeister, J. M., & Krcmar, H. (2012). Towards a competency model for requirements analysts. *Information Systems Journal*, 22(6): 475–503.

Laplante, P. A., Kalinowski, B., & Thornton, M. (2013). A principles and practices exam specification to support software engineering licensure in the United States of America. *Software Quality Professional*, 15(1): 4–15.

Penzenstadler, B., Fernández, D. M., Richardson, D., Callele, D., & Wnuk, K. (2013). The requirements engineering body of knowledge (rebok). In *2013 21st IEEE International Requirements Engineering Conference (RE)*, Rio de Janeiro, Brazil (pp. 377–379). IEEE.

Project Management Institute. (2014). PMI: A Guide to the Project Management Body of Knowledge, Version 4.

Rittel, H., & Webber, M. (1973). Dilemmas in a general theory of planning. *Policy Sciences*, 4: 155–169.

Royce, W. (1975). *Practical Strategies for Developing Large Software Systems*. Addison-Wesley, Boston, MA, p. 59.

Sikora, E., Tenbergen, B., & Pohl, K. (2012). Industry needs and research directions in requirements engineering for embedded systems. *Requirements Engineering*, 17: 57–78.

Software Engineering Body of Knowledge Version 3.0 (SWEBOK). (2014). http://www.computer.org/portal/web/swebok/html/contentsch2#ch2 (accessed January 2017).

Sommerville, I. (2005). *Software Engineering*, 7th edition. Addison-Wesley, Boston, MA.

US Department of Labor: Occupational Outlook Handbook. (2015).

Wang, C., Cui, P., Daneva, M., & Kassab, M. (2018). Understanding what industry wants from requirements engineers: An exploration of RE jobs in Canada. In *Proceedings of the 12th ACM/IEEE International Symposium on Empirical Software Engineering and Measurement* (pp. 1–10), Oulu, Finland.

Wnuk, K., Gorschek, T., & Zahda, S. (2013). Obsolete software requirements. *Information and Software Technology*, 55(6): 921–940.

Wnuk, K., Regnell, B., & Berenbach, B. (2011). Scaling up requirements engineering—Exploring the challenges of increasing size and complexity in market-driven software development. In D. Berry & X. French (Eds.), *Requirements Engineering: Foundation for Software Quality. REFSQ 2011*, LNCS (Vol. 6606, pp. 54–59), Essen, Germany.

Zachman, J. A. (1987). A framework for information systems architecture. *IBM Systems Journal*, 26(3): 276–292.

Zave, P. (1997). Classification of research efforts in requirements engineering. *ACM Computing Surveys*, 29(4): 315–321.

Zave, P., and Jackson, M. (1997). Four dark corners of requirements engineering. *ACM Transactions on Software Engineering Methodology*, 6(1): 1–30.

Chapter 2

Preparing for Requirements Elicitation

Product Business Goals and Mission Statements

Because of the significant costs and time, organizations need to have very compelling reasons when proposing to build or redesign complex systems rather than buying them. The first thing we need to do when undertaking the development of a new system, or redesign of an old one, is to obtain or develop a concise description of these reasons. These could be characterized in a number of different ways (e.g., market intent, product roadmap, business and operational strategy).

Such business reasons are called business goals and will ideally describe high-level goals. The business context of the organizations shall be considered when crafting these, and this context can be thought of in two broad categories:

- The objectives or motivation for building the system
- The environment in which the system is to be built, deployed, and used

For example, there are several types of organizations. These organizational types rely on software/systems in fairly consistent ways. We will look at a couple of examples here:

- Those that generate revenue directly from the construction, sale, or licensing of software-intensive products
- Those that generate revenue from non-software sales and services but rely on software for operations

If the organization is in the business of developing software, it will often have some kind of product roadmap that highlights the business goals of the system under development. This includes things like:

- Market-related information
- The value proposition for the market

DOI: 10.1201/9781003129509-2

- Competitive analysis
- Anticipated number of units
- Target price
- Anticipated revenue
- Associated budgets

The second type of organizations that generate revenue from non-software sales but rely on software for operations often have strategic plans that are focused on their domain and not specifically on software. Consider "Walmart" as an example. Walmart collects data on every transaction that every customer makes in every store (roughly 267 million transactions a day), and it is forever trying to improve their understanding of buying characteristics. This knowledge can influence things like:

- Promotions
- Inventory
- Supply chain process
- Store layout

So the articulated business goals for Walmart's developed software/system should reflect these business needs.

Writing the business goals can be a contentious business, and many people resent or fear doing so because there can be a tendency to get bogged down in minutiae. It is not uncommon for the business goals to be ambiguously understood. They might be understood at a high level without associated tactical approaches, or they could be understood differently across the organization. It is essential, therefore, to establish a consistent view on these goals across an organization as these will serve as a focal point for all involved in the system, and they allow us to weigh the importance of various features by asking the question "how does that functionality serve its intent?". Business goals should be written to be very short, descriptive, compelling, and never detailed.

Business goals are often synonymous with mission objectives, a term typically used by DoD or NASA when creating architectures of weapon or space systems. One of the most widely cited "good" system mission statements is the one associated with the Starship Enterprise from the original Star Trek television series. That mission statement, roughly stated, is:

> to explore strange new worlds, to seek out new life forms and new civilizations, to boldly go where no one[1] has gone before.

This statement is clear, compelling, and inspiring. And it is "useful"—fans of this classic series will recall several episodes in which certain actions to be taken by the starship crew were weighed against the system mission statement.

Some organizations also refer to a concept of operations (Conops) statement, which is similar to the mission statement, though tends to be longer. In some settings, the Conops is a document resembling a very high-level requirements specification.

In agile methodologies, to be discussed later, we could say that the business goals, mission statement, or Conops plays the role of "system metaphor."

For example, Section 1.1 of Appendix A contains a mission statement for the Smith Smart Home, and Section 1 of Appendix B contains a Conops statement for the wastewater pumping control system.

So what might a product or system mission statement for the baggage handling system look like? How about

> To automate all aspects of baggage handling from passenger origin to destination.

For the pet store POS system, consider

> To automate all aspects of customer purchase interaction and inventory control.

These are not necessarily clever or awe-inspiring, but they do convey the essence of the system. And they might be useful later when we need to calibrate the expectations of those involved in its specification. In globally distributed development, in particular, the need for a system metaphor is of paramount importance.

Encounter with a Customer

Suppose your wife (or substitute "husband," "friend," "roommate," or whoever) asks you to go to the store to pick up the following items because she wants to make a cake:

- 5 pounds flour
- 12 large eggs
- 5 pounds sugar
- 1 pound butter

Off you go to the nearest convenience mart (which is close to your home). At the store, you realize that you are not sure if she wants white or brown sugar. So you call her from your cellphone and ask which kind of sugar she wants; you learn she needs brown sugar. You make your purchases and return home.

But your wife is unhappy with your selections. You bought the wrong kind of flour; she informs you that she wanted white and you bought wheat. You bought the wrong kind of butter; she wanted unsalted. You brought the wrong kind of sugar too, dark brown; she wanted light brown. Now you are in trouble.

So you go back to the mart and return the flour, sugar, and butter. You find the white flour and brown sugar, but you could only find the unsalted butter in a tub (not sticks), but you assume a tub is acceptable to her. You make your purchase and return with the items. But now you discover that you made new mistakes. The light brown sugar purchase is fine, but the white flour you brought back is bleached; she wanted unbleached. And the butter in the tub is unacceptable—she points out that unsalted butter can be found in stick form. She is now very angry with you for your ignorance.

So, you go back to the store and sheepishly return the items, and pick up their proper substitutes. To placate your wife's anger, you decide to also buy some of her favorite chocolate candy.

You return home and she is still unhappy. While you finally got the butter, sugar, and flour right, now your wife remembers that she is making omelets for supper and that a dozen eggs won't be enough for the omelets and the cake—she needs 18 eggs. She is also not pleased with the chocolate—she informs you she is on a diet and that she doesn't need the temptation of chocolate lying around.

One more time you visit the mart and return the chocolate and the dozen eggs.

You pick up 18 eggs and return home.

You think you have got the shopping right when she queries: "where did you buy these things?" When you note that you bought the items at the convenience mart, she is livid—she feels the prices there are too high—you should have gone to the supermarket a few miles farther down the road.

We could go on and on with this example—each time your wife discovering a new flaw in your purchases, changing her mind about quantity or brand, adding new items, subtracting others, etc.

But what does this situation have to do with requirements engineering and stakeholders? The situation illustrates many points about requirements engineering. First, you need to understand the application domain. In this case, having a knowledge of baking would have informed you ahead of time that there are different kinds of butter, flour, and sugar and you probably would have asked focusing questions before embarking on your shopping trip. Another point from this scenario—customers don't always know what they want—your wife didn't realize that she needed more eggs until after you made three trips to the store. And there is yet one more lesson in the story: never make assumptions about what customers want—you thought that the tub butter was acceptable; it wasn't. You finally learned that even providing customers with more than they ask for (in this case her favorite chocolate) can sometimes be the wrong thing to do.

But in the larger sense, the most important lesson to be learned from this encounter with a customer is that they can be trouble. They don't always know what they want, and, even when they do, they may communicate their wishes ineffectively. Customers can change their minds, and they may have high expectations about what you know and what you will provide.

Because stakeholder interaction is so important, we are going to devote the rest of this chapter to identifying and understanding the nature of stakeholders, and more especially the stakeholders for whom the system is being built—the customers.

Identifying the System Boundaries

A necessary first step in identifying stakeholders is to create a high-level systems model. This model will identify the set of people and entities involved with the system and define the system boundary—the direct and indirect interactions with other entities (Laplante et al. 2016). Without defining the system boundaries correctly, it is possible that important stakeholders will be overlooked in a potentially disastrous situation. Defining the boundary can be a difficult task. Li et al. (2014) provided a comprehensive discussion on establishing system boundaries.

Context Diagrams

A context diagram is a visual representation diagram for a system of interest showing system boundaries and interactions with other systems, people, and the environment. The context diagram is important for two reasons. First, it provides a basis for dialog about the system for engineers and stakeholders throughout the project life cycle. Second, the context diagram helps combat scope creep, that is, the unchecked growth of functional requirements beyond the intent of the system. The context diagram can also assist in requirements agreement and analysis when discussing ambiguous, complex, and missing requirements.

The context diagram can take many forms. A block diagram may suffice; for example, Figure B.1 is a very simple block diagram depicting the context for the wet-well pumping system. One of the UML/SysML family of diagrams, such as a use-case diagram, can also be used. UML and SysML are discussed in Chapter 4 and Appendices C (UML) and E (use cases, one of the UML models). Rich pictures, from soft systems methodology (SSM) to be described shortly, can also be used for context.

Stakeholders

Stakeholders represent the set of individuals who have some interest (a stake) in the success (or failure) of the system in question. For any system, there are many types of stakeholders, both obvious and sublime. The most obvious stakeholder of a system is the user.

We define the user as the class (consisting of one or more persons) who will use the system. The customer is the class (consisting of one or more persons) who is commissioning the construction of the system. Sometimes the customer is called the client (usually in the case of software systems) or sponsor (in the case where the system is being built not for sale, but internal use). But in many cases, the terms "customer," "client," and "sponsor" are used interchangeably depending on the context. Note that the sponsor and customer can be the same person. And often there is confusion between who the client is and who the sponsor is that can lead to many problems.

In any case, clients, customers, users, and sponsors—however you wish to redefine these terms—are all stakeholders because they have a vested interest in the system. But there are more stakeholders than these. It is said that "the customer is always right, but there are more persons/entities with an interest in the system." In fact, there are many who have a stake in any new system. For example, typical stakeholders for any system might include:

- Customers (clients, users)
- The customers' customers (in the case of a system that will be used by third parties)
- Sponsors (those who have commissioned and/or will pay for the system)
- All responsible engineering and technical persons (e.g., systems, development, test, maintenance)
- Regulators (typically, government agencies at various levels)
- Third parties who have an interest in the system but no direct regulatory authority (e.g., standards organizations, user groups)
- Society (is the system safe?)
- Environment (for physical systems)

And of course, this is an incomplete list. For example, we could go on with a chain of customers' customers' customers…, where the delivered system is augmented by a third party, augmented again, delivered to a fourth party, and so on. This chain of custody has important legal implications too—when a system fails, who is responsible or liable for the failure? In any case, when we use the term "stakeholder," we need to remember that others, not just the customer, are involved.

Negative Stakeholders

Negative stakeholders are those who may be adversely affected by the system. These include competitors, investors, and people whose jobs will be changed, adversely affected, or displaced by the system. There are also internal negative stakeholders—other departments who will take on more workload, jealous rivals, skeptical managers, and more. These internal negative stakeholders can provide passive-aggressive resistance and create political nightmares for all involved. All negative stakeholders have to be recognized and accounted for as much as possible.

Finally, there are always individuals who are not directly affected by systems who are, nonetheless, interested in or opposed to those systems, and because they may wield some power or influence, they must be considered. These interested parties include environmentalists, animal activists, single-issue zealots, advocates of all types, the self-interested, and so on. Some people call these kinds of individuals "gadflies," and they shouldn't be ignored.

Stakeholder Identification

It is very important to accurately and completely identify all possible stakeholders (positive and negative) for any system. Boehm (2003) coined the acronym CRACK to assist in stakeholder identification. That is, a stakeholder is someone who is committed to the project, represents a group of stakeholders, is authorized to make decisions, will collaborate with team members, and is knowledgeable.

Why is stakeholder identification so important? Imagine leaving out a key stakeholder—and discovering them later? Or worse, they discover that a system is being built, in which they have an interest, and they have been ignored. These tardy stakeholders can try to impose all kinds of constraints and requirements changes to the system that can be very costly.

Stakeholder identification is the next step that the requirements engineer must take after the mission statement has been written and the system boundaries identified. In practice, managers are largely responsible for identifying and responding to the interest of stakeholders, and they tend to identify stakeholders from business profitability and operations perspective. But stakeholder identification must also include other perspectives (e.g., moral and social responsibilities). The process of identifying the stakeholders is dynamic in nature. The stakeholder's list should be continuously reassessed within the project life cycle as more stakeholders can be added or removed from the list. A stakeholder who has some social identity can migrate from one economic category to another. In addition, a stakeholder can span several categories (e.g., employees who are also shareholders and/or customers).

Much work has been done in the area of stakeholder theory in terms of determining different types of stakeholders (identification) and managing their interests and responsibilities (analysis). For example, Wang et al. (2015) report on a systemic methodology developed for formally identifying relevant stakeholders throughout the levels of the organization and analyzing their relationship. Crane and Ruebottom (2011) propose an adaption to "stakeholder theory" whereby stakeholders are conceptualized based on their social identity. Gregory et al. (2020) proposed an alternative framework to aid critical reflection in the design and reporting of stakeholder identification and engagement.

Stakeholder Identification Questions

One way to help identify stakeholders is by answering the following set of questions:

- Who is paying for the system?
- Who is going to use the system?
- Who is going to judge the fitness of the system for use?

- What agencies (government) and entities (nongovernment) regulate any aspect of the system?
- What laws govern the construction, deployment, and operation of the system?
- Who is involved in any aspect of the specification, design, construction, testing, maintenance, and retirement of the system?
- Who will be negatively affected if the system is built?
- Who else cares if this system exists or doesn't exist?
- Who have we left out?

Quick Access to stakeholder identification questions
https://phil.laplante.io/requirements/stakeholder/questions.php

Let's try this set of questions on the airline baggage handling system. These answers are not necessarily complete—over time, new stakeholders may be revealed. But by answering these questions as completely as we can now, we reduce the chances of overlooking a very important stakeholder late in the process.

- Who is paying for the system?—airline, grants, passengers, your tax dollars.
- Who is going to use the system?—airline personnel, maintenance personnel, travelers (at the end).
- Who is going to judge the fitness of the system for use?—airline, customers, unions, FAA, OSHA, the press, independent rating agencies.
- What agencies (government) and entities (nongovernment) regulate any aspect of the system?—FAA, OSHA, union contracts, state and local codes.
- What laws govern the construction, deployment, and operation of the system?—various state and local building codes, federal regulations for baggage handling systems, OSHA laws.
- Who is involved in any aspect of the specification, design, construction, testing, maintenance, and retirement of the system?—various engineers, technicians, baggage handlers union, etc.
- Who will be negatively affected if the system is built?—passengers, union personnel.
- Who else cares if this system exists or doesn't exist?—limousine drivers.
- Who have we left out?

And let's try this set of questions on the pet store POS system.

- Who is paying for the system?—pet store, consumers.
- Who is going to use the system?—cashiers, managers, customers (maybe if self-service is provided). Who else?
- Who is going to judge the fitness of the system for use?—company execs, managers, cashiers, customers. Who else?
- What agencies (government) and entities (nongovernment) regulate any aspect of the system?—tax authorities, governing business entities, pet store organizations, better business bureau. Any others?
- What laws govern the construction, deployment, and operation of the system?—tax laws, business, and trade laws. What else?

- Who is involved in any aspect of the specification, design, construction, testing, maintenance, and retirement of the system?—various engineers, CFO, managers, cashiers. We need to know them all.
- Who will be negatively affected if the system is built?—manual cash register makers, inventory clerks. Who else?
- Who else cares if this system exists or doesn't exist?—competitors, vendors of pet products. Who else?
- Who have we left out?

VIGNETTE 2.1   Requirements Engineering in the Movies

Because requirements engineering focuses on both technical aspects of a project and human behavior, valuable lessons, meaning, and purpose can be learned occasionally from observing real life and from film and television. Here is one of our favorite lessons.

In *Harry Potter and the Deathly Hallows: Part 2* (2011), Harry returns to Hogwarts and enlists the help of his friends to find a "horcrux" (cursed object) associated with Rowena Ravenclaw. Here's the key exchange:

Harry: Okay, there's something we need to find, something hidden here in the castle, and it may help us defeat You-Know-Who.
Neville: Right, what is it?
Harry: We don't know.
Dean Thomas: Where is it?
Harry: We don't know that either. I realize that's not much to go on.
Seamus Finnigan: That's nothing to go on.

If only Harry was a Requirements Engineer. Harry's challenge in finding even one horcrux is similar to that of discovering stakeholders and their nature and their desiderata.

And here's a fun exercise—from a favorite movie or television show describe a requirements experience or insight that is depicted.

Watch the *Harry Potter and the Deathly Hallows* clip
https://phil.laplante.io/requirements/harry.php

Rich Pictures

In some systems, the complete set of stakeholders is not easily determined, even with the identifying stakeholder questions. This situation is sometimes the case with novel systems where there is no use history nor impact analysis to inform the identification of stakeholders. In these situations, a more holistic systems-based approach may be appropriate. One such approach is the SSM (Checkland and Scholes 1999).

SSM features a cartoon-like drawing called a "rich picture," which shows various users along with their goals, wants, and needs. Rich pictures resemble annotated use-case diagrams or concept maps, but they are rather informal. Rich pictures can be useful in stakeholder identification.

A first attempt rich picture for the pet store POS system is shown in Figure 2.1. The figure depicts various stakeholders for the pet store and its primary concern. Obvious stakeholders include customers, cashiers, managers, computer maintenance and support people (to upgrade and fix the system), warehouse personnel (to enter inventory data), accountants (to enter and use tax information), and sales clerks (to enter pricing and discounting information). There other many other, less obvious, stakeholders too.

Finally, pets have been included in the rich picture, even though they cannot speak for themselves. Their wants and needs are very important to ensure that the pet store stocks the right products and that all regulatory requirements are met. For example, there are certain guidelines and laws pertaining to the storage of live food for pet snakes and selling animals such as fish and birds. These regulatory requirements are very different from those pertaining to the sale of canned and dry food, grooming supplies, and toys for dogs and cats. Therefore, identifying the set of pets to be served will be an important consideration in the pet store POS system. In general, then, you need to consider voiceless stakeholders when designing any system and you will want to have a mechanism for determining those stakeholders' needs. For example, in the case of pets in the POS system, pet owners or animal welfare advocates could be the proxy for requirements elicitation purposes.

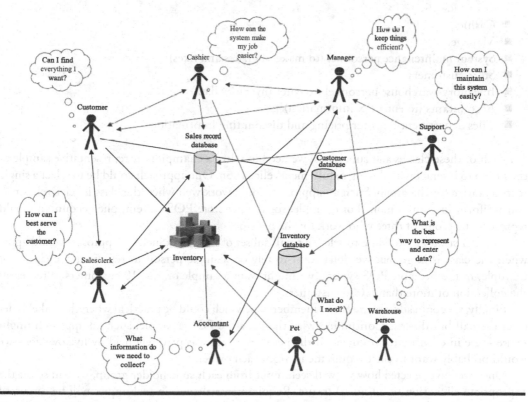

Figure 2.1   Rich picture for pet store POS system.

Developing one or more rich pictures prior to requirements elicitation is an iterative process. With each iteration, missing stakeholders or errors may be identified. This review and revision process should continue until there is agreement that all stakeholders have been correctly identified.

Stakeholder/User Classes

Once the stakeholder (including user) groups have been identified, it may be necessary to divide these groups into classes to adequately address their needs and desires. For example, we already saw in the POS system that the stakeholder "pets" might consist of various kinds of animals with different product needs that could influence system requirements.

A stakeholder/user class subdivision is usually necessary when the classes are large and/or heterogeneous. In many cases, the class subdivision will be needed for the collection of system users. For example, consider the baggage handling system, where users include the following:

■ System maintenance personnel (who will be making upgrades and fixes)
■ Baggage handlers (who interact with the system by turning it on and off, increasing/decreasing speed, and so on)
■ Airline schedulers/dispatchers (who assign flights to baggage claim areas)
■ Airport personnel (who reassign flights to different baggage claim areas)
■ Airport managers and policymakers

For the pet store POS system, user classes would include:

■ Cashiers
■ Managers
■ System maintenance personnel (to make upgrades and fixes)
■ Store customers
■ Inventory/warehouse personnel (to enter inventory data)
■ Accountants (to enter tax information)
■ Sales department (to enter pricing and discounting information)

For each of these classes and subclasses, we need to select a champion or representative sample of the group to interact with during requirements elicitation. One approach could be to select a single representative for the group. Such an approach would work well when the class is relatively small and uniform in composition. For example, for the pet store POS system, one accountant could represent the group of three who work for the pet store chain.

Another strategy would be to select a small subset of the class. Such an approach would apply when the class is large, but we don't want to rely on a single person to represent the class. For example, in the pet store POS system, we might want a sample of, say, 10 customers to represent the collection of more than 5,000 customers.

Finally, we could address the entire membership, which could be practical when the stakeholder class is small but diverse. Continuing with the pet store POS example, such an approach might make sense in eliciting requirements if the pet store chain was privately held by five owners—we would probably want to elicit requirements from each owner.

Once we have selected how we will seek input from each stakeholder group, we can select the appropriate elicitation technique(s) to use. Requirements elicitation techniques will be discussed in Chapter 3.

Stakeholders vs. Use Case Actors

When use cases are used to express requirements, stakeholders that are involved in the use case are referred to as "Actors." Actors are a subcategory of stakeholders. They are external entities that interact with a system in some way. Actors represent a role that can produce and/or consume data, but they do not need to be humans. There are mainly two types of actors that are directly involved with a use case:

- **Primary Actors**: these are the users of the system that will access the system to get a benefit from the services it provides. Primary actors are shown on the left side of the use case diagram.
- **Supporting Actor**: these provide a service (mostly information) to the system. The automated payment authorization service is an example of a supporting actor for an online store system. Supporting actors are shown on the right side of the use case diagram.

On the other hand, stakeholders are entities that care about the project in some way. We can say that actors are always stakeholders, but this relation is not bidirectional (not always a stakeholder is an actor). For example, if the stakeholder has an interest in the system but will not be directly interacting with it, then the stakeholder is not an actor in this case.

A more comprehensive discussion on use cases is provided in Appendix E.

User Characteristics

A requirements engineer should be well aware that individuals, even those within the same user class, may have different characteristics that need to be taken into account. For example, in the baggage handling system, the needs of elderly travelers are different from children or adults. Those with certain disabilities will have different needs. Non-English speakers will have other needs. In many cases, it makes sense to divide user classes into subclasses according to these special needs and characteristics.

Moreover, communicating with individuals or groups within these subclasses probably requires the use of different techniques and most possibly enhanced empathy (e.g., in the case of children or the elderly). Newell et al. (2006) provide a helpful overview of some of these challenges and their solutions.

The differences in cultural characteristics also cannot be ignored when interacting with stakeholders from different countries. Seminal work by sociologist Geert Hofstede (2001) found that there are five dimensions along which cultural differences between countries could be perceived: comfort in dealing with authority (power distance), individualism, the tendency toward a masculine world view (masculinity index), uncertainty avoidance, and long-term orientation. These differences matter to the requirements engineer. For example, an individual from a country with a high power distance may be unwilling to raise an important issue that could embarrass a superior. Or a female individual from a country with a high masculinity index may be afraid to speak up in a meeting when men are present. Therefore, different interaction techniques may be needed in these cases. An entertaining discussion of these issues can be found in Laplante (2010).

Finally, be aware of a conflict of interest or "agency problems" as these can bias and obscure communications with stakeholders. That is, a member of one stakeholder class can be a member of another stakeholder class. For example, a cashier for the pet store POS system may also be a customer. So, in communicating with this individual about how to best conduct a transaction, the requirements engineer cannot know whether the interests of a customer or a cashier are being more favorably represented. The requirements engineer should be aware of these potential biases and control for them in any communications with stakeholders.

Customer Wants and Needs

We mentioned that the primary goal of the requirements engineer is to understand what the customers want. But discovering these wants or desiderata (from the Latin for "desired things") is challenging. You also might think that you, as the requirements engineer, should suggest to the customer what they need, but remember the admonition about substituting your own value system for someone else's—what you think the customer needs might be something that they don't want (remember the encounter with a customer). That being said, it is always helpful to reveal new functionality to the customers so that they can determine if there are features that they need but had not considered. This situation is especially relevant when the customer has less domain knowledge than the requirements engineer. At some point, however, you will need to reconcile wants and needs—requirements prioritization will be very useful then. A very good discussion of stakeholder engagement to identify desiderata is given by Hull et al. (2011).

What Do Customers Want?

The requirements engineer seeks to satisfy customer wants and needs, but it is not always easy to know what these are. Why? Because customers' wants and needs exist on many levels—practical (e.g., minimum functionality of the system), competitive (it should be better than brand X), selfish (they want to show off and tout the system's features), and more. And sometimes the customers want "it all" and they don't want to overpay. Requirements engineers, therefore, have to help customers to set realistic goals for the system to be built.

One way to understand the needs levels of customers is to revisit Maslow's classic hierarchy of self-actualization (Figure 2.2).

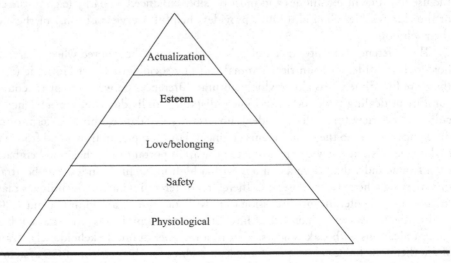

Figure 2.2   Maslow's hierarchy.

Maslow theorized that human beings will seek to satisfy (actualize) their needs starting with the most basic (the bottom of the pyramid) and work their way up to the most elusive, esoteric, and spiritual. But they will never move very far up the pyramid (or stay at a higher level) if lower levels of needs/wants are not satisfied first.

Basic needs include food, water, and sleep. These take precedence over one's physical safety; however—if you were starving, you would risk crossing a very busy street to get food on the other side. People risk jail time by stealing bread. Higher up Maslow's pyramid is the need to be loved and to belong to some group, but he presumes that these needs are fundamentally subordinated to the need for physical safety. You might argue about this one, but the thinking is that some people will sacrifice the chance for love in order to preserve their physical well-being (would you continue belonging to the Sky Diving Club just because you liked one of its members?). Next, one's self-esteem is important, but not as important as the need to belong and be loved (which is why you will humiliate yourself and dress in a Roman costume for your crazy sister-in-law's wedding). Finally, Maslow defined self-actualization as "man's desire for fulfillment, namely to the tendency for him to become actually in what he is potentially: to become everything that one is capable of becoming..." (Maslow 1943).

A variation of Maslow's hierarchy, depicted in Figure 2.3, can help explain the needs and wants of customers.

Here the lowest level is basic functionality. Being a point of sale system implies certain functions must be present—such as create a sale, return an item, update inventory, and so on. At the enabling level, the customer desires features that provide enabling capabilities with respect to other systems (software, hardware, or process) within the organization. So the POS system ties into some management software that allows for sales data to be tracked by managers for forecasting or inventory control purposes. The functionality at the enabling level may not meet or exceed competitors' capabilities. Those functional needs are met at the competitive advantage level. Here the customer wishes for this new system to provide capabilities that exceed those of the competition or otherwise create a business advantage. Finally, groundbreaking desires would imply the development of technology that exceeds current theory or practice and has implications and applications beyond the system in question. For example, some kind of new data-mining technology

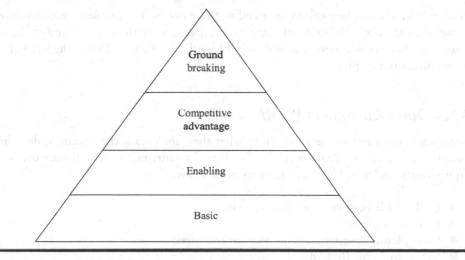

Figure 2.3 Hierarchy of customer needs/wants.

might be desired that exceeds current technologies. As with the Maslow hierarchy, the idea is that the lower-level functionality must not be sacrificed to meet the higher levels of functionality.

While this hierarchy implies four levels of importance of need, it is likely that in any situation there can be more or fewer levels. But the basic idea is to discover and organize customers' needs according to some meaningful hierarchy, which will be most helpful with requirements prioritization later.

This is not the first time that Maslow's theory has been used to illuminate the needs of customers or users. For example, Valacich et al. (2007) introduced a modified four-level Maslow hierarchy to describe user preferences in web-based user interfaces. Hunter (2015) also used a version of the hierarchy to explain requirements types in the Internet of Things (IoT) (see Vignette 2.2).

In any case, let's return to our running examples to consider some of the wants for the baggage handling and pet store POS systems.

For an airline baggage handling system, customers probably want

- Safety
- Speed
- Reliability
- Fault tolerance (no broken luggage!)
- Maintainability
- and so on

For the pet store POS system, customers want

- Speed
- Accuracy
- Clarity (in the printed receipt)
- Efficiency
- Ease of use (especially if self-service provided)
- and more

So we would use our best efforts to attend to these needs. The problem becomes, however, how do we measure the satisfaction of these needs? Because, if these wants and desires cannot be measured, then we will never know if we achieve them. We will discuss the issue of measurable requirements in Chapter 6.

What Don't Customers Want?

Sometimes customers are very explicit in what they don't want the system to do. These specific undesirable features or "do not wants" or "shall not" requirements are frequently overlooked by the requirements engineer. Unwanted features can include

- Undesirable performance characteristics
- Esthetic features
- Gold-plating (excessive and unnecessary features)
- Safety concerns (hazards)

The "shall not" requirements are often the hardest to capture. Sometimes customers don't know what they don't want until they see it. For example, upon seeing the delivered system (or a prototype), they exclaim:

I know I said I wanted it to do that, but I guess I really didn't mean that.

For illustrative purposes, here are some examples of unwanted features of the baggage handling system:

■ The system shall not shut down if main airport power is lost.
■ The system shall not cause a failure in the main airline computer system.
■ The system shall not cause baggage to be destroyed at a rate higher than one bag per minute.

You can see how hard it is to describe what the system is not supposed to do. We will study the issue of unwanted behavior later. In the meantime, to illustrate the point, here are some "shall not" requirements for the pet store POS system:

■ If the register tape runs out, the system shall not crash.
■ If a product code is not found, the system shall not crash.
■ If a problem is found in the inventory reconciliation code, the current transaction shall not be aborted.

These "shall not" requirements can often be rephrased in the positive as "shall" requirements. This topic is discussed more fully in Chapter 4.

Many of the elicitation techniques that we will discuss in the next chapter will tend to uncover unwanted features, but the requirements engineer should always try to discover what the customer does not want as vigorously as what the customer does want.

Why Do Customers Change Their Minds?

One of the greatest challenges in dealing with stakeholders, especially customers, is that they sometimes don't know precisely what they want the system to do. Why is this so? First, no one is omnipotent. A customer can't see every possible desideratum unless they want a complete replica of another system. Helping the customer find these new features as early as possible is the job of the requirements engineer. Another reason why customers change their minds is that, with varying levels of requirements (e.g., features, constraints, business rules, quality attributes), what is important might change as these nonrequirements change during the system life cycle. Sometimes the environment in which the system functions and the customers operate changes (physical changes, economic changes, competition, regulatory changes, etc). For example, while building the pet store POS system, state taxation rules about pet toys might change (maybe only squeaky toys are taxed), and this event might require a significant redesign of the taxation portion of the code. Office politics also play a role in decision-making. When power shifts between stakeholders or in the larger organization, so do project priorities, and changes in requirements often ensue.

Some requirements are so obvious that the customer doesn't think to stipulate them. For example, in what order do you print a list of customers in the pet store POS system? The customer's obvious answer is alphabetically by the last name, but the system designer thought numerically by customer id.

Another reason for changing requirements has to do with the discovery of the return on investment. In one project, for example, it was discovered that what the customer deemed the most important requirement (paperwork reduction) did not add substantial economic value to the business with respect to the cost of implementing the requirement. It turned out that a secondary requirement—the paperwork that should be eliminated had to be nontrivial—was the key economic driver. Prioritization and avoiding late changes are reasons why return on investment data is particularly useful in driving the requirements set.

Sometimes the customer is simply inconsistent or fickle. They change their minds because that is the way they operate. Or they simply don't know what a requirement is (you need to educate them). Other customers are simply, well, stupid. As their requirements engineer, you have to tolerate a certain amount of this changeability (how much is up to you).

Finally, customers will deliberately withhold information for a variety of reasons (e.g., the information is proprietary, they distrust you, they don't like you, they don't think you will understand). The information withheld can rear its ugly head later in the project and require costly changes to the system.

We will discuss how to manage the changes in requirements in Chapter 10.

Stakeholder Prioritization

Until now, we have mostly been referring to the customer as the primary stakeholder, but, of course, there are others. Not all stakeholders are of equal importance. We must prioritize the identified stakeholder roles. For example, the concerns of the Baggage Handlers Union are important but may not be as important as the airport authority (the customer), who is paying for the baggage handling system. On the other hand, federal regulations trump the desires of the customer; for example, the system must comply with all applicable federal standards. Because we have many stakeholders and some of their needs and desires may conflict, we rank or prioritize the stakeholder classes to help resolve these situations.

Glinz and Wieringa (2007) suggest a prioritization scheme based on assessing the risk the system may incur by ignoring or neglecting a stakeholder:

- If neglect might cancel the project or render the system useless, the stakeholder's role is critical.
- If neglect would have a significant negative impact on the system, the stakeholder has a major role.
- If neglect would have a marginal impact on the system, the stakeholder's role is minor.

One may choose another scheme. For example, Table 2.1 contains a partial list of stakeholders for the baggage handling system ranked in a simple high, medium, and low priority scheme. A rationale for the rank assignment is included. Table 2.2 contains a partial list of stakeholders for the pet store POS system ranked using ratings 1–6, where 1 represents the highest importance or priority.

You can certainly argue with this ranking and prioritizations; for example, you may think that the store customer is the most important person in the POS system. But this disagreement highlights an important point—it is early in the requirements engineering process when you want to argue about stakeholder conflicts and prioritization, not later when design decisions may have already been made that need to be undone.

Table 2.1 Partial Ranking of Stakeholders for the Baggage Handling System

Stakeholder Class	Rank	Rationale
System maintenance personnel	Medium	They have a moderate interaction with the system,
Baggage handlers	Medium	They have regular interaction with the system but have an agenda that may run counter to the customer.
Airline schedulers/ dispatchers	Low	They have little interaction with the system.
Airport personnel	Low	Most other airport personnel have little interaction with the system.
Airport managers and policymakers ("the customer")	High	They are paying for the system.

Table 2.2 Partial Ranking of Stakeholders for the Pet Store POS System

Stakeholder Class	Rank	Rationale
Cashiers	2	They have the most interaction with the system.
Managers	1	They are the primary customer/sponsor.
System maintenance personnel	4	They have to fix things when they break.
Store customers	3	They are the most likely to be adversely affected.
Inventory/warehouse personnel	6	They have the least direct interaction with the system.
Accountants/sales personnel	5	They read the reports.

Communicating with Customers and Other Stakeholders

One of the most important activities of the requirements engineer is to communicate with customers and at times with other stakeholders. In many cases, aside from the sales personnel, the requirements engineer is the customer-facing side of the business. Therefore, it is essential that all communications be conducted clearly, ethically, consistently, and in a timely fashion.

The question arises, what is the best format for communications with customers? There are many ways to communicate, and each has specific advantages and disadvantages. For example, in-person meetings are very effective. Verbal information is conveyed via the language used, but also more subtle clues from voice quality, tone, and inflection, and from body language can be conveyed. In fact, agile software methodologies (discussed in Chapter 8) advocate having a customer representative on-site at all times to facilitate continuous in-person communications. But in-person meetings are not economical, and they consume a great deal of time. Furthermore, when you have multiple customers, and geographically distributed customers, how do you meet with them? Damian (2007) provides a comprehensive discussion on lessons learned from practice for dealing with stakeholders in global requirements engineering. This discussion pointed that there are three types of processes that can help stakeholders achieve shared understanding in RE, and these can be challenged in a globally distributed environment:

■ knowledge-acquisition and knowledge-sharing processes that enable the exploration of stakeholders' needs, the application domain, and possible technical solutions;
■ iterative processes that allow the reshaping of this understanding throughout the entire project; and
■ effective communication and coordination processes that support the first two types of processes listed.

Well-planned, intensive group meetings can be an effective form of communication for requirements engineering, and we will be discussing such techniques in the next chapter. But these meetings are expensive and time-consuming and can disrupt the client's business.

Providing periodic status reports to customers during the elicitation and beyond can help to avoid some of these problems. At least from a legal standpoint, the requirements engineer has been making full disclosure of what he knows and what he does not.

Should written communications with the customer take the form of legal contracts and memoranda? The advantage of formal contracts (or change request notices) is that this kind of communication can avoid disputes, or at least provide evidence in the event of a dispute. After all, any communications with the customer can be relevant in court. But formal communications are impersonal, can slow the process of requirements engineering significantly, and can be costly (especially if a lawyer is involved).

Telephone calls and virtual meetings can be used to communicate throughout the requirements engineering process. The informality and speed of this mode are highly desirable. But even with virtual meetings, some of the nuances of co-located communication is lost, and there are always problems of misunderstanding, dropped calls, and interruptions. And the informality of the telephone call is also a liability—every communication with a customer has potential legal implications, but it is usually inconvenient to record every call.

Email can be effective as a means of communication, and its advantages and disadvantages fall somewhere between written memoranda and telephone calls. Email is both spontaneous and informal, but it is persistent—you can save every email transaction. But as with telephone calls and virtual meetings, some of the interpersonal nuanced communication is lost, and as a legal document, email trails are less convincing than formal change request notices.

Wiki technology can be used to communicate requirements information with customers and other stakeholders. The wiki can serve as a kind of whiteboard on which ideas can be shared and refined. Further, with some editing, the wiki can be evolved into the final software requirements specification document. And there are ways to embed executable test cases into the SRS itself using the FitNesse acceptance testing framework. These issues are explored further in Chapter 9.

Finally, there are a number of commercial RE tools that have been useful in supporting distributed team interaction with stakeholders. For example, Rational RequisitePro, Telelogic Doors, and Aha!. We will discuss the tools in RE in Chapter 9.

Managing Expectations

The key to successful communication with customers is managing expectations. Expectations really matter—in all endeavors, not just requirements engineering. If you don't believe this fact, consider the following situations.

Situation A: Imagine you were contracted to do some work as a consultant and you agreed to a fee of $5,000 for the work. You complete the work and the customer is satisfied. But your client pays you $8,000—he has had a successful year and he wants to share the wealth. How do you feel?

Situation B: Now reset the clock—imagine the previous situation didn't happen yet. Imagine now that you agree to do the same work as before, but this time for $10,000. You do exactly the same amount of work as you did in situation A and the customer is satisfied. But now the customer indicates that he had a very bad year and that all he can pay is $8,000, take it or leave it. How do you feel?

In both Situation A and Situation B, you did exactly the same amount of work and you were paid exactly the same amount of money. But you would be ecstatic in Situation A, and upset in Situation B. Why? What is different?

The difference is in your expectations. In Situation A, you expected to be paid $5,000 but the customer surprised you and exceeded your expectations, making you happy. In Situation B, your expectations of receiving $10,000 were not met, making you unhappy.

Some might argue that, in any endeavor, this example illustrates that you should set customers' expectations low deliberately and then exceed them so that you can make the customer extremely happy. But this will not always work, and certainly will not work in the long run—people who get a reputation for padding schedules or otherwise low-balling expectations lose the trust of their customers and clients. Also recognize that the requirements engineer exerts tremendous conscious and unconscious influence on stakeholders. When communicating with customers, a requirements engineer should avoid saying such words as "I would have the system do this …" or "I don't like the system to do that …" These phrases may influence the customer to make a decision that will be regretted later—and potentially blamed on you.

Therefore, our goal as requirements engineers is to carefully manage customers' expectations. That is to understand, adjust, monitor, reset, and then meet customer expectations at all times.

In order to manage customers' expectations, it is important to carefully examine customers' demands too. For example, one of the authors (Kassab) provided consultation to a startup firm that was having difficulty managing requirements. The startup firm was handling change requests (to add, change, or delete new features) from their customer at a very high volume and frequency throughout the life of a project. Upon the author's recommendation, a process to handle these change requests was implemented to manage the customer's expectations.

The process required the customer to document their requests by completing a change request form, similar to the one shown in Figure 2.4 (1) (first form). Upon receiving the change request form, project managers would make an assessment of the impact of the change request. Project managers were required to complete an initial analysis form, similar to the one shown in Figure 2.4 (2) (second form). Once a decision was made, project managers finally would complete a decision form, similar to the one shown in Figure 2.4 (3) (third form), which was communicated back to the customer. If accepted, the request was assigned as a Jira issue with the proper priority.

Quick Access to an editable version of the change requests forms.
https://phil.laplante.io/requirements/change/form.php

46 ■ *Requirements Engineering for Software and Systems*

1.) SUBMITTER – GENERAL INFORMATION				
CR#	[CR001]			
Type of CR	Enhancement	Defect	Other	
Project/Program/Initiative				
Submitter Name	[John Doe]			
Brief Description of Request	[Enter a detailed description of the change being requested]			
Date Submitted	[mm/dd/yyyy]			
Date Required	[mm/dd/yyyy]			
Priority	Low	Medium	High	Mandatory
Reason for Change				
Other Artifacts Impacted				
Assumptions and Notes				
Comments				
Attachments or References	Yes		No	
	Link:			
Approval Signature	[Approval Signature]	Date Signed	[mm/dd/yyyy]	

Figure 2.4 (1) Sample change request forms.

(Continued)

2.) PROJECT MANAGER - INITIAL ANALYSIS			
Hour Impact	*[#hrs]*	*[Enter the hour impact of the requested change]*	
Duration Impact	*[#dys]*	*[Enter the duration impact of the requested change]*	
Schedule Impact	*[WBS]*	*[Detail the impact this change may have on schedules]*	
Cost Impact	*[Cost]*	*[Detail the impact this change may have on cost]*	
Comments	*[Enter additional comments]*		
Recommendations	*[Enter recommendations regarding the requested change]*		
Approval Signature	*[Approval Signature]*	Date Signed	*[mm/dd/xxxx]*

3.) CHANGE CONTROL BOARD – DECISION					
Decision		Approved	Approved with Conditions	Rejected	More Info
Decision Date	*[mm/dd/xxxx]*				
Decision Explanation	*[Document the CCB's decision]*				
Conditions	*[Document and conditions imposed by the CCB]*				
Approval Signature	*[Approval Signature]*		Date Signed		*[mm/dd/xxxx]*

Figure 2.4 (*CONTINUED*) **(2 & 3) Sample change request forms.**

Implementing this change management process helped to build awareness on the customer side of the consequences of their change requests, and hence, their expectations for such requests became noticeably more realistic and reduced in frequency and volume.

Stakeholder Negotiations

It is inevitable that along the way the requirements engineer must negotiate with customers and other stakeholders. Often the negotiations deal with convincing the customer that some desired functionality is impossible or too costly. And expectation setting and management throughout the life cycle of any system project is an exercise in negotiation. While we are not about to embed a crash course in negotiation theory in this book, we wanted to mention a few simple principles that should be remembered.

Set the ground rules upfront. When negotiation is imminent, make sure that the scope and duration of the discussions are agreed upon. If there are to be third parties present, make sure that this is understood. If certain rules need to be followed, make both parties aware. Trying to eliminate unwanted surprises for both sides in the negotiation will lead to success.

Understand people's expectations. Make sure you realize that what matters to you might not matter to the other party. Some people care about money; others care more about their image, reputation, or feelings. When dealing with negotiations surrounding system functionality, understand what is most important to the customer. Ranking requirements will be most helpful in this regard.

Look for early successes. It always helps to build positive momentum if agreement, even on something small, can be reached. Fighting early about the most contentious issues will amplify any bad feelings and make an agreement on those small issues more difficult later.

Be sure to give a little and push back a little. If you give a little in the negotiation, it always demonstrates good faith. But the value of pushing back in a negotiation is somewhat counterintuitive. It turns out that by not pushing back, you leave the other party feeling cheated and empty.

For example, suppose someone advertises a used car for sale at $10,000. You visit the seller, look at the car, and offer $8,000. The seller immediately accepts. How do you feel? You probably feel that the seller accepted too easily and that he had something to hide. Or, you feel that the $10,000 asking price was grossly inflated—why, if you had offered $10,000, the seller would have accepted that—how greedy of him! You would have actually felt better if the seller refused your offer of $8,000 but countered with $9,000 instead. So, push back a little.

Conclude negotiating only when all parties are satisfied. Never end a negotiation with open questions or bad feelings. Everyone needs to feel satisfied and whole at the end of a negotiation. If you do not ensure mutual satisfaction, you will likely not do business together again, and your reputation at large may be damaged (customers talk to one another).

Some of the requirements negotiation theories include: Scrum Win-Win (Khan et al. 2014), TOPSIS (Mairiza et al. 2014), s-CRM (Jeon et al. 2012), and GRNS (Felfernig et al. 2011). Narendhar and Anuradha (2017) provide a good review on which negotiation method in RE can be used under which circumstances. There are many good texts on effective negotiation (e.g., Cohen 2000), and it is advisable that all requirements engineers continuously practice and improve their negotiating skills.

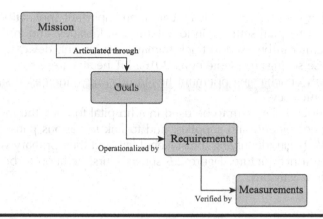

Figure 2.5   Goal-based requirements engineering.

Uncovering Stakeholder Goals

From the encounter with a customer, it should be clear that understanding customer goals is critical to the success of the requirements engineering process. Goals further detail the intentions of the system summarized in the product mission statement.

For example, some goals for the pet store POS:

■ Provide "hassle-free" shopping for all customers
■ Support all coupon and discount processing
■ Support all customer loyalty programs
■ Fully automate inventory entry and maintenance
■ Support all local, state, and federal tax processing

The relationship between the product mission statement, goals, and requirements is depicted in Figure 2.5.

Notice how the goals provide further articulation of the intent contained in the product mission statement. Those goals are operationally described and detailed through the requirements, which must be verified through measurements.

Goal-oriented requirements engineering involves the analysis of stakeholder goals in order to obtain new functional requirements to meet these goals. Existing goal-oriented requirements engineering techniques include the formal methods KAOS, i*, and Tropos (see Chapter 7). These approaches aim at modeling the "who, what, why, where, when, and how" of requirements (Asnar et al. 2011).

Since each stakeholder may have a different set of goals for any system, we must undertake goal understanding after stakeholder identification and prioritization so that we can reconcile differences in goal sets. A simple technique for goal-based understanding using the goal-question-metric paradigm will be discussed in Chapter 3.

VIGNETTE 2.2   Using Rich Pictures to Identify Stakeholders in a Healthcare IoT

The IoT refers to those systems that use the Internet for control, data storage, and analytics. A typical IoT system includes sensors, actuators, and distributed processing power. Healthcare is an important application domain for the IoT in hospital settings, in long-term care facilities, and in the home. These systems can be used to track humans, equipment, devices, and supplies in these settings (Laplante et al. 2016). IoT healthcare systems have the general goals of achieving optimum health outcomes, increased safety, privacy, and efficiency.

Consider an IoT system to be used in a hospital to track the movements of people, equipment, and supplies, and to link to various patient records systems. We begin identifying the stakeholders and their primary wants and needs using a rich picture. Figure 2.6 shows a first iteration to be used for pre-elicitation activities.

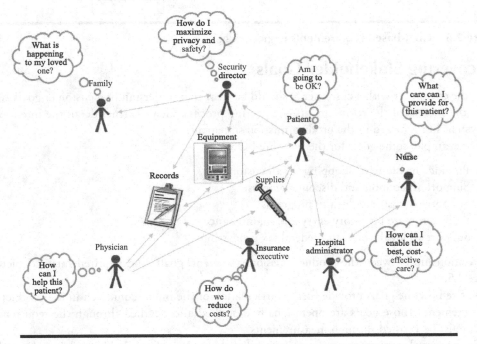

Figure 2.6 First iteration rich picture for a healthcare IoT system. (Adapted from Laplante, N. L., et al., *IEEE Syst. J.,* **2016.)**

We start by noting that hospital administrators are the sponsoring stakeholder, and their primary concern is providing high-quality, cost-effective care. The hospital's security director is concerned with protecting the privacy and safety of the patients, doctors, nurses, staff, and the physical plant. IoT technologies can create new vulnerabilities that must be accounted for in the system requirements.

The patient is a key stakeholder. His or her primary concern is clear if he or she is going to get better. Therefore, IoT requirements would seek to provide significant information and reassurance to the patient. Family members will want to know what is happening to their loved ones, so requirements might focus on status reporting. Physicians and nurses are concerned with helping and caring for the patient.

Insurance company executives are concerned with reducing risk and costs for the system. They will be highly aware of regulatory compliance issues as well as best practices and will seek to ensure that requirements are created along these lines. Other stakeholders will be identified as we refine and improve the rich picture. The primary needs identified for stakeholders will inform the selection and scope of requirements elicitation activities going forward.

Exercises

2.1 Why is it important to have a concept of operation or mission statement at the start of requirements engineering?

2.2 What is the relationship of a system or product mission statement to a company's mission statement?

2.3 When should a domain vocabulary be established?

2.4 At what stage of the requirements development are additions to the requirements considered scope creep?

2.5 Create a rich picture for the
    2.5.1 Baggage handling system
    2.5.2 Smart home system (Appendix A)
    2.5.3 Wet well-pumping control system (Appendix B)

2.6 Using a block diagram or rich picture shows the context for
    2.6.1 The pet store POS system
    2.6.2 The baggage handling system
    2.6.3 The smart home system

2.7 Under what circumstances might the customer's needs and desires be considered secondary?

2.8 Think of an everyday product that you use.
    2.8.1 Try to guess what the mission statement is for the company that manufactures the product.
    2.8.2 Search the Internet for the actual mission statement and compare it to your guess. If it exists, how does the actual mission statement compare to your guess?

2.9 List five goals for the airline baggage handling system.

2.10 List three more stakeholders (other than those already given) and their primary concern for
    2.10.1 The baggage handling system
    2.10.2 The pet store POS system

2.11 Create a comprehensive list of user subclasses for the smart home system described in Appendix A if it were to be built for an arbitrary set of residents and visitors (not just the current user set of two older adults).

2.12 Create a list of subclasses for the following stakeholders:
    2.12.1 Customers for the pet store POS system
    2.12.2 Travelers for the baggage handling system
    2.12.3 Patients in an IoT healthcare system

2.13 For the following, describe whether stakeholder elicitation should use a representative, sample, or exhaustive approach:
    2.13.1 Systems administrators for the pet store POS system
    2.13.2 Maintenance technicians for the baggage handling system
    2.13.3 Homeowners for the smart home system
    2.13.4 Nurses in an IoT healthcare system
2.14 Investigate the KAOS requirements modeling methodology and discuss how goals are modeled in this approach.

Note

1 The original statement used "no man" but was later changed to "no one" to conform to changes in societal norms. This is a brief teaching moment that even mission statements, which are supposed to be immutable, can change.

References

Asnar, Y., Giorgini, P., & Mylopoulos, J. (2011). Goal-driven risk assessment in requirements engineering. *Requirements Engineering*, 16(2): 101–116.

Boehm, B. (2003). *Balancing, Agility and Discipline, A Guide for the Perplexed*. Addison-Wesley, Boston, MA.

Checkland, P., & Scholes, J. (1999). *Soft Systems Methodology in Action*. Wiley, West Sussex.

Cohen, H. (2000). *You Can Negotiate Anything*. Citadel, New York.

Crane, A., & Ruebottom, T. (2011). Stakeholder theory and social identity: Rethinking stakeholder identification. *Journal of Business Ethics*, *102*(1): 77–87.

Damian, D. (2007). Stakeholders in global requirements engineering: Lessons learned from practice. *IEEE Software*, *24*(2): 21–27.

Felfernig, A., Zehentner, C., Ninaus, G., Grabner, H., Maalej, W., Pagano, D., … Reinfrank, F. (2011, July). Group decision support for requirements negotiation. In *International Conference on User Modeling, Adaptation, and Personalization* (pp. 105–116). Springer, Berlin, Heidelberg.

Glinz, M., & Wieringa, R. J. (2007). Guest editors' introduction: Stakeholders in requirements engineering. *IEEE Software*, *24*(2): 18–20.

Gregory, A. J., Atkins, J. P., Midgley, G., & Hodgson, A. M. (2020). Stakeholder identification and engagement in problem structuring interventions. *European Journal of Operational Research*, *283*(1): 321–340.

Hofstede, G. (2001). *Culture's Consequences: Comparing Values, Behaviors, Institutions and Organizations across Nations*. Sage, Thousand Oaks, CA.

Hull, E., Jackson, K., & Dick, J. (2011). *Requirements Engineering in the Problem Domain, Requirements Engineering*, 2nd edition. Springer-Verlag, London, pp. 93–114.

Hunter, J. (2015). The hierarchy of IoT "Thing" needs. *Techcrunch*, 15 September, 2015. https://techcrunch.com/2015/09/05/the-hierarchy-of-iot-thing-needs/ (accessed November 2016).

Jeon, C. K., Kim, N. H., Lee, D. H., Lee, T., & In, H. P. (2012). Stakeholder Conflict Resolution Model (S-CRM) based on supervised learning. *KSII Transactions on Internet and Information Systems (TIIS)*, *6*(11): 2813–2826.

Khan, U. Z., Wahab, F., & Saeed, S. (2014). Integration of scrum with win-win requirements negotiation model. *Middle-East Journal of Scientific Research*, 19(1): 101–104.

Laplante, N. L., Laplante, P. A., & Voas, J. M. (2016). Stakeholder identification and use case representation for Internet-of-things applications in healthcare. *IEEE Systems Journal*. DOI: 10.1109/JSYST.2016.2558449.

Laplante, P. A. (2010). Where in the world is Carmen Sandiego (and is she a software engineer)? *IT Professional*, 12(6): 10–13.

Li, T., Zhang, H., Liu, Z., Ke, Q., & Alting, L. (2014). A system boundary identification method for life cycle assessment. *The International Journal of Life Cycle Assessment*, 19(3): 646–660.

Mairiza, D., Zowghi, D., & Gervasi, V. (2014). Utilizing topsis: A multi criteria decision analysis technique for non-functional requirements conflicts. In *Requirements Engineering* (pp. 31–44). Springer, Berlin, Heidelberg.

Maslow, A. (1943). A theory of human motivation. *Psychological Review*, 50: 370–396.

Newell, A. F., Dickinson, A., Smith, M. J., & Gregor, P. (2006). Designing a portal for older users: A case study of an industrial/academic collaboration. *ACM Transactions on Computer-Human Interaction*, 13(3): 347–375.

Narendhar, M., & Anuradha, K. (2017). Requirement negotiation methods in requirements engineering. *International Journal of Advanced Research in Computer Science*, 8(3): 1–6.

Valacich, J. H., Parboteeah, D. V., & Wells, J. D. (2007). The online consumer's hierarchy of needs. *Communications of the ACM*, 50(9): 84–90.

Wang, W., Liu, W., & Mingers, J. (2015). A systemic method for organizational stakeholder identification and analysis using Soft Systems Methodology (SSM). *European Journal of Operational Research*, 246(-2): 562–574.

Chapter 3

Requirements Elicitation

Introduction

In this chapter, we explore the many ways that requirements can be found, discovered, captured, or coerced. In this context, all of these terms are synonymous with elicitation. But "gathering" is not quite an equivalent term. Requirements are not like fallen fruit to be simply retrieved and placed in a bushel. Requirements are usually not so easy to come by, at least not all of them. Many of the more subtle and complex ones have to be teased out through rigorous, if not dogged processes. The following are common obstacles in the requirements elicitation process:

- **New Project Domain**: when requirements engineer doesn't possess enough knowledge on the industry or the developed solution. Engaging a domain expert can help alleviate this problem.
- **Unclear Project Vision**: when stakeholders don't have a clear understanding of what functionality their system needs. A clear mission statement or ConOps can be very helpful to avoid this problem.
- **Limited Access to Documentation**: when the requirements engineer can't access documentation or when evaluating the current state of the project takes too much time. Consistent and disciplined use of a good document/change management system is important to combat this problem.
- **Focus on the Solution Instead of Requirements**: when the customer focuses more on solutions or architectural tactics instead of requirements themselves. Active expectation management is very important in this regard.
- **Fixation of Specific Functionalities**: when stakeholders insist on designing certain features because they believe they will benefit their business if it is not. Again, expectation management and refocus on the mission statement is the key to addressing this issue.
- **Contradictory Requirements**: when a project includes a wide range of stakeholders, their requirements may contradict each other. We will discuss techniques for addressing this problem in a later chapter.

In addition to the specific suggestions above, there are many general techniques that you can choose to conduct requirements elicitation to overcome the above challenges, and you will probably need

DOI: 10.1201/9781003129509-3

to use more than one and likely different ones for different classes of users/stakeholders. The techniques that we will discuss are as follows:

- Brainstorming
- Card sorting
- Crowdsourcing
- Designer as apprentice
- Domain analysis
- Ethnographic observation
- Goal-based approaches
- Group work
- Interviews
- Introspection
- Joint application development (JAD)
- Laddering
- Protocol analysis
- Prototyping
- Quality function deployment (QFD)
- Questionnaires
- Repertory grids
- Reverse engineering
- Scenarios
- Task analysis
- Use cases
- User stories
- Viewpoints
- Workshops

This list is partially adapted from one suggested by Zowghi and Coulin (1998).

Quick Access to a summary of the elicitation techniques
https://phil.laplante.io/requirements/elicit/summary.php

Requirements Elicitation - First Step

Identifying all customers and stakeholders is the first step in preparing for requirements elicitation. But stakeholder groups, and especially customers, can be nonhomogeneous, and therefore you need to treat each subgroup differently. For example, the different subclasses of users for the pet store point of sale (POS) system include:

- Cashiers
- Managers

- System maintenance personnel
- Store customers
- Inventory/warehouse personnel
- Accountants (to enter tax information)
- Sales department (to enter pricing and discounting information)

Each of these subgroups of users has different desiderata and these need to be determined.
The process, then, to prepare for elicitation is as follows:

- Identify all customers and stakeholders.
- Partition customers and other stakeholders groups into classes according to interests, scope, authorization, or other discriminating factors (some classes may need multiple levels of partitioning).
- Select a champion or representative group for each user class and stakeholder group.
- Select the appropriate technique(s) to solicit initial inputs from each class or stakeholder group.

Here is another example of user class partitioning. There are many different stakeholders for the baggage handling system including:

- Travelers
- System maintenance personnel
- Baggage handlers
- Airline schedulers/dispatchers
- Airport personnel
- Airport managers and policymakers

But there are various kinds of travelers each with different needs. For example, consider the following subclasses:

- Children
- Senior citizens
- Business people
- Casual travelers
- Military personnel
- Civilians
- Casual travelers
- Frequent flyers

Each of these subclasses may need to be approached with different elicitation techniques. For example, surveys may not be appropriate for children, while focus groups may be less useful for military personnel. Many of these subclasses overlap, for example, a person can be a business traveler and also a casual traveler, and these overlaps need to be taken into consideration when analyzing the data from the elicitation activities.

Elicitation Techniques Survey

Now it is time to begin examining the elicitation techniques. We offer these techniques in alphabetical order—no preference is implied. At the end of the chapter, we will discuss the prevalence and suitability of these techniques in different situations.

Brainstorming

Brainstorming consists of informal sessions with customers and other stakeholders to generate overarching goals for the systems. Brainstorming can be formalized to include a set agenda, minute taking, and the use of formal structures (e.g., Robert's Rules of Order). But the formality of a brainstorming meeting is probably inversely proportional to the creative level exhibited at the meeting. These kinds of meetings probably should be informal, even spontaneous, with the only structure embodying some recording of any major discoveries.

During brainstorming sessions, some preliminary requirements may be generated, but this aspect is secondary to the process. The JAD technique incorporates brainstorming (and a whole lot more), and it is likely that most other group-oriented elicitation techniques embody some form of brainstorming implicitly. Brainstorming is also useful for general objective setting, such as mission or vision statement generation. Once brainstorming is selected as a technique, the following guidelines are recommended to get the most out of your brainstorming session:

- Select a facilitator, ideas recorder, and the participants, and reserve the proper place and time for the session.
- Make sure everyone is on the same page regarding the process.
- Give each participant a small amount of time to brainstorm on their own before bringing their ideas to the group.
- Once the brainstorming session has started, keep everyone on topic.
- Do not limit creativity, free association, or the number of ideas.
- If the session is long, build in some coffee break times.
- Write all ideas down in plain view of the entire group.

Once the brainstorming session is over, begin the refining process utilizing other elicitation techniques.

Card Sorting

This technique involves having stakeholders complete a set of cards that includes key information about functionality for the system/software product. It is also a good idea for the stakeholders/customers to include a ranking and rationale for each of the functionalities.

The time period to allow customers and stakeholders to complete the cards is an important decision. While the exercise of card sorting can be completed in a few hours, rushing the stakeholders will likely lead to important, missing functionalities. Giving stakeholders too much time, however, can slow the process unnecessarily. It is recommended that a minimum of 1 week (and no more than 2 weeks) be allowed for the completion of the cards. Another alternative is to have the customers complete the cards in a 2-hour session and then return 1 week later for another session of card completion and review.

In any case, after each session of card generation, the requirements engineer organizes these cards in some manner, generally clustering the functionalities logically. These clusters form the basis of the requirements set. The sorted cards can also be used as an input to the process to develop CRC (capability, responsibility, collaboration) cards to determine program classes in the eventual code. Another technique to be discussed shortly, QFD, includes a card-sorting activity.

To illustrate the process, Figure 3.1 depicts a tiny subset of cards generated by the customer for the pet store POS system, lying in an unsorted pile. In this case, each card contains only a

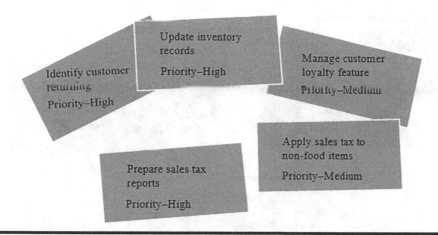

Figure 3.1 A tiny subset of the unsorted cards generated by customers for the pet store POS system.

brief description of the functionality and a priority rating is included (for brevity, no rationale is shown).

The requirements engineer analyzes this pile of cards and decides that two of the cards pertain to "customer management" functions, two cards to "tax functions," and one card to "inventory features" or functions, and arranges the cards in appropriate piles as shown in Figure 3.2.

The customer can be shown this sorted list of functionalities for correction or missing features. Then, a new round of cards can be generated if necessary. The process continues until the requirements engineer and customer are satisfied that the system features are substantially captured.

Crowdsourcing

Crowdsourcing is a business model that harnesses the power of a large and diverse number of people to contribute knowledge and solve problems. The term "crowdsourcing" is a combination of crowd and outsourcing and was coined in 2006 by Wired magazine author Jeff Howe in his article "The Rise of Crowdsourcing" (Howe 2006).

Capturing the requirements directly from the crowd (the bigger the crowd, the better) gives the requirements engineers access to a wide diversity of actual and potential users which has the potential to increase the comprehensiveness and quality of the captured requirements. Crowd-based requirements engineering as a term was coined by Groen et al. (2015) to be a requirements engineering approach for acquiring and analyzing any kind of users' feedback from the crowd, with the aim of seeking validated user requirements. It was further elaborated by Groen and Koch (2016) be "the combined set of techniques for analyzing data from the crowd using text and usage mining, motivational techniques for stimulating the further generation of data, and crowdsourcing to validate requirements."

Crowdsourcing in requirements elicitation has the potential to increase the comprehensiveness and quality of the captured requirements by giving the requirements engineers access to a wide diversity of actual and potential users. The captured requirements from the crowd (the bigger the crowd, the better) will almost always be superior to products developed only with input from experts. But who is "the crowd" for your system? It could be system users in the company, it could be the sales representatives, or it could be millions of customers worldwide, it all depends on the system that is being developed.

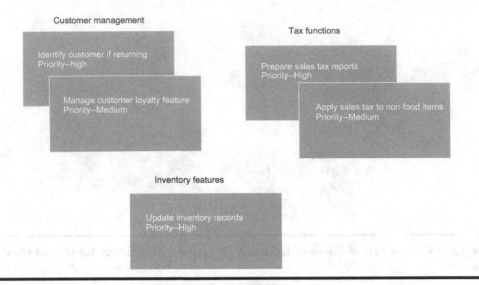

Figure 3.2 **Sorted cards for the pet store POS system.**

Social media (e.g., LinkedIn, Facebook, Twitter) provide the best-developed mechanisms to efficiently get feedback from crowds. The more efficient the communication mechanism to the crowd, the more people can be included in the process. A typical process to capture the requirements via crowdsourcing includes the following steps:

1. Choosing the potential crowd to be targeted.
2. Choosing the proper social media tool that is both popular with that crowd and that will allow the requirements engineer to ask questions of the crowd.
3. Creating a community with the crowd (perhaps by sharing useful information and news that's relevant to the crowd).
4. Once the "crowdsourced elicitation process" is set up, it comes down to asking the right questions when opportunities arise. Asking questions to the crowd will be only efficient if the right questions are asked. That's where expertise comes in, along with other requirements elicitation discussed in this chapter. Since most people don't know what they want, it is the requirements engineer's job to present the right questions in an effective manner in order to trigger useful insights from the crowd. The same types of questions that don't work in one-on-one elicitation will not work in crowd elicitation as well (see the questionnaires technique).

The elicitation process via crowdsourcing is often carried on with the support of tools. Some developed crowd-based tools for requirements elicitation include CrowdREquire (Adepetu et al. 2012), Refine (Snijders et al. 2015), StakeRare (Lim and Finkelstein 2011), and Requirements Bazaar (Renzel and Klamma 2014).

Groen et al. (2015) discussed the concept of crowd-based requirements engineering and its landscape and challenges to emphasize its use. The highlighted challenges are as follows:

■ Challenges related to largeness

- Challenges related to diversity
- Challenges related to anonymity
- Challenges related to competence
- Challenges related to collaboration
- Challenges related to intrinsic motivations
- Challenges related to volunteering
- Challenges related to extrinsic incentives
- Challenges related to opt-out opportunity
- Challenges related to feedback

In most cases, crowdsourcing every aspect of the system is not necessary or beneficial, but in almost all cases the feedback from the crowd on a myriad of questions about the requirements can be of great benefit.

Designer as Apprentice[1]

Designer as apprentice is a requirements discovery technique in which the requirements engineer "looks over the shoulder" of the customer in order to learn enough about the customer's work to understand their needs. The relationship between customer and designer is like that between a master craftsman and apprentice. That is, the apprentice learns a skill from the master just as we want the requirements engineer (the designer) to learn about the customer's work from the customer. The apprentice is there to learn whatever the master knows (and therefore must guide the customer in talking about and demonstrating those parts of the work). The designer is there to address specific needs.

It might seem that the customer needs to have some kind of teaching ability for this technique to work, but that is not true. Some customers cannot talk about their work effectively but can talk about it as it unfolds. Moreover, customers don't have to work out the best way to present it, or the motives; they just explain what they're doing.

Seeing the work also reveals what matters. For example, people are not aware of everything they do and sometimes why they do it. Some actions are the result of years of experience and are too subtle to express. Think about an expert system that automates cake decoration based on customer preferences. Other actions are just habits with no valid justification. The presence of an apprentice provides the opportunity for the master (customer) to think about the activities and how they came about.

Seeing the work reveals details since, unless we are performing a task, it is difficult to be detailed in describing it. Finally, seeing the work reveals structure. Patterns of working are not always obvious to the worker. An apprentice learns the strategies and techniques of work by observing multiple instances of a task and forming an understanding of how to do it themselves, incorporating the variations.

In order for this technique to work, the requirements engineer must understand the structure and implication of the work, including:

- The strategy to get work done
- Constraints that get in the way
- The structure of the physical environment as it supports work
- The way work is divided

- Recurring patterns of activity
- The implications these have on any potential system

The designer must demonstrate an understanding of the work to the customer so that any misunderstandings can be corrected.

Finally, using the designer as apprentice approach provides other project benefits beyond requirements discovery. For example, using this technique can help improve the process that is being modeled.

Both customer and designer learn during this process—the customer learns what may be possible and the designer expands their understanding of the work. If the designer has an idea for improving the process, however, this must be fed back to the customer immediately (at the time).

Domain Analysis

We have already emphasized the importance of having domain knowledge (whether it is had by the requirements engineer and/or the customer) in requirements engineering. Domain analysis involves any general approach to assessing the "landscape" of related and competing applications to the system being designed. Such an approach can be useful in identifying essential functionality and, later, missing functionality. Domain analysis can also be used later for identifying reusable components (such as open-source software elements that can be incorporated into the final design). The QFD elicitation approach explicitly incorporates domain analysis, and we will discuss this technique shortly.

Ethnographic Observation

Ethnographic observation refers to any technique in which observation of indirect and direct factors inform the work of the requirements engineer. Ethnographic observation is a technique borrowed from social science in which observations of human activity and the environment in which the work occurs are used to inform the scientist in the study of some phenomenon. In the strictest sense, ethnographic observation involves long periods of observation (hence, an objection to its use as a requirements elicitation technique).

To illustrate ethnographic observation, imagine the societal immersion of an anthropologist studying different cultures. The anthropologist lives among the culture being studied, but in a way that is minimally intrusive. While eating, sleeping, hunting, celebrating, mourning, and so on within the culture, all kinds of direct and indirect evidence of how that society functions and its belief systems are collected.

As another example, imagine that you are leading the requirements elicitation activities for a new home security system that includes alarms, window and door sensors, cameras, and motion sensors. The system interacts with the security monitoring company via a dedicated telephone line and with the home residents via one or more keypads (as well as through the sensors). You may choose ethnographic observation to capture the requirements related to how children (age 5–12) will interact directly or indirectly with the system.

In applying ethnographic observation to requirements elicitation, the requirements engineer immerses himself in the workplace culture of the customer. Here, in addition to observing work or activity to be automated, the requirements engineer is also in a position to collect evidence of customer needs derived from the surroundings that may not be communicated directly. Designer as apprentice is one requirements elicitation technique that includes the activity of ethnographic observation.

To illustrate this technique in practice, consider this situation in which ethnographic observation occurs:

- You are gathering requirements for a smart home for a customer.
- You spend long periods of time interviewing the customer about what they want.
- You spend time interacting with the customer as they go about their day and ask questions ("why are you running the dishwasher at night, why not in the morning?").
- You spend long periods of time passively observing the customer "in action" in the current home to get nonverbal clues about wants and desires.
- You gain other information from the home itself—the books on the bookshelf, paintings on the wall, furniture styles, evidence of hobbies, signs of wear and tear on various appliances, etc.

Ethnographic observation can be very time-consuming and requires substantial training of the observer. There is another objection based on the intrusiveness of the process. There is a well-known principle in physics known as the Heisenberg uncertainty principle, which, in layperson's terms, means that you can't precisely measure something without affecting that which you are measuring. So, for example, when you are observing the work environment for a client, processes and behaviors change because everyone is out to impress—so an incorrect picture of the situation is formed, leading to flawed decisions.

Goal-Based Approaches

Goal-based approaches comprise any elicitation techniques in which requirements are recognized to emanate from the mission statement, through a set of goals that lead to requirements. That is, looking at the mission statement, a set of goals that fulfill that mission is generated. These goals may be subdivided one or more times to obtain lower-level goals. Then, the lower-level goals are branched out into specific high-level requirements. Finally, the high-level requirements are used to generate lower level ones.

For example, consider the baggage handling system mission statement:

To automate all aspects of baggage handling from passenger origin to destination.

The following goals might be considered to fulfill this mission:

- **Goal 1**: To completely automate the tracking of baggage from check-in to pick-up.
- **Goal 2**: To completely automate the routing of baggage from check-in counter to plane.
- **Goal 3**: To reduce the amount of lost luggage to 1%.

These goals can then be decomposed into requirements using a structured approach such as goal-question-metric (GQM). GQM is an important technique used in many aspects of systems engineering such as requirements engineering, architectural design, systems design, and project management. GQM incorporates three steps: state the system's objectives or goals; derive from each goal the questions that must be answered to determine if the goal is being met; and decide what must be measured in order to be able to answer the questions (Basili and Weiss 1984).

For example, in the case of the baggage handling system, consider goal 3. Here the related question is "what percentage of luggage is lost for a given (airport/airline/ flight/time period/etc.)?" This question suggests a requirement of the form:

The percentage of luggage lost for a given [airport/airline/flight/time period/etc.] shall be not greater than 1%.

The associated metric for this requirement, then, is simply the percentage of luggage lost for a particular (airport/airline/flight/time period/etc.). Of course, we really need a definition for lost luggage, since so-called lost luggage often reappears days or even months after it is declared lost. Also, reasonable assumptions need to be made in framing this requirement in terms of an airport's reported luggage losses over some time period, or for a particular airline at some terminal, and so forth.

In any case, we deliberately picked a simple example here—the appropriate question for some goal (requirement) is not always so obvious, nor is the associated metric so easily derived from the question. Here is where GQM really shows its strength.

For example, it is common to see requirements that are a variation of the following:

The system shall be user friendly

The problem with such a requirement is that there is no way to demonstrate its satisfaction—any person on the acceptance testing team can declare that the system is not user-friendly. But user-friendliness is a reasonable goal for the system. So, following GQM, we create a series of questions that pertain to the goal of user-friendliness, for example:

1. How easy is the system to learn to use?
2. How much help does a new user need?
3. How many errors does a user get?

Next, we define one or more metrics for each of these questions. Let's generate one for each.

1. The time it takes a user to learn how to perform certain functions
2. The number of times a user has to use the help feature over some period of time
3. The number of times a user sees an error message during certain operations over a period of time.

Finally, we work with the customer to set acceptable ranges for these metrics. After the system is built, requirements satisfaction can be demonstrated through testing trials with real users. The parameters of this testing, such as the characteristics and number of users, duration of testing, and so on, can be defined later and incorporated into the system test plan. In this way, we can define an acceptable level of user-friendliness.

Group Work

Group work is a general term for any kind of group meetings that are used during the requirements discovery, analysis, and follow-up processes. The most celebrated group-oriented work for requirements elicitation is JAD, which we will discuss shortly.

Group activities can be very productive in terms of bringing together many stakeholders but risk the potential for conflict and divisiveness. The key to success in any kind of group work is in the planning and execution of the group meetings. Here are the most important things to remember about group meetings.

- Do your homework—research all aspects of the organization, problems, politics, environment, and so on.
- Publish an agenda (with time allotted for each item) several days before the meeting occurs.
- Stay on the agenda throughout the meeting (no meeting scope creep).
- Have a dedicated note-taker (scribe) on hand.
- Do not allow personal issues to creep in.
- Allow all to have their voices heard.
- Look for consensus at the earliest opportunity.
- Do not leave until all items on the agenda have received sufficient discussion.
- Publish the minutes of the meeting within a couple of days of meeting close and allow attendees to suggest changes.

These principles will come into play for the JAD approach to requirements elicitation. Group work of any kind has many drawbacks. First, group meetings can be difficult to organize and get the many stakeholders involved to focus on issues. Problems of openness and candor can occur as well because people are not always eager to express their true feelings in a public forum. Because everyone has a different personality, certain individuals can dominate the meeting (and these may not be the most "important" individuals). Allowing a few to own the meeting can lead to feelings of being "left out" for many of the other attendees.

Running effective meetings, and hence using group work, requires highly developed leadership, organizational, and interpersonal skills. Therefore, the requirements engineer should seek to develop these skills whenever possible.

Interviews

Elicitation through interviews involves in-person communication between two individual stakeholders or a small group of stakeholders (sometimes called a focus group). Interviews are an easy-to-use technique to extract system-level requirements from stakeholders, especially usability requirements.

Three kinds of interviews can be used in elicitation activities, and they can be applied to individuals or focus groups:

- Unstructured
- Structured
- Semi-structured

Unstructured interviews, which are probably the most common type, are conversational in nature and serve to relax the participants. Like a spontaneous "confession," these can occur any time and any place whenever the requirements engineer and stakeholder are together, and the opportunity to capture information this way should never be lost. But depending on the skill of the interviewer, unstructured interviews can be hit or miss. Therefore, structured or semi-structured interviews are preferred.

Structured interviews are much more formal in nature, and they use predefined questions that have been rigorously planned. Templates are very helpful when interviewing using the structured style. The main drawback to structured interviews is that some customers may withhold information because the format is too controlled.

Semi-structured interviews combine the best of structured and unstructured interviews. That is, the requirements engineer prepares a carefully thought-out list of questions, but then allows for spontaneous unstructured questions to creep in during the course of the interview.

While structured interviews are preferred, the choice of which one to use is very much an opportunistic decision. For example, when the client's corporate culture is very informal and relaxed, and trust is high, then unstructured interviews might be preferred. In a stodgier, process-oriented organization, structured and semi-structured interviews are probably more desirable.

Hickey and Davis (2003) conducted in-depth interviews with some of the world's most experienced analysts in requirements elicitations. Their findings on using the "interviews" technique in practice revealed that experts use it in the following situations:

■ Gathering initial background information when working on new projects in new domains.
■ Whenever heavy politics are present to ensure that the group session does not self-destruct.
■ When there is a need to isolate and show conflicts among stakeholders.
■ When senior management has an idea, but the employees consider this idea to be unreasonable, the problem can be addressed by interviewing subject matter experts and visionaries.
■ Using it with subject matter experts is essential when the users and customers are inaccessible.

Here are some sample interview questions that can be used in any of the three interview types.

■ Name an essential feature of the system.
■ Why is this feature important?
■ On a scale of one to five, five being most important, how would you rate this feature?
■ How important is this feature with respect to other features?
■ What other features are dependent on this feature?
■ What other features must be independent of this feature?
■ What other observations can you make about this feature?

Whatever interview technique is used, care must be taken to ensure that all of the right questions are asked. That is, leave out no important questions, and include no extraneous, offensive, or redundant questions. When absolutely necessary, interviews can be done via telephone, videoconference, or email, but be aware that in these modes of communication, certain important nuanced aspects to the responses may be lost.

Introspection

When a requirements engineer develops requirements based on what he thinks the customer wants, then he is conducting the process of introspection. In essence, the requirements engineer puts himself in the place of the customer and opines "if I were the customer I would want the system to do this ..."

An introspective approach is useful when the requirements engineer's domain knowledge far exceeds that of the customer. Occasionally, the customer will ask the engineer questions similar to the following: "if you were me, what would you want?" While introspection will inform every aspect of the requirements engineer's interactions, remember our admonition about not telling a customer what he ought to want. Introspection is also a great way to gather requirements when users are too busy to be involved in interviews, group sessions, or questionnaires. Introspection can be used to assess political and power relationships when working in new organizations.

Joint Application Design

JAD involves highly structured group meetings (sometimes called "mini-retreats") with system users, system owners, and analysts focused on a specific set of problems for an extended period of time. These meetings occur 4–8 hours per day and over a period lasting 1 day to a couple of weeks. JAD has even been adapted for multisite implementation when participants are not colocated (Cleland-Huang and Laurent 2014). While traditionally associated with large, government systems projects, the technique can be used in industrial settings on systems of all sizes.

JAD and JAD-like techniques are commonly used in systems planning and systems analysis activities to obtain group consensus on problems, objectives, and requirements. Specifically, the requirements engineer can use JAD sessions for the concept of operation definition, system goal definition, requirements elicitation, requirements analysis, requirements document review, and more.

Planning for a JAD review or audit session involves three steps:

1. Selecting participants
2. Preparing the agenda
3. Selecting a location

Great care must be taken in preparing each of these steps.

Reviews and audits may include some or all of the following participants:

■ Sponsors (e.g., senior management)
■ A team leader (facilitator, independent)
■ Users and managers who have ownership of requirements and business rules
■ Scribes (i.e., meeting minutes and note-takers)
■ Engineering staff

The sponsor, analysts, and managers select a leader. The leader may be in-house or a consultant. One or more scribes (note-takers) are selected, normally from the software development team. The analyst and managers must select individuals from the user community. These individuals should be knowledgeable and articulate in their business area.

Before planning a session, the analyst and sponsor must determine the scope of the project and set the high-level requirements and expectations of each session. The session leader must also ensure that the sponsor is willing to commit people, time, and other resources to the effort. The agenda depends greatly on the type of review to be conducted and should be constructed to allow for sufficient time. The agenda, code, and documentation must also be sent to all participants well in advance of the meeting so that they have sufficient time to review them, make comments, and prepare to ask questions.

The following are some rules for conducting software requirements, design audits, or code walkthroughs. The session leader must make every effort to ensure these practices are implemented.

■ Stick to the agenda.
■ Stay on schedule (agenda topics are allotted specific time).

- Ensure that the scribe is able to take notes.
- Avoid technical jargon (if the review involves nontechnical personnel).
- Resolve conflicts (try not to defer them).
- Encourage group consensus.
- Encourage user and management participation without allowing individuals to dominate the session.
- Keep the meeting impersonal.
- Allow the meetings to take as long as necessary.

The end product of any review session is typically a formal written document providing a summary of the items (specifications, design changes, code changes, and action items) agreed upon during the session. The content and organization of the document obviously depend on the nature and objectives of the session. In the case of requirements elicitation, however, the main artifact could be a first draft of the SRS.

Laddering

In laddering, the requirements engineer asks the customer short prompting questions (probes) to elicit requirements. Follow-up questions are then posed to dig deeper below the surface. The resultant information from the responses is then organized into a tree-like structure.

To illustrate the technique, consider the following sequence of laddering questions and responses for the pet store POS system. "RE" refers to the requirements engineer.

RE: Name a key feature of the system.
Customer: Customer identification.
RE: How do you identify a customer?
Customer: They can swipe their loyalty card.
RE: What if a customer forgets their card?
Customer: They can be looked up by phone number.
RE: When do you get the customer's phone number?
Customer: When customers complete the application for the loyalty card.
RE: How do customers complete the applications? ...
    And so on.

Figure 3.3 shows how the responses to the questions are then organized in a ladder or hierarchical diagram.

The laddering technique assumes that information can be arranged in a hierarchical fashion, or, at least, it causes the information to be arranged hierarchically.

Protocol Analysis

A protocol analysis is a process where customers, together with the requirements engineers, walk through the procedures that they are going to automate. During such a walkthrough, the customers explicitly state the rationale for each step that is being taken.

For example, for the major package delivery company discussed in the section of domain vocabulary understanding in Chapter 1, it was the practice to have engineers and other support professionals ride with the regular delivery personnel during the busy winter holiday season. This

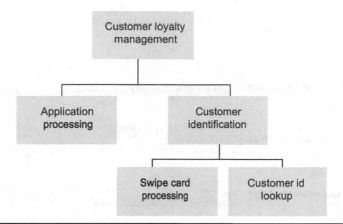

Figure 3.3   Laddering diagram for the pet store POS system.

practice not only addressed the surge in packages to be delivered but also reacquainted the engineers with the processes and procedures associated with the company's services as they were actually applied. Observations made by the engineers in the field often led to processes optimization and other innovations.

While you will see shortly that protocol analysis is very similar to designer as apprentice, there are subtle differences. These differences lie in the role of the requirements engineer who is more passive in protocol analysis than in designer as apprentice.

Prototyping

Prototyping involves the construction of models of the system in order to discover new features, particularly usability requirements. Prototyping is a particularly important technique for requirements elicitation. It is used extensively, for example, in the spiral software development model, and agile methodologies consist essentially of a series of increasingly functional non-throwaway prototypes.

Suppose a company is developing a new online commerce website for their products. A new customer support application is proposed to allow customers to view the status of their orders online. The company needs to determine the best "look and feel" of the user interface. Developing different prototypes of the user interface and showing it to users followed by interviewing them about these interfaces can be the appropriate approach to follow.

Prototypes can involve working models and non-working models. Working models can include executable code in the case of software systems and simulations, or temporary or to-scale prototypes for non-software systems. Non-working models can include storyboards and mock-ups of user interfaces. Building architects use prototypes regularly (e.g., scale drawings, cardboard models, 3-D computer animations) to help uncover and confirm customer requirements. Systems engineers use prototypes for the same reasons.

In the case of working software prototypes, the code can be deliberately designed to be throwaway or it can be deliberately designed to be reused (non- throwaway). For example, graphical user interface code mock-ups can be useful for requirements elicitation and the code can be reused. And agile software development methodologies incorporate a process of continuously evolving non-throwaway prototypes.

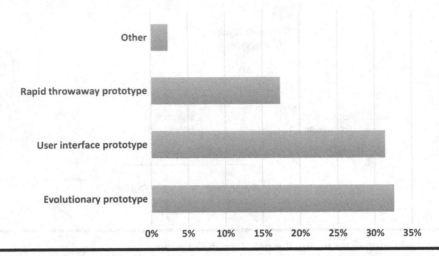

Figure 3.4   Prototype methods selection across software development life cycle methodology. (Kassab and Laplante 2022).

Recently, 3-D printing has become an important tool in building physical models of certain systems. 3-D printing has two important advantages over other rapid prototyping technologies. The first is cost—industrial quality 3-D printers can be purchased for a few thousand dollars, while rapid prototyping machines using traditional computer numerical control (CNC) can cost several hundreds of thousands. The second advantage is that 3-D printers can take as input standard format files produced by commonly used computer-aided design (CAD) programs (Berman 2012).

There are a number of different ways to use prototyping—for example, within a fourth-generation environment (i.e., a simulator), throwaway prototyping, evolutionary prototyping (where the prototype evolves into the final system), or user interface prototyping. Some organizations may use more than one type of prototyping. The results from the requirements engineering state of practices conducted in 2020 (Kassab and Laplante 2022) pertaining to the selection frequency for these different prototyping techniques in practice are shown in Figure 3.4.

There are at least three dangers to consider when using prototyping for requirements elicitation.

Stay Updated: For the up-to-date data from the RE state of practice survey. https://phil.laplante.io/requirements/updates/survey.php

First, in some cases, software prototypes that were not intended to be kept are kept because of schedule pressures. This situation is potentially dangerous since the code was likely not designed using the most rigorous techniques. The unintended reuse of throwaway prototypes occurs often in the industry.

The second problem is that prototyping is not always effective in discovering certain nonfunctional requirements (NFRs). Suppose that you are conducting requirements elicitation activities for a new smart washing machine/dryer combination. While prototyping can be good for understanding existing functionalities, revealing missing functionalities, and identifying unwanted functionalities, many NFRs (e.g., security, safety) won't be easy to identify. Prototyping is particularly not suitable for those requirements that can only be derived by an analysis of prevailing standards and laws (Kassab and Ormandjieva 2014).

Finally, problems can occur when using prototypes to discover the ways in which users interact with the system. The main concern is that users interact differently with a prototype (in which the consequences of behavior are not real) vs. the actual system. Consider, for example, how users might drive in a vehicle simulator, where there is no real injury or damage from a crash. The drivers may behave much more aggressively in the simulator than they would in a real vehicle, leading to possibly erroneous requirements discovery.

Quality Function Deployment

Quality function deployment (QFD) is a technique for discovering customer requirements and defining major quality assurance points to be used throughout the production phase. QFD provides a structure for ensuring that customers' needs and desires are carefully heard, then directly translated into a company's internal technical requirements—from analysis through implementation to deployment. The basic idea of QFD is to construct relationship matrices between customer needs, technical requirements, priorities, and (if needed) competitor assessment. In essence, QFD incorporates card sorting, laddering, and domain analysis.

Because these relationship matrices are often represented as the roof, ceiling, and sides of a house, QFD is sometimes referred to as the "house of quality" (Figure 3.5: Akao 1990).

QFD was introduced by Yoji Akao in 1966 for use in manufacturing, heavy industry, and systems engineering. It has also been applied to software systems by IBM, DEC, HP, AT&T, Texas Instruments, and others.

When we refer to the "voice of the customer," we mean that the requirements engineer must empathically listen to customers to understand what they need from the product, as expressed by the customer in their words. The voice of the customer forms the basis for all analysis, design, and development activities, to ensure that products are not developed from only "the voice of the engineer." This approach embodies the essence of requirements elicitation.

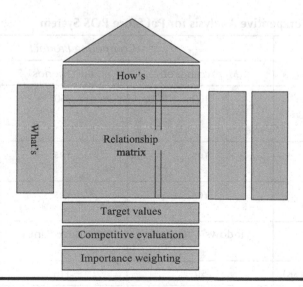

Figure 3.5   QFD's "house of quality" (Akao 1990).

The following requirements engineering process is prescribed by QFD:

- Identify stakeholder's attributes or requirements.
- Identify technical features of the requirements.
- Relate the requirements to the technical features.
- Conduct an evaluation of competing products.
- Evaluate technical features and specify a target value for each feature.
- Prioritize technical features for the development effort.

QFD uses a structured approach to competitive analysis. That is, a feature list is created from the union of all relevant features for competitive products. These features comprise the columns of a competition matrix. The rows represent the competing products, and the corresponding cells are populated for those features included in each product. The matrix can be then used to formulate a starter set of requirements for the new or revised product. The matrix also helps to ensure that key features are not omitted from the new system and can contribute to improving the desirable quality of requirements completeness.

To illustrate, a partial competitive analysis for the pet store POS system is shown in Table 3.1.

Notice that only very high-level features are shown, though we could drill down to whatever level of detail is desired, greatly expanding the matrix. The matrix gives us a starter set of mandatory and optional features. For example, noting that wireless support is found in all these products might indicate that such a feature is mandatory in the new pet store POS system.

Since it incorporates a total life cycle approach to requirements engineering, QFD has several advantages over other stand-alone elicitation techniques. QFD improves the involvement of users and managers. It shortens the development lifecycle and improves overall project development. QFD supports team involvement by structuring communication processes. Finally, it provides a preventive tool that avoids the loss of information.

There are some drawbacks to QFD, however. For example, there may be difficulties in expressing temporal requirements. And QFD is difficult to use with an entirely new project type—how do you discover customer requirements for something that does not exist, and how do you build

Table 3.1   Partial Competitive Analysis for Pet Store POS System

Feature	Competing Product		
	MyFavoritePet	*BestFriends*	*Fido-2.0*
Maximum simultaneous users supported	100	250	Unlimited
Wireless device support	Yes	Yes	Yes
Business analytics features	Yes	Yes	No
Operating system support	Windows/Mac/Linux	Windows/Linux	Windows/Mac
Cost (base system) ($K)	50	110	75

and analyze the competitive products? In these cases, the solution is to look at similar or related products, but still there is apt to be a cognitive gap.

Sometimes it is hard to find measurements for certain functions and to keep the level of abstraction uniform. And, the less we know, the less we document. Finally, as the feature list grows uncontrollably, the house of quality can become a "mansion."

Even if QFD is not used as the primary requirements elicitation approach, its approach to competitive systems analysis should be employed wherever possible. The structured nature of the QFD competitive analysis is an effective way to ensure that no important requirements are missing, leading to more complete requirements set.

Questionnaires/Surveys

Requirements engineers often use questionnaires and other survey instruments to reach large groups of stakeholders. Surveys are generally used at the early stages of the elicitation process to quickly define the scope boundaries.

Survey questions of any type can be used. For example, questions can be closed (e.g., multiple-choice, true-false) or open-ended—involving free-form responses. Closed questions have the advantage of easier coding for analysis, and they help to bind the scope of the system. Open questions allow for more freedom and innovation, but can be harder to analyze and can encourage scope creep.

For example, some possible survey questions for the pet store POS system are as follows:

■ How many unique products (SKUs) do you carry in your inventory? (a) 0–1,000 (b) 1,001–10,000 (c) 10,001–100,000 (d) >100,000
■ How many different warehouse sites do you have? _____
■ How many different store locations do you have? _____
■ How many unique customers do you currently have? _____

There is a danger in overscoring and underscoring if questions are not adequately framed, even for closed-ended questions. Therefore, survey elicitation techniques are most useful when the domain is very well understood by both stakeholders and requirements engineers.

Before undertaking large-scale surveys, it is important to conduct a pilot study with a small subset of the intended survey population. The results are analyzed, and the survey participants are interviewed for the purpose of identifying confusing, missing, or extraneous questions. Then the instrument can be refined before administering the survey to the greater population.

In analyzing survey data, particularly when asking participants to identify and rank desirable features, be careful of the following effect.

> When given a set of choices which do not have to be realized, a person will tend to desire a much larger number of options then if the decision were actually to be made.

We call this effect the "ice cream store effect" because of the following example. Consider an entrepreneur who decides to open a handmade ice cream shop. As part of their product research, they survey a number of people with an instrument in which the respondents check off the flavors of ice cream they would purchase. They find that of the 30 different flavors listed in the survey, 20 of the flavors are selected by 50% of the respondents or more. Thus, they decide to produce and stock these 20 flavors roughly in proportion to the demand indicated by the survey results. Yet, after 1 week of opening their ice cream store, they discover that 90% of business is due to the top three flavors—chocolate, vanilla, and strawberry. Of the 17 other flavors in the survey they keep in inventory, several have never even been purchased. They realize that even though customers said they would buy these flavors in the survey when it came time to exercise their choice, they behaved differently. Therefore, remember the ice cream store effect when giving customers choices about feature sets—they will say one thing and do another.

Surveys can be conducted via telephone, email, in person, and using web-based technologies. There are a variety of commercial tools and open-source solutions that are available to simplify the process of building surveys and collecting and analyzing results that should be employed.

Repertory Grids

Repertory grids incorporate a structured ranking system for various features of the different entities in the system and are typically used when the customers are domain experts. Repertory grids are particularly useful for the identification of agreement and disagreement within stakeholder groups.

The grids look like a feature or quality matrix in which rows represent system entities and desirable qualities, and columns represent rankings based on each of the stakeholders. While the grids can incorporate both qualities and features, it is usually the case that the grids have all features or all qualities to provide for consistency of analysis and dispute resolution.

To illustrate the technique, Figure 3.6 represents a repertory grid for various qualities of the baggage handling system. Here, we see that for the airport operations manager, all qualities are essentially of highest importance (safety is rated as slightly lower, at 4). But for the Airline Worker's Union representative, safety is the most important (after all, his union membership has to interact with the system on a daily basis). In essence, these ratings reflect the agendas or differing viewpoints of the stakeholders. Therefore, it is easy to see why the use of repertory grids can be very helpful in confronting disputes involving stakeholder objectives early. In addition, the grids can provide valuable documentation for dealing with disagreements later in the development of the system because they capture the attitudes of the stakeholders about qualities and features in a way that is hard to dismiss. Still, when using repertory grids, remember the ice cream store effect—stakeholders will say one thing in a public setting and then act differently later.

Reverse Engineering

If an existing system has outdated documentation (or even nonexistent documentation), then reverse engineering can be applied to the system to extract the requirements from the system to

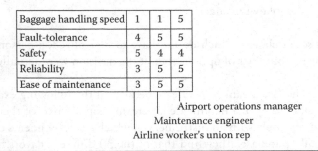

Figure 3.6   Partial repertory grid for the baggage handling system.

understand what the system does. This elicitation technique can be particularly useful for migration projects when dealing with legacy systems. Generally, there are two types of reverse engineering techniques:

- **Black Box Reverse Engineering**: the system is studied without examining its internal structure (function and composition of software).
- **White-Box Reverse Engineering**: The inner workings of the system are studied (analyzing and understanding of software code).

Several methods are proposed in the literature for reverse engineering as requirements elicitation techniques. For example, Yu et al. (2005) proposed a methodology to extract stakeholder goal models from both structured and unstructured legacy code. The methodology consists of the four major steps: (i) refactoring source code by extracting methods based on comments; (ii) converting the refactored code into an abstract structured program through state chart refactoring and hammock graph construction; (iii) extracting a goal model from the structured programs abstract syntax tree; and (iv) identifying NFRs and deriving soft goals based on the traceability between the code and the goal model. Other existing studies include Hassan et al. (2015), Fahmi and Choi (2007), and Alderson and Liu (2012).

Scenarios

Scenarios are informal descriptions of the system in use that provide a high-level description of system operation, classes of users, and exceptional situations.

Here is a sample scenario for the pet store POS system.

> A customer walks into the pet store and fills the cart with a variety of items. When checking out, the cashier asks if the customer has a loyalty card. If so, the cashier swipes the card, authenticating the customer. If not, then the cashier offers to complete one on the spot.
>
> After the loyalty card activity, the cashier scans products using a bar code reader. As each item is scanned, the sale is totaled and the inventory is appropriately updated. Upon completion of product scanning a subtotal is computed. Then any coupons and discounts are entered. A new subtotal is computed and applicable taxes are added. A receipt is printed and the customer pays using cash, credit card, debit card, or check. All appropriate totals (sales, tax, discounts, rebates, etc.) are computed and recorded.

Scenarios are quite useful when the domain is novel (consider a scenario for the space station, for example). User stories are, in fact, a form of scenario.

Task Analysis

Like many of the hierarchically oriented techniques that we have studied already, task analysis involves a functional decomposition of tasks to be performed by the system. That is, starting at the highest level of abstraction, the designer and customers elicit further levels of detail. This detailed decomposition continues until the lowest level of functionality (single task) is achieved.

As an example, consider the partial task analysis for the pet store POS system shown in Figure 3.7.

Here, the overarching pet store POS system is deemed to consist of three main tasks: inventory control, sales, and customer management. Drilling down under the sales functions, we see that

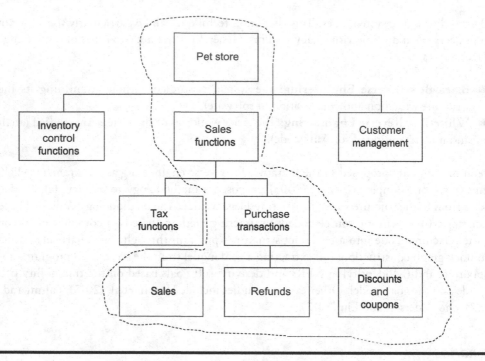

Figure 3.7   Partial task analysis for the pet store POS system.

these consist of the following tasks: tax functions and purchase transactions. Next, proceeding to the purchase transaction function, we decompose these tasks into sales, refunds, discounts, and coupons tasks.

The task analysis and decomposition continue until a sufficient level of granularity is reached (typically, to the level of a method or nondecomposable procedure) and the diagram is completed.

Use Cases[2]

Use cases are a way for more sophisticated customers and stakeholders to describe their desiderata. Use cases depict the interactions between the system and the environment around the system, in particular, human users and other systems. They can be used to model the behavior of pure software or hybrid hardware-software systems.

Use cases describe scenarios of operation of the system from the designer's (as opposed to customers') perspective. Use cases are typically represented using a use case diagram, which depicts the interactions of the system with its external environment. In a use case diagram, the box represents the system itself. The stick figures represent "actors" that designate external entities that interact with the system. The actors can be humans, other systems, or device inputs. Internal ellipses represent each activity of use for each of the actors (use cases). The solid lines associate actors with each use. Figure 3.8 shows a use case diagram for the baggage inspection system.

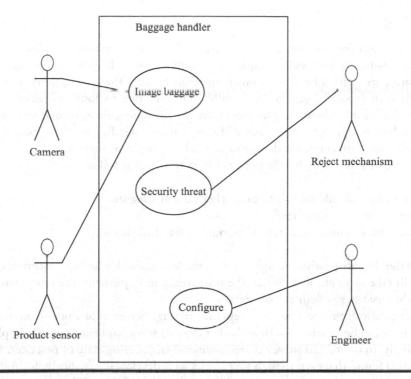

Figure 3.8 Use case diagram of baggage inspection system.

Three uses are shown—capturing an image of the baggage ("image baggage"), the detection of a security threat (in which case the bag is rejected from the conveyor for offline processing), and then configuration by the systems engineer. Notice that the imaging camera, product sensor, and reject mechanism are represented by a human-like stick figure—this is typical—the stick figure represents a system "actor" whether human or not.

Appendix B, Figure B.4 provides another example—a use case diagram for the wet well.

Each use case is a form of documentation that describes scenarios of operation of the system under consideration as well as pre- and postconditions and exceptions. In an iterative development life cycle, these use cases will become increasingly refined and detailed as the analysis and design workflows progress.

Interaction diagrams are then created to describe the behaviors defined by each use case. In the first iteration, these diagrams depict the system as a "black box," but once domain modeling has been completed, the black box is transformed into a collaboration of objects as will be seen later.

If well developed, sometimes, the use cases can be used to form a pattern language, and these patterns and the derived design elements can be reused in related systems (Issa and Al-Ali 2010). Using patterns when specifying requirements ensures a greater level of consistency and can reduce errors in automated measurement of certain requirements properties.

Because they have become such an important tool in requirements discovery and in modeling requirements in the requirements specification document, a comprehensive discussion of use cases, with many examples, can be found in Appendix E.

User Stories

User stories are short conversational texts that are used for initial requirements discovery and project planning. User stories are widely employed in conjunction with agile methodologies.

User stories are written by the customers in terms of what the system needs to do for them and in their own "voice." User stories usually consist of two to four sentences written on a three-by-five-inch card. About 80 user stories are usually appropriate for one system increment or evolution, but the appropriate number will vary widely depending on the application size and scope, and the development methodology to be used (e.g., agile vs. incremental).

An example of a user story for the pet store POS system is as follows:

- Each customer should be able to easily check out at a register.
- Self-service shall be supported.
- All coupons, discounts, and refunds should be handled this way.

User stories should only provide enough detail to make a reasonably low-risk estimate of how long the story will take to implement. When the time comes to implement, the story developers will meet with the customer to flesh out the details.

User stories also form the basis of acceptance testing. For example, one or more automated acceptance tests can be created to verify whether the user story has been correctly implemented.

Surprisingly, in their 2020 survey on requirements engineering state of practices, Kassab and Lapante (2022) found that even though user stories were developed specifically for Agile methodologies, the technique is being used with the Waterfall model—28% of Waterfall projects reported using it. According to that survey, "User stories" witnessed the biggest jump in overall usage from 2013 when a similar survey was conducted (43% in 2020 from 14% in 2013 (Kassab et al. 2014)).

User stories are discussed further in Chapter 8 and in Appendix D.

Viewpoints

Viewpoints are a way to organize information from the (point of view of) different constituencies. For example, in the baggage handling system, there are different perspectives of the system for each of the following stakeholders:

- Baggage handling personnel
- Travelers
- Maintenance engineers
- Airport managers
- Regulatory agencies

By recognizing the needs of each of these stakeholders and the contradictions raised by these viewpoints, conflicts can be reconciled using various approaches.

The actual viewpoints incorporate a variety of information from the business domain, process models, functional requirements specifications, organizational models, etc.

Sommerville and Sawyer (1997) suggested the following components should be in each viewpoint:

- A representation style, which defines the notation used in the specification
- A domain, which is defined as "the area of concern addressed by the viewpoint"

- A specification, which is a model of a system expressed in the defined style
- A work plan, with a process model, which defines how to build and check the specification
- A work record, which is a trace of the actions taken in building, checking, and modifying the specification

Viewpoint analysis is typically used for prioritization, agreement, and ordering of requirements.

Workshops

On a most general level, workshops are any gathering of stakeholders to resolve requirements issues. We can distinguish workshops as being of two types—formal and informal.

Formal workshops are well-planned meetings and are often "deliverable" events that are mandated by contract. For example, DOD-MIL-STD-2167 incorporated multiple required and optional workshops (critical reviews). A good example of a formal workshop style is embodied in JAD.

Informal workshops are usually less boring than highly structured meetings. But informal meetings tend to be too sloppy and may lead to a sense of false security and lost information. If some form of the workshop is needed, it is recommended that a formal one be held using the parameters for successful meetings previously discussed.

Eliciting Nonfunctional Requirements

Nonfunctional requirement (NFR) elicitation techniques differ from functional requirements elicitation techniques. NFRs are generally stated informally during the requirements analysis, are often contradictory, and are difficult to enforce and validate during the software development process. Borg et al. (2003) carried out interviews aimed at identifying the roots of NFR-related problems in different organizations. The conclusion was that NFR-related problems occur at four stages of the development process: elicitation, documentation, management, and test; elicitation being flagged as the main source of potential NFR-related problems as NFR omission at the elicitation stage propagates through the entire development process. The reasons are, for instance, that (i) certain constraints are unknown at the requirements stage, (ii) NFRs tend to conflict with each other, and (iii) separating FRs and NFRs makes it difficult to trace dependencies between them, whereas functional and nonfunctional considerations are difficult to separate if all requirements are mixed together.

It is not easy to choose a method for eliciting, detailing, and documenting NFRs among the variety of existing methods. The following are the desirable characteristics of NFR elicitation methods (Herrmann et al. 2007):

1. A guided process to ease the method usage by less experienced personnel and to support repeatability of the results
2. Derivation of measurable NFRs to ease quality assurance
3. Reuse of artifacts to support completeness of the derived NFRs to support learning and to avoid rework
4. Intuitive and creative elicitation of quality to capture also the hidden requirements and thus support completeness
5. Focused effort for efficient elicitation and NFR prioritization to support trade-off decisions

6. Handling dependencies between NFRs to support trade-off decisions
7. Integration of NFRs with functional requirements

Common ways to discover NFRs include competitive analysis of system qualities: NFRs can be discovered by analyzing the qualities for competing products in the market. For example, what is the response time for a competing product? And do we need to do better?

Another technique for discovering NFS is to use a pre-established questionnaire where a requirements engineer develops a questionnaire to be asked of the stakeholders and development team. For example: "How should the system respond to input errors? What parts of the system are likely candidates for later modification? What data of the system must be secure?" These questions can be prepared while following some template or standard (e.g., ISO 9126) in order to focus and ask questions about each type of NFR in the standard.

Elicitation Summary

This tour has included many elicitation techniques, and each has its advantages and disadvantages, which were discussed along the way. Clearly, some of these techniques are too general, some too specific, some rely too much on stakeholder knowledge, some not enough, etc. Therefore, it is clear that some combination of techniques is needed to successfully address the requirements elicitation challenge.

Which Combination of Requirements Elicitation Techniques Should Be Used?

There is scant research to provide guidance on selecting an appropriate mix of requirements elicitation techniques. One notable exception is the knowledge-based approach for the selection of requirements engineering techniques (KASRET), which guides users to select a combination of elicitation from a library (Eberlein and Jiang 2011). The library of techniques and assignment algorithms are based on a literature review and survey of industrial and academic experts. The technique has not yet been widely used, however.

In order to provide some guidance on appropriate elicitation techniques, we first cluster the techniques previously discussed into categories or equivalence classes based on the kinds of information the techniques are likely to uncover. The classes (interviews, domain-oriented, group work, ethnography, prototyping, goals, scenarios, viewpoints) and the included elicitation techniques are shown in Table 3.2.

Now we can summarize how effective various techniques are in dealing with various aspects of the elicitation process as shown in Table 3.3 (based on work by Zowghi and Coulin 1998).

For example, interview-based techniques are useful for all aspects of requirements elicitation (but are very time-consuming). On the other hand, prototyping techniques are best used to analyze stakeholders and to elicit the requirements. Ethnographic techniques are good for understanding the problem domain, analyzing stakeholders, soliciting requirements, and so on.

Finally, there is a clear overlap between these elicitation techniques (clusters) in that some accomplish the same thing and, hence, are alternatives to each other. In other cases, these techniques complement one another. In Table 3.4, alternative (A) and complementary (C) elicitation groupings are shown.

You can use Tables 3.2–3.4 to guide you through selecting an appropriate set of elicitation techniques. In selecting a set of techniques to be used, you would select a set of complementary techniques. For example, a combination of viewpoint analysis and some form of prototyping would be desirable. Conversely, using both viewpoint analysis and scenario generation would probably yield excessively redundant information.

As an example, consider the case of an IoT healthcare system to track humans in a hospital. Here it would be appropriate to start with an initiative to define system overall goals and desired outcomes. Next, we would use a domain analysis to explore the laws, regulations, and standards that apply to the system. User stories, scenarios, and interviews would be appropriate for elicitation of requirements from users of the system. There would undoubtedly be some prototyping throughout the requirements discovery and refinement process. Each of these techniques would move from informal to more rigorous as requirements discovery proceeded. This approach is often used for federal, state, and municipal governments, and even for large industrial projects.

There is no "silver bullet" combination of elicitation techniques, however. The right mix will depend on the application domain, the culture of the customer organization and that of the requirements engineer, the size of the project, and many other factors. Learning from experience is very important in this regard. Finally, Hickey and Davis (2003) provide further insights on selecting the appropriate combinations of elicitation techniques from the viewpoint of a number of experts who were presented with various requirements elicitation scenarios and asked which approaches they would use.

Table 3.2 Organizing Various Elicitation Techniques Roughly by Type

Technique Type	Techniques
Domain oriented	Card sorting Designer as apprentice Domain analysis Laddering Protocol analysis Task analysis
Ethnography	Ethnographic observation
Goals	Goal-based approaches QFD
Group work	Brainstorming Group work JAD Workshops
Interviews	Interviews Introspection Questionnaires
Prototyping	Prototyping
Scenarios	Scenarios Use cases User stories
Viewpoints	Viewpoints Repertory grids

Source: Zowghi, D., & Coulin, C., Requirements elicitation: A survey of techniques, approaches, and tools, in A. Aurum & C. Wohlin (Eds.), (2005). *Engineering and Managing Software Requirements*, pp. 19–46. Springer, 1998.

Table 3.3 Techniques and Approaches for Elicitation Activities

	Interviews	Domain	Group Work	Ethnography	Prototyping	Goals	Scenarios	Viewpoints
Understanding the domain	●	●	●			●	●	●
Identifying sources of requirements	●	●	●			●	●	●
Analyzing the stakeholders	●	●	●	●	●	●	●	●
Selecting techniques and approaches	●	●	●					
Eliciting the requirements	●	●	●	●	●	●	●	●

Source: Zowghi, D., & Coulin, C., Requirements elicitation: A survey of techniques, approaches, and tools, in A. Aurum & C. Wohlin (Eds.), (2005). *Engineering and Managing Software Requirements*, pp. 19–46, Springer, 1998.

Table 3.4 Complementary and Alternative Techniques

	Interviews	Domain	Group Work	Ethnography	Prototyping	Goals	Scenarios	Viewpoints
Interviews		C	A	A	A	C	C	C
Domain	C		C	A	A	A	A	A
Groupwork	A	C		A	C	C	C	C
Ethnography	A	A	A		C	C	A	A
Prototyping	A	A	C	C		C	C	C
Goals	C	A	C	C			C	C
Scenarios	C	A	C	A	C	C		A
Viewpoints	C	A	C	A	C	C	A	

Source: Zowghi, D., & Coulin, C., Requirements elicitation: A survey of techniques, approaches, and tools, in A. Aurum & C. Wohlin (Eds.), (2005). *Engineering and Managing Software Requirements*, pp. 19–46, Springer, 1998.

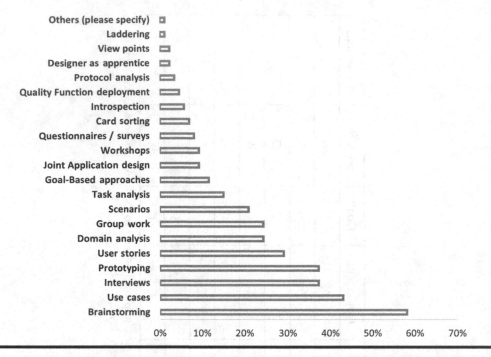

Figure 3.9   Summary of answers to the question "which requirements elicitation technique(s) do you use?" (Kassab and Laplante 2022).

Prevalence of Requirements Elicitation Techniques

Before we conclude this discussion, let's get an idea of how various elicitation techniques are commonly used in the industry. Hickey and Davis (2003) found that less experienced analysts often select a technique based on one of two reasons: (i) it is the only one they know, or (ii) they think that a technique that worked well last time must surely be appropriate this time. If we return to the 2020 survey of requirements engineering state of practices (Kassab and Laplante 2022), then a summary of the answers to the question "which requirements elicitation technique(s) do you use?" is shown in Figure 3.9. The responses revealed that brainstorming, use cases, interviews, prototyping, and user stories were the top five frequently used elicitation techniques. On average, a participant selected three elicitation techniques.

Stay Updated: For the up-to-date data from the RE state of practice survey. https://phil.laplante.io/requirements/updates/survey.php

Eliciting Hazards

We previously noted that "shall not" behaviors are the set of output behaviors that are undesired and that hazards were a subset of those behaviors that tended to cause serious or catastrophic failures.

The terms "serious" and "catastrophic" are subjective but generally involve loss of life, serious injury, major infrastructure damage, or great financial loss.

For example, some "shall not" requirements for the pet store POS include the following:

- The system shall not expose customer information to external systems.
- The system shall not allow unauthorized access.
- The system shall not allow customers to overdraw store credit.

In these examples, the first two requirements might be considered hazards because the potential for financial damage to the company is far greater than for the third requirement.

Hazards are a function of input anomalies that are either naturally occurring (such as hardware failures) or artificially occurring (such as attacks from intruders) (Voas and Laplante 2010). These anomalous input events need to be identified, and their resultant failure modes and criticality need to be determined during the requirements elicitation phase in order to develop an appropriate set of "shall not" requirements for the system. As with any other requirements, "shall not" requirements need to be prioritized.

Typical techniques for hazard determination include the traditional development of misuse cases, anti-modeling, and formal methods (Robinson 2010). Checklists of unwanted behavior that have been created from previous versions of the system or related systems are also helpful in identifying unwanted behavior. Prevailing standards and regulations may also include specific "shall not" requirements; for example, in the United States, the Health Insurance Portability and Accountability Act (HIPPA 1996) prohibits the release of certain personal information to unauthorized parties, and standard construction codes in all jurisdictions include numerous prohibitions against certain construction practices.

Misuse Cases

Use cases are structured, brief descriptions of desired behavior. Just as there are use cases describing desired behavior, there are misuse cases (or abuse cases) describing undesired behavior. Typical misuses for most systems include security breaches and other malicious behaviors as well as abuse by untrained, disoriented, or incapable users. Cleland-Huang et al. (2016) describe several ways of identifying misuse cases based on threat modeling and brainstorming.

An easy way to create use cases is to assume the role of a persona non grata, that is, an unwanted user of the system, and then model the behaviors of such a person (Cleland-Huang 2014). Personae non gratae can include hackers, intruders, spies, and even well-meaning, but bumbling users.

To see how identifying these persons helps in creating hazard requirements, consider the following examples. In the pet store POS system, it would be appropriate to consider how a hacker would infiltrate this system and then create requirements that would thwart the hacker's intentions. In the baggage handling system, a requirements engineer could assume the role of a clumsy or absent-minded traveler and then prescribe requirements that would ensure the safety of such persons. The need to create misuse cases is a reason to completely identify all negative stakeholders since these persons are likely to comprise large many personae non gratae.

Antimodels

Another way of deriving unwanted behavior is to create antimodels for the system. Antimodels are related to fault trees, that is, the model is derived by creating a cause and effect hierarchy for unwanted behaviors leading to system failure. Then, the causes of the system failure are used to create the "shall not" requirements. For example, consider the security functionality for the baggage handling system involving the unwanted outcome of damaged baggage shown in Figure 3.10.

The figure leads us to write the following raw requirements:

- If a baggage jam is sensed, then the conveyor shall not move.
- If the baggage feeder is stuck, then the conveyor shall not move.
- If the baggage door is stuck, then the conveyor shall not move.
- If the conveyor is stuck, then the baggage feeder should not move.

These requirements need further analysis and possibly simplification, but the anti-model helped us to derive these raw requirements in a systematic way.

Formal Methods

To be discussed in Chapter 7, mathematical formalisms can be used to create a model of the system and its environment as related to their goals, operations, requirements, and constraints. These formalisms can then be used in conjunction with automated model checkers to examine various properties of the system and ensure that unwanted ones are not present. For example, if using UML/SysML to model behavior, formal activity diagrams could be annotated with security concerns. Activity diagrams are described in Appendix C, UML.

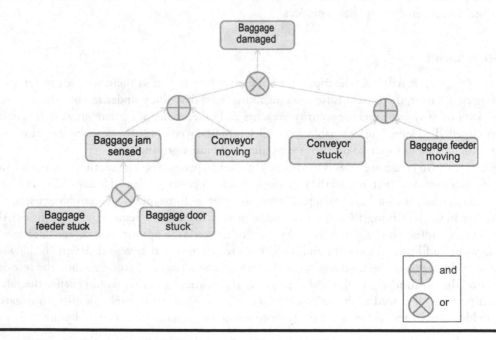

Figure 3.10 Partial antimodel for baggage handling system.

VIGNETTE 3.1 Requirements Engineering for Safety-Critical Systems

Safety-critical systems are those in which failure can result in loss of human life or significant injury. These kinds of systems include public infrastructure for power and water, medical systems, transportation systems, and much more. With the growth of smart and autonomous systems, for example, smart highways, driverless vehicles, and robotic surgery, the focus on safety requirements has become paramount.

There are some domain-specific standards (e.g., nuclear, medical devices, aviation) and general standards that provide guidance for developing requirements for safety-critical systems. One example of a general standard is NASA-STD-8719.13, which provides guidance for safety considerations in software-intensive systems (NASA 2013).

This standard uses a risk assessment matrix table to show the probability and severity of a particular hazard (Figure 3.11).

Systems severity levels	System hazard likelihood of occurrence				
	Improbable	Unlikely	Possible	Probable	Likely
Catastrophic	4	3	2	1	1
Critical	5	4	3	2	1
Moderate	6	5	4	3	2
Negligible	7	6	5	4	3

1= Highest priority (highest system risk), 7= Lowest priority (lowes system risk).

Figure 3.11 The system risk index matrix of NASA-STD-8719.13C. (From NASA [National Aeronautics and Space Administration], NASA-STD 8719.13, Software Safety Standard, 2013, http://NASA.gov [accessed January 2017].)

The standard stipulates that safety requirements be tagged for traceability through the requirements engineering life cycle needed in order to enable the assessment of impacts and changes to the requirements. The uniquely identified hazards can be listed in a special section in the requirements document, or be designated by a flag beside the requirement, or be tagged within a requirements management tool.

Software safety requirements can do more than protect against unsafe system behavior. The software can be used proactively to monitor the system, analyze critical data, look for trends, and signal or act when events occur that may be precursors to a hazardous state. Once such an indicator is detected, the software can be used to avoid or mitigate the effects of that hazard. Avoidance or mitigation could include restoring the system fully or partially or putting the system into a safe state.

Share your Opinion: What kinds of requirements engineering elicitation techniques does your organization use?
https://phil.laplante.io/requirements/opinion.php

Exercises

3.1 What are some different user classes for the smart home system described in Appendix A?

3.2 What are some difficulties that may be encountered in attempting to elicit requirements without face-to-face interaction?

3.3 Does the Heisenberg uncertainty principle apply to techniques other than ethnographic observation? What are some of the ways to alleviate the Heisenberg uncertainty principle?

3.4 During ethnographic observation, what is the purpose of recording the time and day of the observation made?

3.5 Should requirements account for future scalability and enhancements?

3.6 Which subset of the techniques described in this chapter would be appropriate for a setting where the customers are geographically distributed?

3.7 Investigate the concept of "active listening." How would this technique assist in requirements elicitation?

3.8 Which elicitation techniques would you use to elicit system requirements from the following stakeholder groups?
    3.8.1 Passengers in the baggage handling system
    3.8.2 Cashiers in the pet store POS system
    3.8.3 Doctors in the IoT healthcare system described at the end of Chapter 2

3.9 If you are working on a course project, list the elicitation techniques that you would you use to elicit system requirements from each of the stakeholder groups.

3.10 There are several "shall not" requirements in the SRS of Appendix A. Which, if any, of these would you consider being hazards?

3.11 Speculate as to why there are no "shall not" requirements in the SRS in Appendix B.

*3.12 For the pet store point of sale system, develop an antimodel pertaining to inventory control. For example, the system should not record negative inventory. Write the corresponding "shall not" requirements for this antimodel.

Note

1, 2 This discussion is adapted from one found in Laplante (2006), with permission.

References

Adepetu, A., Khaja, A. A., Al Abd, Y., Al Zaabi, A., & Svetinovic, D. (2012, March). CrowdREquire: A requirements engineering crowdsourcing platform. In *2012 AAAI Spring Symposium Series*, Palo Alto, California.

Akao, Y. (1990). *Quality Function Deployment: Integrating Customer Requirements into Product Design.* Productivity Press, Cambridge, MA.

Alderson, A., & Liu, K. (2012). Reverse requirements engineering: The AMBOLS approach. *Systems Engineering for Business Process Change: Collected Papers from the EPSRC Research Programme* (pp. 196–208). Springer, London.

Basili, V. R., & Weiss, D. (1984). A methodology for collecting valid software engineering data. *IEEE Transactions on Software Engineering*, 10: 728–738.

Berman, B. (2012). 3-D printing: The new industrial revolution. *Business Horizons*, 55(2): 155–162.

Borg, A., Yong, A., Carlshamre, P., & Sandahl, K. (2003). The bad conscience of requirements engineering: An investigation in real-world treatment of non-functional requirements. In *Third Conference on Software Engineering Research and Practice in Sweden (SERPS'03)* (pp. 1–8). Lund.

Centers for Medicare & Medicaid Services. (1996). The Health Insurance Portability and Accountability Act of 1996 (HIPAA). Online at http://www.hhs.gov/hipaa (accessed June 2017).

Cleland-Huang, J. (2014). How well do you know your personae non gratae? *IEEE Software*, 31(4): 28–31.

Cleland-Huang, J., Denning, T., Kohno, T., Shull, F., & Weber, S. (2016). Keeping ahead of our adversaries. *IEEE Software*, 33(3): 24–28.

Cleland-Huang, J., & Laurent, P. (2014). Requirements in a global world. *IEEE Software*, 31(6): 34–37.

Eberlein, A., & Jiang, L. (2011). Selecting requirements engineering techniques. In P. Laplante (Ed.), *Encyclopedia of Software Engineering* (pp. 962–978). Taylor & Francis. Boca Raton, FL, Published online.

Fahmi, S. A., & Choi, H. J. (2007, November). Software reverse engineering to requirements. In *2007 International Conference on Convergence Information Technology (ICCIT 2007)*, (pp. 2199–2204). IEEE, Gwangju, South Korea.

Groen, E. C., Doerr, J., & Adam, S. (2015, March). Towards crowd-based requirements engineering a research preview. In *International Working Conference on Requirements Engineering: Foundation for Software Quality* (pp. 247–253). Springer, Cham.

Groen, E. C., & Koch, M. (2016). How requirements engineering can benefit from crowds. *Requirements Engineering Magazine*, 8: 10.

Hassan, S., Qamar, U., Hassan, T., & Waqas, M. (2015, August). Software reverse engineering to requirement engineering for evolution of legacy system. In *2015 5th International Conference on IT Convergence and Security (ICITCS)* (pp. 1–4). IEEE, Kuala Lumpur, Malaysia.

Herrmann, A., Kerkow, D., & Doerr, J. (2007). Exploring the characteristics of NFR methods – A dialogue about two approaches. In *REFSQ 2007*, LNCS 4542 (pp. 320–334). Trondheim, Norway.

Hickey, A. M., & Davis, A. M. (2003). Elicitation technique selection: How do experts do it? In *Proceedings 11th IEEE International Requirements Engineering Conference, 2003* (pp. 169–178). IEEE, Monterey Bay, CA, USA.

Howe, J. (2006). The rise of crowdsourcing. *Wired Magazine*, 14(6): 1–4.

Issa, A. A., & Al-Ali, A. (2010). Use case patterns driven requirements engineering. In *Proceedings Second International Conference on Computer Research and Development* (pp. 307–313). Kuala Lumpur, Malaysia.

Kassab, M., & Laplante, P. (2022). The current and evolving landscape of requirements engineering state of practice. *IEEE Software.* DOI: 10.1109/MS.2022.3147692.

Kassab, M., Neill, C., & Laplante, P. (2014). State of practice in requirements engineering: Contemporary data. *Innovations in Systems and Software Engineering*, 10(4): 235–241.

Kassab, M., & Ormandjieva, O. (2014). Non-functional requirements in process-oriented approaches. In Phillip A. Laplante (Ed.), *Encyclopedia of Software Engineering* (pp. 1–11). Taylor & Francis, Boca Raton, FL.

Laplante, P.A. (2006). *What Every Engineer Needs to Know About Software Engineering.* CRC/Taylor & Francis, Boca Raton, FL.

Lim, S. L., & Finkelstein, A. (2011). StakeRare: Using social networks and collaborative filtering for large-scale requirements elicitation. *IEEE Transactions on Software Engineering*, 38(3): 707–735.

NASA (National Aeronautics and Space Administration). (2013). NASA-STD 8719.13, Software Safety Standard. http://NASA.gov (accessed January 2017).

Renzel, D., & Klamma, R. (2014). Requirements bazaar: Open-source large scale social requirements engineering in the long tail. *IEEE Computer Society Special Technical Community on Social Networking E-Letter*, Vol. 2 No. 3.

Robinson, W. N. (2010). A roadmap for comprehensive requirements modeling. *Computer*, 43(5): 64–72.

Snijders, R., Dalpiaz, F., Brinkkemper, S., Hosseini, M., Ali, R., & Ozum, A. (2015, August). REfine: A gamified platform for participatory requirements engineering. In *2015 IEEE 1st International Workshop on Crowd-Based Requirements Engineering (CrowdRE)* (pp. 1–6). IEEE, Ottawa, Canada.

Sommerville, I., & Sawyer, P. (1997). Viewpoints for requirements engineering. *Software Quality Journal*, 3: 101–130.

Voas, J., & Laplante, P. (2010). Effectively defining shall not requirements. *IT Professional*, 12(3): 46–53.

Yu, Y., Wang, Y., Mylopoulos, J., Liaskos, S., Lapouchnian, A., & do Prado Leite, J. C. S. (2005, September). Reverse engineering goal models from legacy code. In *13th IEEE International Conference on Requirements Engineering (RE'05)* (pp. 363–372). IEEE, Paris, France.

Zowghi, D., & Coulin, C. (1998). Requirements elicitation: A survey of techniques, approaches, and tools. In A. Aurum & C. Wohlin (Eds.), *Engineering and Managing Software Requirements* (pp. 19–46). Springer, Berlin.

Chapter 4

Writing the Requirements Document

Requirements Agreement and Analysis

Throughout the elicitation activity and especially before finalizing the systems requirements specification, raw requirements should be analyzed for problems and these problems reconciled. Requirements analysis is the activity of analyzing requirements for problems. Problematic requirements include those that are:

- Confusing
- Extraneous
- Duplicated
- Conflicting
- Missing

We make these concepts more rigorous in Chapter 6.

Requirements analysis is usually informal; for example, customers and stakeholder representatives are shown elicited requirements in a format that they can understand to ensure that the requirements engineer's interpretation of all customers' desiderata is correct. Formal methods provide more rigorous techniques for requirements analysis.

Requirements agreement is the process of reconciling differences in the same requirement derived from different sources. Requirements agreement is generally an informal process, though a systematic process should be used. Use cases and user stories are often effective tools for this purpose (Hull et al. 2011). Screen mockups are another useful technique, especially when used in conjunction with functional prototypes (Ricca et al. 2010). Others suggest using scenarios, storyboards, paper prototypes, scripted concept generators, and functional prototypes for requirements agreement (Sutcliffe et al. 2011).

As part of the customer agreement process, Hull et al. (2011) suggest asking the following questions:

- Is the requirement complete?
- Is the requirement clear?

DOI: 10.1201/9781003129509-4

- Is the requirement implementable?
- Is the qualification plan clear and acceptable?

You will note in Chapter 6 how these questions relate to the desirable qualities of a requirements specification, and we will look at more precise ways to achieve these qualities throughout the requirements engineering life cycle in Chapters 6 and 7.

Requirements Representation

Various techniques can be used to describe functionality in any system, including use cases, user stories, natural languages, formal languages, stylized (restricted natural) languages, and a variety of formal, semiformal, and informal diagrams. Eberlein and Jiang (2011) provide a focused discussion of selecting various requirements techniques including specification approaches. When making such selections, they suggest considering the following factors:

- Technical issues such as the maturity and effectiveness of the technique being considered
- The complexity of the techniques
- Social and cultural issues such as organization's opposition to change
- The level of education and knowledge of developers

Certain characteristics of the software project such as time and cost constraint the complexity of the project, and the structure and competence of the software team.

Selecting the right representation style for a single project or organization-wide use is a significant challenge, and in a sense, a theme of this book.

Approaches to Requirements Representation

Generally, there are three forms of requirements representation: formal, informal, and semiformal. Requirements specifications can strictly adhere to one or another of these approaches, but usually they contain elements of at least two of these approaches (informal and one other).

- **Informal Representation** techniques cannot be completely transliterated into a rigorous mathematical notation. Informal techniques include natural language (i.e., human languages), flowcharts, ad hoc diagrams, and most of the elements that you may be used to seeing in Software/Systems Requirements Specifications Document (SRS). Natural language is in fact the dominant form of requirements for projects in the industry. All SRS documents will have some informal elements. We can state this fact with confidence because even the most formal requirements specification documents have to use natural language, even if it is just to introduce a formal element. Kassab and Laplante (2022) found that 69% of respondents of the RE survey reported that requirements were expressed in terms of natural language, that is, informally (Figure 4.1). As natural language is notoriously imprecise, using natural language may face fundamental challenges in constructing software and systems. Some approaches bypass the challenges by using a constrained form of natural language, ensuring some degree of precision without going as far as the formal notations. Examples of these approaches include Stimulus (Jeannet and Gaucher 2015), Relax (Whittle et al. 2009), and Requirements Grammar (Scott and Cook 2004).

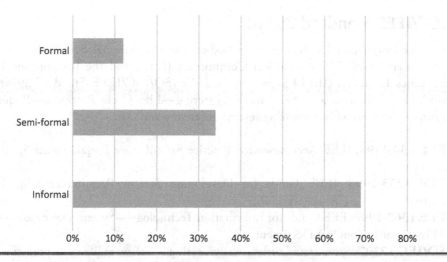

Figure 4.1   Requirement specification notation prevalence (Kassab and Laplantel 2022).

- **Semiformal Representation** techniques include those that, while appearing informal, have at least a partial formal basis. For example, many of the diagrams in the Unified Modeling Language (UML) or Systems Modeling Language (SysML) family of metamodeling languages including the use case diagram are semiformal or can be formalized. UML and SysML are generally considered semiformal modeling techniques; however, they can be made entirely formal with the addition of appropriate mechanisms (Laplante 2006). Other semiformal techniques include KAOS (Dardenne et al. 1993), Rectify (Dassault Systems 2016), and User Requirements Notations (URN) (Berenbach et al. 2012).
- **Formal Representations** such as Z and VDM rely on mathematical formalisms to represent requirements. We will explore these further in Chapter 7. But even formal requirements specifications documents will have elements of the informal or semiformal specification.

In this chapter, we will focus our attention mostly on requirements representation using informal techniques. While the majority of respondents of the 2020 survey on requirements engineering state of practice (Kassab and Laplante 2022) stated that they encountered or used informal specifications in the industry (Figure 4.1), it is probably the case that even the most informal of specification documents have some semiformal and formal elements. Likewise, even the most formal of specifications will have some informal elements (e.g., an introduction in natural language). In practice, it is recommended to have an appropriate blend of informal (natural language), semiformal (e.g., UML/SysML models), and formal elements.

Stay Updated: For the up-to-date data from the RE state of practice survey.
https://phil.laplante.io/requirements/updates/survey.php

ISO/IEC/IEEE Standard 29148

In 2011, three international standardization bodies, the International Standards Organization (ISO), the International Electrotechnical Commission (IEC), and the Institute for Electrical and Electronics Engineers (IEEE) jointly released the ISO/IEC/IEEE Standard 29148 (2011). This standard for Systems and Software Engineering—Life Cycle Processes—Requirements Engineering is the result of harmonizing several different standards:

- **IEEE 830-1998**: IEEE Recommended Practice for Software Requirements Specifications (retired).
- **IEEE 1233-1998**: IEEE Guide for Developing System Requirements Specifications (retired).
- **IEEE 1362-1998**: EEE Guide for Information Technology—System Definition—Concept of Operations (ConOps) Document.
- **ISO/IEC 12207**: Systems and software engineering—Software life cycle processes.
- **ISO/IEC 15288**: Systems and software engineering—System life cycle processes.

In 2018, the second edition of ISO/IEC/IEEE Standard was released in response to the revision of ISO/IEC/ IEEE 15288 and ISO/IEC/IEEE 12207. The purpose of the revision was to accomplish the harmonization of the structures and contents of the two documents while supporting the requirements of the assessment community.

IEEE Standard 830-1998, which was retired, was frequently referenced in previous editions of this text. Standard 29148 expanded on IEEE 830 to incorporate the full system life cycle, recognize advances in agile and lean engineering, and focus on software-intensive systems as a general case of software. For simplicity, from here on, ISO/IEC/IEEE 29148 will be referred to as IEEE 29148.

IEEE 29148 is based on a model that produces a document that helps:

- to establish the basis for agreement between the acquirers or suppliers on what the product is to do;
- to force a rigorous assessment of requirements before design can begin and reduce later redesign;
- to provide a realistic basis for estimating product costs and risks, and schedules organizations to develop validation and verification plans;
- to provide an informed basis for deploying a product to new users or new operational environments; and
- to provide a basis for product enhancement (IEEE 29148 2018).

There are several benefits to observing the IEEE 29148 guidelines. First, they provide a simple framework for the organization of the requirements document. Moreover, the IEEE 29148 outline is particularly beneficial to the requirements engineer because it has been widely deployed across a broad range of application domains (Figure 4.2).

Finally, and perhaps more importantly, IEEE 29148 provides guidance for organizing the functional and nonfunctional requirements of the SRS.

The standard also describes ways to represent functional and nonfunctional requirements under Section 3, Specific Requirements. These requirements are the next subject for discussion.

1. Introduction
1.1 Purpose
1.2 Scope
1.3 Product overview
1.3.1 Product perspective
1.3.2 Product functions
1.3.3 User characteristics
1.3.4 Limitations
1.4 Definitions
2. References
3. Requirements
3.1 Functions
3.2 Performance requirements
3.3 Usability Requirements
3.4 Interface Requirements
3.5 Logical database requirements
3.6 Design constraints
3.7 Software system attributes
3.8 Supporting information
4. Verification
        (parallel to subsections in Section 3)
5. Appendices
5.1 Assumptions and dependencies
5.2 Acronyms and abbreviations

Figure 4.2   Table of contents for an SRS document as recommended by IEEE Standard 29148. (From ISO/IEC/IEEE Standard 29148, 2nd edition, *Systems and Software Engineering—Life Cycle Processes—Requirements Engineering*, 2018.)

Quick Access the latest version of the standard
https://phil.laplante.io/requirements/standard.php

Recommendations on Representing Nonfunctional Requirements

IEEE 29148 describes several types of nonfunctional requirements in Sections 3.2–3.7 involving:

■ Performance requirements
■ Usability requirements
■ Interface requirements
■ Logical database requirements
■ Design constraints
■ Software system attribute requirements

Performance requirements specify both the static and the dynamic numerical requirements placed on the software or on human interaction with the software as a whole. Typical performance requirements might include the number of simultaneous users to be supported, the number of transactions and tasks, and the amount of data to be processed within certain periods for both normal and peak workload conditions. The performance requirements should be stated in measurable terms, for example:

90% of the transactions shall be processed in less than 2 s.

Rather than

An operator shall not have to wait too long for the transaction to complete.

Usability requirements define the requirements measuring the effectiveness in using the system's functionalities, efficiency, satisfaction criteria, and avoidance of harm that could arise from use in specific contexts of use. These kinds of requirements are hard to discover in novel systems or concerning new features. Prototyping can be helpful in discovering such requirements.

Interface requirements define all inputs into and outputs from the system. Each defined interface should include the following content:

- Name of item
- Description of purpose
- Source of input or destination of output
- Valid range, accuracy, and/or tolerance
- Units of measure
- Timing
- Relationships to other inputs/outputs
- Data formats
- Command Formats
- Data items or information included in the input and output

Logical database requirements are types of information used by various functions such as:

- Types of information used by various functions
- Frequency of use
- Accessing capabilities
- Data entities and their relationships
- Integrity constraints
- Security
- Data retention requirements

Design constraint requirements are related to standards compliance and hardware limitations. We will see how to organize constraints in the requirements statements shortly.

Finally, software or system attributes often can include reliability, availability, security, maintainability, portability, and many others. In general, any such attribute is termed an "ility" because most end with an "ility." But other attributes, such as safety, timeliness, security, and others, are also ilities.

Recommendations on Representing Functional Requirements

The functional requirements should capture all system inputs and the exact sequence of operations and responses (outputs) to normal and abnormal situations for every input possibility. Functional requirements may use case-by-case descriptions or other general forms of description.

IEEE 29148 provides a number of organizational options for the functional requirements. Functional requirements can be organized by one or more the following organizational schemes:

- System mode (suitable if the system behaves differently under different modes of operations, e.g., navigation, combat, diagnostic)
- User class (suitable if the system provides different functions to different user classes, e.g., user, supervisor, diagnostic)
- Object (by defining classes/objects, attributes, functions/methods, and messages)
- Feature (describes what the system provides to the user)
- Stimulus (e.g., sensor 1, sensor 2, actuator 1, ...)
- Response (suitable when the system is best described by all the functions in support of the generation of a response)
- Functional hierarchy (e.g., using a top-down decomposition of tasks)

A combination of these organizational techniques can be used within one SRS document. For example, the smart home description given in Appendix A is a *feature-driven description* of functionality. One clue that it is feature-driven is the mantra "the system shall" for many of the requirements. The wet-well pumping system specification in Appendix B is also largely described by functionality.

Alternatively, consider a system organized by functional *mode*—the NASA WIRE (wide-field infrared explorer) system (1996). This system was part of the Small Explorer program and describes a submillimeter-wave astronomy satellite incorporating standardized space-to-ground communications. Information was to be transmitted to the ground and commands received from the ground, according to the Consultative Committee for Space Data Systems (CCSDS) and Goddard Space Flight Center standards (NASA WIRE 1996). The flight software requirements are organized as follows:

- System management
- Command management
- Telemetry management
- Payload management
- Health and safety management
- Software management
- Performance requirements

Reviewing the document under the system management mode, we see the operating system functionality described as follows:

Operating System

01 The operating system shall provide a common set of mechanisms necessary to support real-time systems such as multitasking support, CPU scheduling, basic communication, and memory management.

    01.1 The operating system shall provide multitasking capabilities.

    01.2 The operating system shall provide event-driven, priority-based scheduling.

        01.2.1 Task execution shall be controlled by a task's priority and the availability of resources required for its execution.

    01.3 The operating system shall provide support for intertask communication and synchronization.

Table 4.1 Object-Oriented vs. Structured Representation

	Structured	Object-Oriented
System components	Functions	Objects
Data and control specification	Separated through internal decomposition	Encapsulated within objects
Characteristics	Hierarchical structure	Inheritance relationship of objects
	Functional description of system	Behavioral description of system
	Encapsulation of knowledge within functions	Encapsulation of knowledge within objects

01.4 The operating system shall provide real-time clock support.

01.5 The operating system software shall provide task-level context switching for the 80387 math coprocessor.

Also under system functionality is command validation, defined as follows:

Command Validation

211 The flight software shall perform CCSDS command structure validation.

211.1 The flight software shall implement CCSDS Command Operations Procedure number 1 (COP-1) to validate that CCSDS transfer frames were received correctly and in order.

211.2 The flight software shall support a fixed-size frame acceptance and reporting mechanism (FARM) sliding window of 127 and a fixed FARM negative edge of 63.

211.3 The flight software shall telemeter status and discard the real-time command packet if any of the following errors occur:

Checksum fails validation prior to being issued

An invalid length is detected.

An invalid application ID is detected.

211.4 The flight software shall generate and maintain a command link control word (CLCW). Each time an update is made to the CLCW, a CLCW packet is formatted and routed for possible downlink (NASA WIRE 1996).

It is also very common in software-based systems to take an ***object-oriented approach*** to describe the system behavior. This is particularly the case when the software is expected to be built using a pure object-oriented language such as Java.

Object-oriented representations involve highly abstract system components called objects, and their encapsulated attributes and behavior. The differences between traditional structured descriptions of systems and object-oriented descriptions of systems are summarized in Table 4.1.

When dealing with requirements organized in an object-oriented fashion, it is very typical to use user stories (especially in conjunction with agile software development methodologies, which we will discuss in Chapter 8) or use cases and use case diagrams to describe behavior.

VIGNETTE 4.1   Security Requirements

The SRS for any system (no matter how trivial) should consist of a section on "Security Requirements." Even non-critical software systems (e.g. an Internet of Things capable toy) can sometimes be used as an entry point to the critical systems to which they may be connected (either deliberately or unintentionally), threatening the safety and well-being of the public. Even if the system is somehow unable to connect to other systems or is non-critical in nature, the system itself needs to be protected from the insertion of malware and the theft of the intellectual property (e.g., a proprietary design or confidential personal or corporate information) contained within.

Many customers (but not all) will require security requirements in the system, but it is up to the requirements engineer to ensure that security requirements are written. The requirements engineer should help identify and minimize system vulnerabilities and protect against any attacks, such as those that steal passwords for login and network access and those that insert malware. These requirements should be presented in a section of the SRS labeled "Security Requirements" or similar terminology.

Many governments and international agencies, such as the US National Institute of Standards, publish guidelines on the identification and creation of security requirements. For example, Enhanced Security Requirements for Protecting Controlled Unclassified Information: A Supplement to NIST Special Publication 800-171.

REFERENCES

Ron Ross, Victoria Pillitteri, Gary Guissanie, Ryan Wagner, Richard Graubart and Deborah Bodeau, NIST SP 800-172 Enhanced Security Requirements for Protecting Controlled Unclassified Information: A Supplement to NIST Special Publication 800-171, February 2021, available at https://csrc.nist.gov/publications/detail/sp/800-172/final

UML/SysML

The UML is a set of modeling notations that are used in software and systems engineering requirements specification, design, systems analysis, and more. The SysML (System Modeling Language) is an extension of UML dedicated to systems engineering and provides requirements diagrams, allowing users to express requirements in a textual representation and cover non-functional requirements. SysML diagrams can also express traceability links between requirements or between requirements and implementation elements as well as other modeling artifacts.

Both UML and SysML are quite similar. For the reader's convenience, a primer on UML by a major contributor to the UML standard, James Rumbaugh, is found in Appendix C. Use cases are one of the models found in UML and are very important to requirements engineers. Therefore, an extensive discussion of use cases is found in Appendix E.

The Requirements Document

The system (or software) requirements specification document is the official statement of what is required of the system developers. Hay (2003) likens the SRS to the score for a great opera—without a score there is chaos, but the finished score ensures that the activities and contributions of the various performers in the opera are coordinated.

It is also important to remember that, under those circumstances where there is a customer-vendor relationship between the sponsor and builders of the system, the SRS is a contract and is, therefore, enforceable under civil contract law (or criminal law if certain types of fraud or negligence can be demonstrated).

Users of a Requirements Document

There are several users of the requirements document, each with a unique perspective, needs, and concerns. Typical users of the document include:

- Customers
- Managers
- Developers
- Test engineers
- Maintenance engineers
- Stakeholders

Customers specify the requirements and are supposed to review them to ensure that they meet their needs. Customers also specify changes to the requirements. Because customers are involved, and they are likely not engineers, SRS documents should be accessible to the layperson (formal methodologies excepted).

Managers at all levels will use the requirements document to plan a bid for the system and to plan for managing the system development process. Managers, therefore, are looking for strong indicators of cost and time taken in the SRS.

And of course, developers use the requirements specification document to understand what system is to be developed. At the same time, test engineers use the requirements to develop validation tests for the system. Later, maintenance engineers will use the requirements to help understand the system and the relationship between its parts so the system can be upgraded or fixed. Other stakeholders who will use the SRS include all of the direct and indirect beneficiaries (or adversaries) of the system in question, as well as lawyers, judges, plaintiffs, juries, district attorneys, arbiters, mediators, etc., who will view the SRS as a legal document in the event of disputes.

It is not uncommon to have multiple specialized versions of the requirements document for a project that can be intended for different audiences. For example, a project can have a variation of:

- **Product Requirements Document**: It is written from a user's point of view to understand what a product *should* do.
- **A User-Interface Requirements Document**: describes the look and feel of the User Interface (UI) of the system. Most often it includes mockup screenshots and wireframes to give readers an idea of what the finished system will look like.

- **Technical Requirements Document**: includes requirements like the programming language the system should be developed in and the processor speed required to run the system. It might also consider the limitations of the system and its performance.
- **Quality Requirements Document**: outlines the expectations of the customer for the quality of the final product. It consists of various criteria, factors, and metrics that must be satisfied.

Requirements Document Requirements

What is the correct format for an SRS? There are many acceptable formats. For example, the IEEE 29148 standard provides a general format for a requirements specification document. Other standards bodies and professional organizations, such as the Project Management Institute (PMI), have standard requirements formats and templates. Most requirements management software will produce documents in customizable formats. But the "right" format depends on what the sponsor, customer, employer, and other stakeholders require.

That the SRS document should be easy to change is evident for the many reasons we have discussed so far. Furthermore, since the SRS document serves as a reference tool for maintenance, it should record forethought about the lifecycle of the system, that is, to predict changes.

In terms of general organization, writing approach, and discourse, best practices include:

- Using consistent modeling approaches and techniques throughout the specification, for example, a top-down decomposition, structured, or object-oriented approaches
- Separating operational specification from descriptive behavior
- Using consistent levels of abstraction within models and conformance between levels of refinement across models
- Modeling nonfunctional requirements as a part of the specification models—in particular, timing properties
- Omitting hardware and software assignments in the specification (another aspect of design rather than the specification)

Following these rules will always lead to a better SRS document. Finally, IEEE 29148 describes certain desirable qualities for requirements specification documents, and these will be discussed in Chapter 5.

Preferred Writing Style

Engineers (of all types) have acquired an unfair reputation for poor communications skills, particularly writing. In any case, it should be clear now that requirements documents should be very well written.

There are excellent texts on technical writing that should be studied by every requirements engineer (we highly recommend Laplante 2019). You are urged to improve your writing through practice and by studying well-written SRS documents to learn from them. You should also read literature, poetry, and news, as much can be learned about the economy and clarity of presentation from these writings. Some have even suggested screenwriting as an appropriate paradigm for writing requirements documents (or for user stories and use cases) (Norden 2007).

In any case, approach the requirements document like any writing—be prepared to write and rewrite, again and again. Have the requirements document reviewed by several other stakeholders (and by non-stakeholders who write well).

Text Structure Analysis

Metrics can be helpful in policing basic writing features. Most word processing tools calculate average word, sentence, and paragraph length, and these are valuable to collect because they can highlight writing that is too simplistic or too difficult to understand. One important feature of writing that is not computed by standard word processors, however, is the numbering structure depth.

Numbering structure depth is a metric that counts the numbered statements at each level of the source document. For example, first-level requirements numbered 1.0, 2.0, 3.0, and so forth are expected to be very high-level (abstract) requirements. Second-level requirements numbered 1.1, 1.2, 1.3, …

2.1, 2.2, 2.3, … 3.1, 3.2, etc. are subordinate requirements at a lower level of detail. Even more detailed requirements will be found at the third level, numbered as 1.1.1, 1.1.2, and so on. A specification can continue to fourth or even fifth-level requirements, but normally third or fourth levels of detail should be sufficient. In any case, the counts of requirements at levels 1, 2, 3, and so on provide an indication of the document's organization, consistency, and level of detail.

A well-organized SRS document should have a consistent level of detail, and if you were to list out the requirements at each level, the resultant shape should look like a pyramid in that there should be a few numbered statements at level 1 and each lower level should have increasingly more numbered statements than the level above it (Figure 4.3a).

On the other hand, requirements documents whose requirements counts at each level resemble an hourglass shape (Figure 4.3b) are usually those that contain a large amount of introductory and administrative information. Finally, diamond-shaped documents represented by a pyramid followed by decreasing statement counts at lower levels (Figure 4.3c) indicate an inconsistent level of detail representation (Hammer et al. 1998).

The NASA ARM tool introduced in Chapter 6 computes text structure for a given SRS document in an appropriate format.

Requirement Format

Each requirement should be in a form that is clear, concise, and consistent in the use of language. A requirement can be written in the form of a natural language or some other form of language. If expressed in the form of a natural language, the statement should include a subject and a verb, together with other elements necessary to adequately express the information content of the requirement. Using consistent language assists users of the requirements documents (Hull et al. 2011) and renders analysis of the SRS document by software tools much easier.

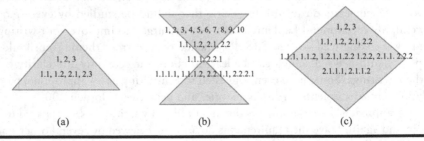

Figure 4.3   (a) Pyramid, (b) hourglass, and (c) diamond-shaped configurations for requirements text structure.

A simplified standard requirement form is:

■ The [noun phrase] shall (not) [verb phrase]

where [noun phrase] describes the main subject of the requirement, the shall (or shall not) statement indicates a required or prohibited behavior, and [verb phrase] indicates the actions of the requirement.

Consider these two requirements for the baggage handling system:

■ The system shall reject untagged baggage
■ The system shall not reject damaged baggage

We assume that the terms "baggage," "reject," "tagged," and "damage" have been defined somewhere in the SRS document.

It is not uncommon for requirements to be written in a nonstandard form. For example, the first requirement shown could be rewritten as:

■ Untagged baggage shall be rejected by the system

Both versions of this requirement are equivalent, but representing requirements in the standard form is desirable for the reasons already noted.

It is also not uncommon to omit the article in requirements, so the two example requirements would be:

■ System shall reject untagged baggage
■ System shall not reject damaged baggage

There is nothing wrong with this formulation of requirements.

It is important to identify requirements in some way for ease of reference, usually with hierarchical numbering. Hence the standard form for a requirement extends to the following:

■ [identifier] The [noun phrase] shall (not) [verb phrase]

Returning to the previous two requirements, and inventing some identifier numbers, we have the following two examples:

1.2.1 The system shall reject untagged baggage.
1.2.2 The system shall not reject damaged baggage.

Finally, it is desirable to place measurable constraints on performance for functional requirements whenever possible, enriching the standard requirement form to:

■ [identifier] The [noun phrase] shall (not) [verb phrase] [constraint phrase]

As an example of a requirement that includes a measurable target, let's return to the baggage handling system. For requirement number 1.2.1, a realistic constraint might be incorporated to yield the following requirement:

- *Attribute:* Average time that a scanner unit needs to scan a piece of luggage.
- *Scale:* Seconds (type: ratio scale).
- *Procedure:* Measure time required to scan a package for forbidden contents: take the average over 1,000 pieces of various types.
- *Planned value:* 50% less than reference value.
- *Lowest acceptable value:* 30% less than reference value.
- *Reference value:* Average time needed by competing or similar products to scan a piece of luggage.

Figure 4.4 Sample requirements attribute metrics for baggage handling system.

1.2.1 The system shall reject untagged baggage within 5 seconds of identification.

There may be variances and tolerances on the measurable targets of a requirement. For example, consider a hypothetical requirement 3.4.2 in the baggage handling system: 3.4.2. Each baggage scanner unit shall process, on average, 10 pieces of luggage per minute.

This target in this requirement can be augmented with certain constraint attributes in the format shown in Figure 4.4 (Glinz 2008).

There are other variations of the "final" individual requirement standard form, for example, those given by Hull et al. (2011) and in IEEE 29148. These variations, however, can be shown to be reducible to the format previously introduced.

Use of Imperatives

"Shall" is a command word or imperative that is frequently used in requirements specifications. Other command words and phrases that are frequently used include "should," "must," "will," "responsible for", and there are more. Wilson et al. (1997) describe the subtle differences in these imperatives, which are summarized in Table 4.2.

It is considered good practice to use "shall" as the imperative when requirements are mandatory and to omit the use of weak imperatives such as should for optional requirements. When requirements are ranked for importance, as they should be, a requirement in standard form with a low rank is, essentially, one that can be deemed an optional requirement.

Shall or Shall Not?

When writing requirements, it is generally preferred to use the positive form, rather than the negative form. That is, the requirement should be written using "shall" statements rather than "shall not" statements. IEEE 29148 advises against writing "shall not" requirements altogether. But it is possible, and sometimes desirable, for there to be exceptions particularly for those requirements pertaining to hazards. Consider the following example. The requirement

- The system shall permit access only to authorized users.

can be rewritten equivalently as

- The system shall not permit access to unauthorized users.

In this case, the first form is easier to understand because it does not involve a double negative. But consider the problem of designing test cases to prove that the requirement has been met.

Table 4.2 Imperative Words and Phrases

Imperative	Most Common Use
Are applicable	To include, by reference, standards, or other documentation as an addition to the requirements being specified
Is required to	As an imperative in specifications statements written in the passive voice
Must	To establish performance requirements or constraints
Responsible for	In requirements documents that are written for systems whose architectures are predefined
Shall	To dictate the provision of a functional capability
Should	Not frequently used as an imperative in requirement specification statements
Will	To cite things that the operational or development environment is to provide to the capability being specified

Source: Wilson, W.M., Rosenberg, L.H., & Hyatt, L.E., Automated analysis of requirement specifications, in *Proceedings of the 19th International Conference on Software Engineering*, pp. 161–171, 1997.

In the first form, the test cases have to involve the subsets of authorized and unauthorized users. Then we need to decide if we are to use exhaustive testing, or equivalence class testing, or pairwise testing, and so on for both authorized and unauthorized users. Conversely, in the second form, only a set of test cases involving one or more unauthorized users need to be created in order to show satisfaction of the requirement. We still have to decide on how to select a subset of the universe of unauthorized users, but the second form of the requirement simplifies testing significantly.

In some cases, there may be simply a psychological impact on stakeholders (particularly lawyers) in using the "shall not" form instead of the "shall" form of the requirement. Consider this example:

■ The system shall not harm humans.

   This requirement seems more straightforward than the requirement

■ The system shall cause humans no harm.

In any case, you should write requirements using shall statements whenever possible, and when clarity is not sacrificed. But "shall not" requirements may occasionally be preferred to their "shall" equivalent form.

Avoiding Imprecision in Requirements

All requirements in a specification document should be precise. That is by using measurable quantities and avoiding vague modifiers such as countless, some, approximately, huge, tiny, microscopic, and so on. These vague words need to be replaced with measurements. Assigning measurements to requirements is also a very powerful form of validation.

For example, consider the following requirement for the baggage handling system:

4.3.1 The number of pieces of baggage lost by the system shall be minimal.

This statement is vague and gives designers no real guidance. The following requirement is an improvement:

> 4.3.1 The number of pieces of baggage lost by the system shall be less than 0.001% of pieces entered.

This requirement is now measurable, and thus, more precise.

More examples of imprecise words/phrases are:

- Vague pronouns (such as "it," "this,", "that")
- Ambiguous terms such as adverbs and adjectives (such as "almost always," "many," "few," "significant," "minimal") and ambiguous logical statements (such as "or," "and/or")
- Open-ended, non-verifiable terms (such as "provide support," "but not limited to," "as a minimum")
- Comparative phrases (such as "better than," "higher quality")
- Loopholes (such as "if possible," "as appropriate," "as applicable")
- Terms that imply totality (such as "all," "always," "never," and "every")

In a requirement, ambiguous terms should be replaced with fractions or ranges. For example, the fractions 1/100, 1/20, 1/2, and 9/10 could replace the aforementioned words, respectively. Appropriate ranges could be "<10," "10–20," "21–50," and "51–99," respectively.

Other forms of imprecision using fuzzy words can lead to ambiguity in requirements. For example, specifying something to happen "frequently" in a requirement should be replaced by a range or an approximate rate. For example, instead of "frequently" use "more than 75% of the time." Providing a hard number or range allows for a reference point for future refinement of the information from a fuzzy number (or range) to a finer range or exact number.

You can't always replace hedging words with precision. Sometimes only an approximation is possible. For example, consider a discussion of the likelihood of some event for which you have no prior statistics, and not even a probability model for constructing predictive statistics. This would be a case where stakeholder negotiated probabilities would be needed to agree to some acceptance criteria.

To address the problem of ambiguous and incomplete nonfunctional requirements, consultant Tom Gilb (2005) developed *Planguage*, a language with a rich set of keywords that permits precise statements of quality attributes and other project goals. We will discuss *Planguage* further in the next chapter.

A more recent approach to specify requirements is presented by Meyer et al. (2019) based on the "PEGS" view of requirements where the four parts cover Project, Environment, Goals, and System. This approach treats requirements as a project activity, not necessarily a lifecycle step. The starting point, which gives its name to the approach, is that requirements should cover the four aspects mentioned, the four "PEGS," defined as follows:

- A goal is a result desired by an organization.
- A system is a set of related artifacts, devised to help meet certain goals.
- A project is the set of human processes involved in the planning, construction, revision, and operation of a system.
- An environment is the set of entities (such as people, organizations, devices, other material objects, regulations, and other systems) external to the project and system but with the potential to affect the goals, project, or system or to be affected by them.

Requirements Document Size

There are few studies covering the statistics of requirements specification document size in terms of number of requirements, page length, and other measures. Yet, the question is often asked—how long should the SRS document be? In a 2003 survey of software professionals concerning software requirements engineering practices, respondents were asked about SRS document length (Neill and Laplante 2003). Respondents reported that about 40% of documents comprised fewer than 25 pages and about 30% were between 25 and 50 pages. Roughly 25% of SRS documents were reported to be between 51 and 100 pages, 17% between 101 and 250 pages, and the balance was of length 251 or more pages.

The aforementioned numbers are consistent with those found in a study of 56 NASA requirements specifications documents developed in the 1990s through the early 2000s (Wilson et al. 1997). For the projects studied, the SRS document sizes ranged from about 140–28,000 lines of text, with a median of 2,200 and an average of 4,700 lines of text. Assuming 60 lines of text per page, these figures convert to a range of 2–470 pages, a median of 37, and an average of 78 pages.

Based on our own observations and discussions with many practitioners, we suspect that if these studies were repeated today the numbers would not differ significantly. Clearly then, the industrial norm is for relatively short requirements specification documents. No judgment can be rendered as to whether this finding is good or bad since each individual SRS document has to be judged on qualities other than length.

Behavioral Specifications

In some cases, the requirements engineer may be asked to reverse engineer requirements for an existing system when the requirements do not exist, are incomplete, are out of date, or are incorrect. It may also be necessary to generate requirements for open-source software (software that is free for use and/or redistribution under the terms of a license) for the purposes of generating test specifications. In these cases, a form of SRS, the behavioral specification, needs to be generated.

The behavioral specification is identical in all aspects to the requirements specification, except that the former represents the engineers' best understanding of what the users intended the system to do, while the latter represents the users' best understanding of what the system should do. The behavioral specification has an additional layer of inference and, therefore, can be expected to be even less complete than a requirements specification.

Fortunately, there is an approach to generating the behavioral specification. The technique involves assembling a collection of as many artifacts as possible that could explain the system's intent. Then, these artifacts are used to reconstruct the system's intended behavior as it is best understood. The foregoing description is adapted from the original paper by Elcock and Laplante (2006).

Artifacts that may be used to derive a behavioral specification include (but should not be limited to):

- Any existing requirements specification, even if it is out of date, incomplete, or known to be incorrect
- User manuals and help information (available when running the program)
- Release notes
- Bug reports and support requests

- Application forums
- Relevant versions of the software under consideration

Some of these artifacts may have to be scavenged from various sources, such as customer files, emails, open-source community repositories, or archives. A brief description of how these artifacts are used in generating the specification is given below.

Starting with user manuals, statements about the behavior of an application in response to user input can be directly related to the behavioral specification. User manuals are particularly well suited for this purpose since describing the response of software to user stimulus is germane to their existence. Help information, such as support websites and application help menus, is also rich in information that can either be used directly or abstracted to define behavioral requirements.

Next, release notes can be consulted. Release notes are typically of limited benefit to the process of developing test cases since they tend to focus on describing which features are implemented in a given release. There is usually no description of how those supported features should function. However, release notes are particularly important in resolving the conflict that arises when an application does not respond as expected for features that were partially implemented in or removed from future implementation.

Defect reports can also be a great source for extracting behavioral responses because they are typically written by users, they identify troublesome areas of the software, and they often clarify a developer's intention for the particular behavior. Defect reports, at least for open-source systems, are readily found in open repositories such as Bugzilla. While it is true that in some cases bug reports contain too much implementation detail to be useful for extracting behavioral responses, they can be discarded if behavior cannot be extrapolated from the details.

In many ways, the content of support requests is similar to bug reports in that they identify unexpected behavior. Unlike bug reports, however, support requests can sometimes be helpful in identifying features that are not fully implemented as well as providing information that illuminates the expectations of users. It is important to investigate both since, in addition to providing the insights mentioned, they may also aid in rationalizing unexpected behavior.

For many open-source projects, and some closed-source projects, there are web-based forums associated with the project. Within these forums, various amounts of behavioral information can be extracted. Ranging from useless to relevant, open-discussion postings need to be carefully filtered and applied only when other development artifacts are lacking. As with other artifacts, these postings can also be used to clarify behavior.

Finally, in the absence of any other artifacts, the software being tested could itself be an input to developing structural (glass-box) test cases. Assuming that this approach is necessary, the reality is that it will really be the tester's characterization of correct behavior that will largely prevail in defining the test cases.

Once the discovery process has concluded, the behavioral specification can be written. The format of the behavioral specification is identical to that for the requirements specification (Elcock and Laplante 2006) and all of the IEEE 29148 rules should be applied.

Best Practices and Recommendations

Writing effective requirements specifications can be very difficult even for trivial systems because of the many challenges that we have noted. Some of the more common dangers in writing poor SRS documents include:

- Mixing of operational and descriptive specifications
- Combining low-level hardware functionality and high-level systems and software functionality in the same functional level
- Omission of timing information

Other bad practices arise from failing to use language that can be verified. For example, consider this set of requirements:

- The system shall be completely reliable.
- The system shall be modular.
- The system will be fast.
- Errors shall be less than 99%.

What is wrong with these? The problem is they are completely vague and immeasurable, and therefore, their satisfaction cannot be demonstrated. For example, what does "completely reliable" mean? Any arbitrary person will have a different meaning for reliability for a given system. Modularity (in software) has a specific meaning, but how is it measurable? What does "fast" mean? Fast as a train? Faster than a speeding bullet? This requirement is just so vague. And finally "errors shall be less than 99%" is a recipe for a lawsuit. 99% of what? Over what period of time?

For the above set of requirements, the improved version is:

- Response times for all level one actions will be less than 100 ms.
- The cyclomatic complexity of each module shall be in the range of 10–40.
- 95% of the transactions shall be processed in less than 1 s.
- Meantime before the first failure shall be 100 hours of continuous operation.

But even this set is imperfect, because there may be some important details missing—we really can't know what, if anything, is missing outside the context of the rest of the SRS document. Using Glintz's attribute template (Figure 4.4) is a good solution to this problem.

Some final recommendations for the writing of specification documents are:

- Invent and use a standard format and use it for all requirements.
- Use language in a consistent way.
- Use "shall" for mandatory requirements.
- Use "should" for desirable requirements.
- Use text highlighting to identify key parts of the requirement.
- Avoid the use of technical language unless it is warranted.

Dick (2010) also suggests avoiding the following when writing the requirements document:

- Rambling phrases
- Hedging clauses, such as "if necessary"
- Conjunctions such as "and," "or," and "but," since these
- Can lead to multiple requirements in a single statement
- Speculative terms such as "perhaps" and "sometimes"
- Vague terms such as "user-friendly" and "generally" since these cannot be verified

But we are not done yet with our treatment of writing the requirements specification. Chapters 5 and 6 are devoted to perfecting the writing of specific requirements.

VIGNETTE 4.2   Product Line Engineering

It is often the case that requirements specifications for product line systems (see Exercise 1.14) consist of a baseline or template of shared requirements. Then, different requirements are added for the diverse instances of related products in the line. Purposeful requirements engineering for a product line can mean significant cost savings over haphazard approaches. Consider the following example. A very large (multibillion dollar) multinational firm produces engines of all kinds—from small gas-powered to very large jet turbines. The company has more than 150 major projects with over 50,000 requirements in each. Many of these projects have very similar requirements sets. Yet the leveraging effect of product line requirements engineering is not realized.

Tool use is inconsistent across projects and engineering sites, so the power of reusability, change tracking, and identification of missed or excessive requirements is lost. One would expect that a large organization would do things better, but part of the problem for this company is that there were no tools available in the early days of the company and the older engineers didn't realize the potential for product line engineering and the need to change the way things were done. Further problems arose as the company grew over the decades by acquiring companies that used many different tools and had far-ranging engineering practices.

Of course, standardizing tool use and engineering practices across the company is a big part of the solution, but so too is training, so that all engineers use the tool in the same correct way. A corporate and project "metaphor" needs to be created so that engineers are all "on the same page." Solid product line engineering practices would then be possible.

The situation just described is not unique to large companies.

Exercises

4.1   Under what circumstances is it appropriate to represent an SRS using informal techniques only?

4.2   What can the behavioral specification provide that a requirements document cannot?

4.3   If the customer requests that future growth and enhancement ideas be kept, where can these ideas be placed?

4.4   What are some items to be included under "data retention" in the SRS?

4.5   Here are some more examples of vague and ambiguous requirements that have actually appeared in real requirements specifications. Discuss why they are vague, incomplete, or ambiguous. Provide improved versions of these requirements (make necessary assumptions).

   4.5.1   The tool will allow for expedited data entry of important data fields needed for subsequent reports.

   4.5.2   The system will provide an effective means for identifying and eliminating undesirable failure modes and/or performance degradation below acceptable limits.

4.5.3 The database creates an automated incident report that immediately alerts the necessary individuals.

4.5.4 The engineer shall manage the system activity, including the database.

4.5.5 The report will consist of data, in sufficient quantity and detail, to meet the requirements.

4.5.6 The data provided will allow for a sound determination of the events and conditions that precipitated the failure.

4.5.7 The documented analysis report will include, as appropriate, investigation findings, engineering analysis, and laboratory analysis.

4.6 In Section 9.4 of the SRS in the appendix, which requirements are suitable for representing using the "measurable targets" in the format shown in Figure 4.4?

4.7 Many of the requirements in the appendix can be improved in various ways. Select ten requirements listed and rewrite them in an improved form. Discuss why your rewritten requirements are superior using the vocabulary of IEEE 29148.

4.8 Convert the following "shall not" requirements in Appendix A to "shall" format: 4.1.1, 5.2.2, 6.1.8, 8.1.17.

4.9 Convert the following "shall not" requirements in Appendix A to "shall" format: 9.1.2.6, 9.3.3, 9.3.5, 9.3.13, 9.5.10.

*4.10 Research different techniques for creating context diagrams and prepare a report highlighting the strengths and weaknesses of each.

*4.11 The requirements specifications in Appendices A and B use imperatives inconsistently. Using a textual analysis tool such as ARM or via manual analysis, assess the use of imperatives in these documents and suggest improvements.

References

Berenbach, B., Schneider, F., & Naughton, H. (2012, September). The use of a requirements modeling language for industrial applications. In *2012 20th IEEE International Requirements Engineering Conference (RE)* (pp. 285–290). IEEE, Chicago, IL, USA.

Dardenne, A., Van Lamsweerde, A., & Fickas, S. (1993). Goal-directed requirements acquisition. *Science of Computer Programming*, 20(1–2): 3–50.

DassaultSystems.(2016).CATIAReqtify.Retrievedfromhttps://www.3ds.com/products-services/catia/products/reqtify

Dick, J. (2010). Requirements engineering: Principles and practice. In P. Laplante (Ed.), *Encyclopedia of Software Engineering* (pp. 949–961). Taylor & Francis, Boca Raton, FL, Published online.

Eberlein, A., & Jiang, L. (2011). Requirements engineering: Technique selection. In P. Laplante (Ed.), *Encyclopedia of Software Engineering* (pp. 962–978). Taylor & Francis, Boca Raton, FL, Published online.

Elcock, A., & Laplante, P. A. (2006). Testing without requirements. *Innovations in Systems and Software Engineering: A NASA Journal*, 2: 137–145.

Gilb, T. (2005). *Competitive Engineering: A Handbook for Systems Engineering, Requirements Engineering, and Software Engineering Using Planguage.* 1 edition. Butterworth-Heinemann, UK.

Glinz, M. (2008). A risk-based, value-oriented approach to quality requirements. *Computer*, 25(2): 34–41.

Hammer, T. F., Huffman, L. L., Rosenberg, L. H., Wilson, W., & Hyatt, L. (1998). Doing requirements right the first time. *CROSSTALK The Journal of Defense Software Engineering*, 1: 20–25.

Hay, D. (2003). *Requirements Analysis: From Business Views to Architecture.* Prentice Hall PTR, Upper Saddle River, NJ.

Hull, E., Jackson, K., & Dick, J. (2011). Requirements engineering in the problem domain. In E. Hull, K. Jackson, & J. Dick. (Eds.), *Requirements Engineering* (pp. 93–114). Springer-Verlag, London.

IEEE Standard 830-1998. (1998). *Recommend Practice for Software Requirements Specifications*. IEEE Standards Press, Piscataway, NJ.

ISO/IEC/IEEE Standard 29148. (2011). *Systems and Software Engineering—Life Cycle Processes—Requirements Engineering*. IEEE, Piscataway, NJ.

Jeannet, B., & Gaucher, F. (2015). Debugging real-time systems requirements: simulate the "what" before the "how". In *Embedded World Conference*, Nürnberg, Germany.

Kassab, M., & Laplante, P. (2022). The current and evolving landscape of requirements engineering state of practice. *IEEE Software*. DOI: 10.1109/MS.2022.3147692.

Laplante, P. A. (2006). *What Every Engineer Should Know about Software Engineering*. CRC Press, Boca Raton, FL.

Laplante, P. A. (2019). *Technical Writing: A Practical Guide for Engineers and Scientists*, 2nd edition. CRC Press, Taylor & Francis, Boca Raton, FL.

Meyer, B., Bruel, J. M., Ebersold, S., Galinier, F., & Naumchev, A. (2019, October). Towards an anatomy of software requirements. In *International Conference on Objects, Components, Models and Patterns* (pp. 10–40). Springer, Cham.

NASA WIRE (Wide-Field Infrared Explorer) Software Requirements Specification. (1996). http://science.nasa.gov/missions/wire/ (accessed January 2017).

Neill, C. J., & Laplante, P. A. (2003). Requirements engineering: The state of the practice. *Software*, 20(6): 40–45.

Norden, B. (2007). Screenwriting for requirements engineers. *Software*, 24: 26–27.

Ricca, F., Scanniello, G., Torchiano, M., Reggio, G., & Astesiano, E. (2010). On the effectiveness of screen mockups in requirements engineering: Results from an internal replication. In *Proceedings of the 2010 ACM-IEEE International Symposium on Empirical Software Engineering and Measurement* (Vol. 17, p. 26), Bolzano, Italy.

Scott, W., & Cook, S. C. (2004). *A Context-Free Requirements Grammar to Facilitate Automatic Assessment*, AWRE'04 9th Australian Workshop on Requirements Engineering, Adelaide, Australia.

Sutcliffe, A., Thew, S., & Jarvis, P. (2011). Experience with user-centred requirements engineering. *Requirements Engineering*, 16(4): 267–280.

Whittle, J., Sawyer, P., Bencomo, N., Cheng, B. H., & Bruel, J. M. (2009, August). Relax: Incorporating uncertainty into the specification of self-adaptive systems. In *2009 17th IEEE International Requirements Engineering Conference* (pp. 79–88). IEEE, Atlanta, GA, USA.

Wilson, W. M., Rosenberg, L. H., & Hyatt, L. E. (1997). Automated analysis of requirement specifications. In *Proceedings of the 19th International Conference on Software Engineering* (pp. 161–171). Boston, Massachusetts, USA.

Chapter 5

On Nonfunctional Requirements

Motivation to Consider NFRs Earlier in Development

Systems are characterized both by their functional behavior (what the system does) and by their nonfunctional behavior (how the system behaves concerning some observable attributes like reliability, reusability, and maintainability).

In the software/system marketplace, in which functionally equivalent products compete for the same customer, nonfunctional requirements (NFRs) become more important in distinguishing between the competing products. However, in practice, NFRs may receive little attention relative to functional requirements (FRs) (Weber and Wesbrot 2003). This is mainly true because of the nature of NFRs which poses a challenge when taking the choice of treating them at an early stage of the development process. NFRs are characterized to be:

1. **Subjective and Relative**: where the satisfaction of one NFR can be based on or influenced by personal feelings, opinions, or tastes. For example, "maintaining a system with 98% availability" can be considered acceptable by some stakeholders and for some systems, but not by others.
2. **Scattered**: where NFRs tend to become scattered among multiple modules when they are mapped from the requirements domain to the solution space. For example, a "high-modifiability" related requirement can only be satisfied with multiple implementations and configurations that are spread at multiple modules of the system.
3. **Interacting**: in the sense that attempts to achieve one NFR can help or hinder the achievement of other NFRs at particular functionality. For example, the attempt to achieve security through implementing authentication may come at the expense of performance by increasing latency. Such interaction creates an extensive network of interdependencies and tradeoffs among NFRs which is not easy to trace or estimate (Chung et al. 2000).

Once a system has been deployed, it is typically straightforward to observe whether a certain FR has been met, as the areas of success or failure in their context can be rigidly defined.

DOI: 10.1201/9781003129509-5

However, the same is not true for NFRs as these can refer to concepts that can be interdependent and difficult to measure. The problem of lacking any early NFR integration within the specified system is likely to cause an increase in the effort and maintenance overhead (Seffah et al. 2005), or even a catastrophic project failure. The following list provides valid examples:

- **Target Data Breach (O'Neil 2015)**: Between November 27 and December 18, 2013, the Target Corporation's network was breached, which became the second-largest credit and debit card breach after the TJX breach in 2007. In the Target incident, 40 million credit and debit card numbers were stolen in only 2 weeks and a half. Target's systems were compromised by a stolen password. Upon analyzing the "security" breach, it was found that an HVAC company had access to their networks to monitor the environmental conditions of the servers. The perpetrators were then able to legitimately gain access to Target's systems through the available access that was allowed for the HVAC company. They eventually installed malware on the point of sale (POS) systems.

- **Siemens**: Possible Hearing Damage in Some Cell Phones (Siemens 2004): In 2004, Siemens issued a "safety" warning that some of its cell phones may have a software problem that could cause them to emit a loud noise, possibly causing a hearing loss for the phone user. The malfunction happens only if, while the phone is in use, the battery runs down to the point that the phone automatically disconnects the call and begins to shut down.

- **Mars Climate Orbiter (Breitman et al. 1999)**: This was one of two NASA spacecraft in the Mars Surveyor '98 program. The mission failed because of a software "interoperability" issue. The craft drifted off course during its voyage and entered a much lower orbit than planned and was destroyed by atmospheric friction. The metric/imperial mix-up which destroyed the craft was caused by a software error back on Earth. The thrusters on the spacecraft which were intended to control its rate of rotation were controlled by a computer that underestimated the effect of the thrusters by a factor of 4.45. This is the ratio between a pound-force—the standard unit of force in the imperial system—and a Newton—the standard unit in the metric system. The software on Earth was working in pounds-force, while the spacecraft expected figures in Newton.

- **London Ambulance System (LAS) (Finkelstein and Dowell 1996)**: In 1992, The London Ambulance Service introduced a new computer-aided dispatch system that was intended to automate the system that dispatched ambulances in response to calls from the public and the emergency services. This new system was extremely inefficient and ambulance response times increased markedly. Shortly after its introduction, it failed and LAS reverted to the previous manual system. The failure of the system was mainly due to a failure to consider "human and organizational factors" in the design of the system.

- **The National Library of Medicine MEDLARS II system (Boehm and In 1996)**: The project was initially developed with many layers of abstraction to support a wide range of future publication systems. The initial focus of the system was toward improving "portability" and "evolvability" qualities. The system was scrapped after two expensive hardware upgrades due to "performance" problems.

- **The New Jersey Department of Motor Vehicles' licensing system (Babcock 1985)**: This system was written in the fourth-generation programming language, ideal to save development time. When implemented, the system was so slow that at one point more than a million

New Jersey vehicles roamed the streets with unprocessed license renewals. The project aimed at satisfying "affordability" and "timeliness" objectives but failed due to "performance scalability" problems.

Despite this obvious importance and relevance of NFRs, they are almost always left to be verified after the implementation is finished, which means NFRs are not mapped directly and explicitly from requirements engineering to implementation. This is mainly due to the enormous pressure toward "urgent" deployment. This may leave software/system development with potential exacerbation of the age-old problem of requirements errors that are not detected until very late in the process. Some of the well-known problems of the software development due to the NFRs omission include: (i) cost and schedule overruns, (ii) software systems discontinuation, and (iii) dissatisfaction of software systems users.

The importance of software/system compliance with the imposed NFRs requires management of their scope (Kassab et al. 2007), which brings up the importance of clearly defining, tracing, and effort estimating the complex and frequently ill-defined NFRs and their interrelations in an increasingly complex large-scale software system.

Watch a short video from Dr. Kassab discussing NFRs in software engineering.
https://phil.laplante.io/requirements/NFRs.php

What Is an NFR?

In general, and because of their diverse nature, NFRs have been (at best) specified in loose, fuzzy terms that are open to wide-ranging and subjective interpretation. As such, they provide little guidance to architects and engineers as they make the already tough tradeoffs necessary to meet schedule pressures and functionality goals. For instance, most software engineering approaches (IEEE 29148), *The Unified Software Development Process* (Grady 1992), and industrial practices specify NFRs separately from FRs of a system. This is mainly because the early integration of NFRs is difficult to achieve and usually accomplished at the later phases of the software development process. However, since the integration is not supported from the requirements phase to the implementation phase, some of the software engineering principles, such as abstraction, localization, modularization, uniformity, and reusability, can be compromised. Furthermore, the resulting system is more difficult to maintain and evolve.

Instead, NFRs need to be made precise and clear right from the requirements phase. But to be able to specify the NFRs in precise terms, there must be a general understanding of what the term NFR stands for, and what are the relations that the NFR may be exposed to during the life cycle of the project. Although the term "nonfunctional requirement" has been in use for more than 40 years, there is still no consensus in the requirements engineering community on what NFRs (Kassab 2009) are and what are relations that an individual NFR may participate in. Table 5.1 gives an overview of selected definitions from the literature or the web which are representative of the definitions that exist. We provided our definition in the last row of the table derived from experience and knowledge of the existing definitions.

Table 5.1 Definitions of the Term "Non-Functional Requirement(s)".

Source	Definition
Antón (1997)	Requirements describe the nonbehavioral aspects of a system, capturing the properties and constraints under which a system must operate.
Davis (1993)	Requirements that represent the required overall attributes of the system, including portability, reliability, efficiency, human engineering, testability, understandability, and modifiability.
IEEE 610.12	The term is not defined. The standard distinguishes design requirements, implementation requirements, interface requirements, performance requirements, and physical requirements.
ISO/IEC/IEEE Standard 29148	The term is not defined. The standard defines the categories functionality, external interfaces, performance, attributes (portability, security, etc.), and design constraints. Project requirements (such as schedule, cost, or development requirements) are explicitly excluded.
Jacobson et al. (1999)	A requirement that specifies system properties, such as environmental and implementation constraints, performance, platform dependencies, maintainability, extensibility, and reliability. A requirement that specifies physical constraints on a functional requirement.
Kotonya and Sommerville (1998)	Requirements that are not specifically concerned with the functionality of a system. They place restrictions on the product being developed and the development process, and they specify external constraints that the product must meet.
Mylopoulos et al. (1992)	"... global requirements on its development or operational cost, performance, reliability, maintainability, portability, robustness, and the like. (...) There is not a formal definition or a complete list of nonfunctional requirements."
Ncube (2000)	The behavioral properties that the specified functions must have, such as performance and usability.
Robertson and Robertson (1999)	A property, or quality, that the product must have, such as an appearance, or a speed or accuracy property.

(*Continued*)

Table 5.1 (*Continued*)   Definitions of the Term "Non-Functional Requirement(s)".

Source	Definition
Wiegers (2003)	A description of a property or characteristic that a software system must exhibit or a constraint that it must respect other than an observable system behavior.
Wikipedia: Nonfunctional Requirements	Requirements specify criteria that can be used to judge the operation of a system rather than specific behaviors.
Our definition	Umbrella term to cover all those requirements which are not explicitly defined as functional.

Employing our definition then and the example of the brick presented in Vignette 1.1, one can clearly see that while the color and texture of the brick might be important to the homeowner, it has no role in the function of the home. Therefore, here color and texture are NFRs.

What Are the Different Types of NFRs?

In general, NFRs can be of two main types: quality requirements or constraints.

Quality Requirements: is the totality of characteristics of an entity that bear on its ability to satisfy stated and implied needs (ISO 25000). Software quality is an essential and distinguishing attribute of the final product. Evaluation of software products to satisfy software quality needs is one of the processes in the software development life cycle.

Software product quality can be evaluated by measuring internal attributes (typically static measures of intermediate products which specify internal quality from the internal view of the product), or by measuring external attributes (typically by measuring the behavior of the code when executed to specify the required level of quality from the external view), or by measuring quality in use attributes (which represents the user's view of the quality of the software product when it is used in a specific environment and a specific context of use). Figure 5.1 presents the three views of the product quality at different stages in the software life cycle.

Many approaches (Boehm (1976), Chung et al. (2000), and (ISO 25000)) classify software quality in a structured set of characteristics which are further decomposed into sub-characteristics. For example, "Confidentiality" quality is a sub-characteristic of "Security" which is a sub-characteristic of "Reliability."

We will discuss the quality requirements in more detail shortly in this chapter.

Constraints: are not subject to negotiations and, unlike qualities, are off-limits during design tradeoffs. Constraints are defined in Leffingwell and Widrig (2003) as restrictions on the design of the system, or the process by which a system is developed, that does not affect the external behavior of the system but that must be fulfilled to meet technical, business, or contractual obligations. A key property of a constraint is that a penalty or loss of some kind applies if the constraint is not respected. There are different types of constraints:

■ **Design Implementation Constraints**: These are premade design/implementation decisions that we have to live with and limit our technical choices. They become load-bearing walls on the design space. Examples of these types of constraints include:

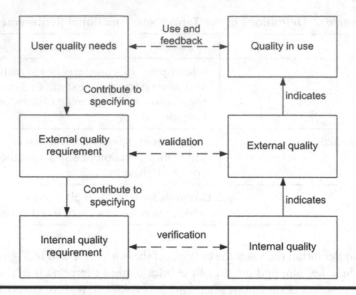

Figure 5.1 Quality in the software/systems life cycle (ISO12601).

- – Utilization of and/or integration with legacy systems
- – Mandated technologies, architectural patterns, languages, protocols, standards, patterns, and so forth
- ■ **Business Constraints**: conditions that exist within the business that cannot be changed and impact design. These constraints may include:
  - – Cost and benefit expectation
  - – Rollout (incremental delivery) schedule
  - – Targeted market and time to market
  - – Projected lifetime of the system
  - – Alignment of team structure, available expertise, and software structures
  - – Product lines or another reuse strategy to amortize investments

Operating Constraints: These are constraints that include physical constraints, personnel availability, skill-level considerations, and system accessibility for maintenance.

Political/Cultural Constraints: These are constraints that include policy and legal issues (e.g., what laws and standards apply to the product).

While the origin of the constraints may differ, all will impact early design decisions to one degree or another. This is why it is important to document all constraints. We can simply write them down including any supporting information that is relevant or available including:

- ■ Rationale
- ■ Any associated flexibility in the constraint
- ■ Any suitable alternatives
- ■ Originating stakeholder(s)

While there is no need to prioritize constraints according to importance, as by definition they must be adhered to, it can still be helpful to assign difficulty rankings.

Quality Requirements in Practice

Quality requirements play a significant role in the creation of an architecture for a given system. If ignored or not fully understood, you risk creating a system that may not be fit for its intended purpose. In practice, quality requirements are not stand-alone requirements, as their existence is always dependent on other concepts or requirements in the project context. For example, if a functional requirement is "when the user presses the red button the Options dialog appears," then:

■ A *performance* quality requirement annotation might describe how quickly the dialog will appear;
■ An *availability* quality annotation might describe how often this function will fail, and how quickly it will be repaired;
■ A *usability* quality annotation might describe how easy it is to learn this function.

In general, there are three problems when dealing with quality requirements in practice:

Problem 1: In most of the time, the definitions provided for a quality requirement are not testable.

It is meaningless to say that a system shall be "modifiable." Single-word descriptions of quality attribute properties are meaningless and wide open to interpretation. Such a vague description lacks quantifiable measures which are impossible to establish without a specific context.

Problem 2: Endless time is wasted on arguing over which quality a concern belongs to. Is a system failure due to a denial of service attack an aspect of availability, performance, security, or usability?

Many approaches classify software quality in a structured set of characteristics which are further decomposed into sub-characteristics. But there is no widely accepted standard and utilized standard for quality requirements taxonomy. It is impossible to provide an exhaustive or precise definition for any quality attribute as the specific meaning will depend on the project context.

Problem 3: Each attribute community has developed its own vocabulary. While classic quality attributes (e.g., availability, modifiability, performance, security) are well discussed in books and literature, in practice, there are other qualities attributes that are are domain-specific or may emerge in the context of disruptive technology. Here are some examples:

Iowability: In the book "Software Architecture in Practice" (Bass et al. 2021), the authors refer to a system architecture that was designed with the conscious goal of retaining key employees and attracting talented new hires to relocate to a quiet region of the American Midwest. That system's architects spoke of imbuing the system with "Iowability." This quality attribute was achieved for that system by bringing in state-of-the-art technology and giving their development teams wide creative latitude. While it is unlikely to find this quality attribute in any standard list of quality attributes, it is still important to that system in that specific context. Quality attribute requirements are derived from the context, not a textbook.

Caring: This quality could mean different things to different people and different systems. Consider, for example, a robotic surgery system. These systems are now used extensively for many types of procedures including heart, cancer, and prostate surgery. While current systems are robotic in the sense that the machine mimics the movements of a human surgeon, fully autonomous robot surgical systems are envisioned in the near future, replacing surgeons and nurses in the operating room. While we expect the human surgeon and nurses to care about the patient, as systems engineers, what should we require of a fully autonomous robot surgeon? Caring likely encompasses elements of the qualities of trust, reliability, privacy, and more, but none of these, by themselves, capture the full essence of caring. Instead, caring is a super-ility resulting as some composite of other ilities (and the system quality called empathy). One possible hierarchical representation for caring in terms of these other ilities is given in Figure 5.2.

Consider, for example, the constituent qualities of caring in the robotic surgery system. The surgeon wants the system to be safe and reliable, likely as the primary concern. Both the safe and trustworthy operation of the system contribute to a sense of reliability in the system and are of concern to the systems engineers. The patient shares these concerns but also wants the system to preserve his privacy (e.g., by not exposing medical records or embarrassing images). If the actors in the operating room were humans, the patient would probably also expect a sense of empathy from the surgeons or nurses. Of course, robot surgeons look nothing like human surgeons; therefore, there would need to be a means by which the robots could emote empathy via speech or facial expression generation on some display device. These diverse concerns, with respect to the qualities related to caring, will inform the specific system requirements discovery and representation process. Laplante et al. discuss the caring quality in the context of IoT in the healthcare domain (Laplante et al. 2017).

■ **Humanization**: There are may be questions on the moral role that an IoT system may play in human lives, particularly concerning personal control. For example, applications in the IoT involve more than computers interacting with other computers. Fundamentally, the success of the IoT will depend less on how far the technologies are connected and more on the humanization of the technologies that are connected. IoT technology may reduce people's autonomy, shift them toward habits, and then shift power to corporations focused on financial gain. For the education domain, many studies indicate that face-to-face interaction between students will not only benefit a child's social skills but also positively contribute to character building. The issue that may arise from increased IoT technologies in the education domain is the partial loss of the social aspect of going to school. Hence, considering "Humanization" as a quality for any such systems is worth the research.

Dehumanization of humans in interacting with machines is a valid concern, and it is discussed in many studies (e.g., Kassab et al. 2018, 2020).

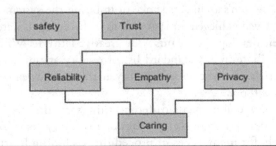

Figure 5.2 Taxonomy for caring quality requirement (Laplante et al. 2017).

Specifying with Quality Attribute Scenarios

The Software Engineering Institute (SEI) provided a solution to the above problems in the form of using quality attribute scenarios as a means of characterizing quality attributes. Quality Attribute Scenarios are a means to better characterize quality attributes. While "Use case scenarios" are descriptions of the expected response given specific functional input, the "Quality Attribute Scenarios (QAS)" are descriptions of a quality attribute response given a stimulus. The SEI has developed a six-part framework to help ensure that quality attribute requirements are actionable and to some extent "testable." The six parts of the QAS are as follows:

- **Stimulus**: A condition that affects the system.
- **Source of the Stimulus**: The entity that generated the stimulus.
- **Environment**: The condition under which the stimulus occurred.
- **Artifact Stimulated**: The artifact that was stimulated by the stimulus.
- **Response**: The activity that results because of the stimulus.
- **Response Measure**: The measure by which the system's response will be evaluated.

So an alternative way to specify a security quality requirement through a quality scenario is:

> A disgruntled employee from a remote location attempts to modify the pay rate table during normal operations. The system maintains an audit trail and the correct data is restored within 12 hours.

If we analyze the above scenario, we can detect the six elements:

- **Stimulus**: attempting to modify
- **Source of the Stimulus**: a disgruntled employee
- **Environment**: normal operation
- **Artifact Stimulated**: the pay rate table
- **Response**: maintaining an audit trail and restoring the correct data
- **Response Measure**: 12 hours.

With using the QAS, the focus is not on the naming of the quality attribute but on its underlying concerns—to illustrate the concepts that are fundamental to the domain. It is not important what term we use to specify a quality requirement as long as it is characterized in a QAS form.

Nevertheless, early in a project six-part, QAS may be impractical. For example, it may not be possible to know the exact "artifact stimulated" until we do some design. Also, stakeholders may not know specific responses or response measures as realistic response measures may need to be determined via experimentation. While it may take too long to develop six-part scenarios in real-time requirements elicitations, it is recommended to start by getting the essentials (source, stimulus, response) and then to develop the details through analysis and iteration with stakeholders.

Specifying with Planguage

Tom Gilb developed *Planguage*, a language with a rich set of keywords that permits precise specifications of requirements including qualities. For a comprehensive discussion of Planguage, see Gilb's book *Competitive Engineering: A Handbook for Systems Engineering, Requirements Engineering, and Software Engineering Using Planguage* (Butterworth-Heinemann 2005).

Planguage offers a very large number of keywords to permit expressing different types of requirements, but you need only a few of them to get started. Following is an example of how to express a performance requirement by using Planguage. Expressed in traditional form, the requirement might read: "The system shall take at most 5 seconds to display search results 99% of the time."

To avoid ambiguity, a quality requirement shall be more specific by highlighting the following elements:

Tag: Performance.Results.response time
Ambition: Fast response time to generate results on the base user platform.
Scale: Seconds of elapsed time between pressing the Enter key or clicking OK to request the search results and the beginning of the display of the report.
Meter: Stopwatch testing performed on a predefined number of test searches that represent a defined usage operational profile for a field office accountant.
Goal: No more than 5 seconds for 99% of the time a search is conducted.
Stretch: No more than 2 seconds for predefined reports, 5 seconds for all reports.
Wish: No more than 1 second for all searches.
Base User Platform Defined: Quad-core processor, 8GB RAM, Windows 10, single-user, at least 60% of system RAM and 80% of system CPU capacity free, the network connection speed of at least 50 Mbps.

Popular Quality Requirements

Security

To be able to argue about security, one has to understand what it means. Let's examine security in the light of three contexts:

■ Think about "Fort Knox US Army post in Kentucky." In addition to housing various US Army functions, it is also the home to a gold bullion depository with 5,000+ tons of gold housed there. So, what is the business asset that needs protection in this context? And what does protect mean here?

■ Think also of the "The Central Intelligence Agency (CIA)" which is a US civilian intelligence organization with the primary purpose is to collect information about foreign governments, corporations, and individuals. It uses this information to influence public policymakers. What is the business asset that needs protection? And what does protect mean in this case?

■ Now, think of an "electrical distribution grid" system. Again, what is the asset that needs protection? And what protects mean in this case?

To define security in a specific context, we need to understand the business asset that requires protection. We then want to ensure the system protects that business asset, and we do this by imagining how the business asset is at risk and then articulate the desired response to that risk.

So, security is defined as a measure of the system's ability to protect data and information from unauthorized access while still providing access to people and systems that are authorized.

An action taken against a computer system with the intention of doing harm is called an attack and can take a number of forms. It may be an unauthorized attempt to access data or services or to modify data, or it may be intended to deny services to legitimate users. The related concerns that are refined security are typically classified as "security" concerns. These concerns are typically:

- **Confidentiality:** The property that reflects the extent to which data and services are only available to those that are authorized to access them.
- **Integrity:** It reflects the extent to which data or services can be delivered as intended. This property can also refer to data or services.
- **Non-repudiation:** It refers to the ability to guarantee that the sender cannot later repudiate or deny having sent the message. It can also refer to the guarantee that the recipient cannot later deny having received the message.
- **Availability:** This is the property that reflects the extent to which the system will be available for legitimate use. A denial-of-service attack is meant to disrupt the availability of a system, and it is a security concern. Availability builds on reliability by adding the notion of recovery (repair). Fundamentally, availability is about minimizing service outage time by mitigating faults.

Not all of these four security concerns are of the same importance to every system. For example, privacy concern is not necessary of the same importance for a security system in museum as for a managing accounts system in a financial institution.

People often have trouble articulating the security concerns, and it is common to articulate the solution (e.g., "System must support login") instead of the concern itself. "Supporting login" is a functional requirement that may emerge as a solution to satisfy a "confidentiality" concern.

> **Example:** Here it is interesting to note that even a brick can have a security requirement. This sort of NFR is essential for embassy and government security agency buildings. For any bricks used in a high-security system, there is a security requirement on those bricks—they should not contain or be easily able to accept any sort of eavesdropping device, and provisions must be made to ensure that this requirement is met, for example, through secure sourcing and testing.

Modifiability

Modifiability is about change, and our interest in it is in the cost and risk of making changes. To plan for modifiability, a requirement engineer must consider three questions:

- What can change?
- What is the likelihood of the change?
- When is the change made and who makes it?

An example of a quality scenario for "Modifiability" can be written as:

> The developer wishes to change the user interface by modifying the code at design time. The modifications are made with no side effects within three hours.

Interoperability

Interoperability is about the degree to which two or more systems can usefully exchange meaningful information. Like all quality attributes, interoperability is not a yes-or-no proposition but has shades of meaning. An example of a quality scenario for "Interoperability" can be written as:

> Our vehicle information system sends our current location to the traffic monitoring system. The traffic monitoring system combines our location with other information, overlays this information on a Google Map, and broadcasts it. Our location information is correctly included with a probability of 99.9%.

Performance

Performance is about time and the system's ability to meet timing requirements. When events occur—interrupts, messages, requests from users or other systems, or clock events marking the passage of time—the system, or some element of the system, must respond to them in time. Characterizing the events that can occur (and when they can occur) and the system or element's time-based response to those events is the essence in discussing performance. An example of a quality scenario for "Performance" can be written as:

> Users initiate transactions under normal operations. The system processes the transactions with an average latency of two seconds.

Testability

Testability refers to the ease with which software or system can be made to demonstrate its faults through (typically execution-based) testing. Specifically, testability refers to the probability, assuming that the software or system has at least one fault, that it will fail on its next test execution. If a fault is present in a system, then we want it to fail during testing as quickly as possible. For a system to be properly testable, it must be possible to control each component's inputs (and possibly manipulate its internal state) and then to observe its outputs (and possibly its internal state). An example of a quality scenario for "testability" can be written as:

> The unit tester completes a code unit during development and performs a test sequence whose results are captured and that gives 85% path coverage within 3 hours of testing.

Usability

Usability is concerned with how easy it is for the user to accomplish the desired task and the kind of user support the system provides. Over the years, a focus on usability has shown itself to be one of the cheapest and easiest ways to improve a system's quality (or, more precisely, the user's perception of quality). Usability comprises the following areas:

- Learning system features
- Using a system efficiently
- Minimizing the impact of errors
- Adapting the system to user needs
- Increasing confidence and satisfaction

An example of a quality scenario for "testability" can be written as:

> The user downloads a new application and is using it productively after two minutes of experimentation.

Other Approaches to Handle NFRs in RE

Most of the early work on NFRs focused on measuring how much a software system is following the set of NFRs that it should satisfy, using some form of quantitative analysis (Fenton and Pfleeger 1997, Keller 1990, and Lyu 1996) offering predefined metrics to assess the degree to which a given software object meets a particular NFR. Those approaches that are concerned with measuring how much software complies with NFRs are called product-oriented approaches.

On the contrary, process-oriented approaches focus on the software development process. It aims to help software engineers search for alternatives to sufficiently meet NFRs while developing the software. Two worth mentioning: (i) NFRs framework and (ii) incorporating NFRs into UML models.

NFR Framework

The NFR framework (Chung et al. 2000) is a process-oriented and goal-oriented approach that is aimed at making NFRs explicit and putting them at the forefront in the stakeholder's mind. It requires the following interleaved tasks, which are iterative:

Task 1: Acquiring knowledge about the system's domain, FRs, and the particular kinds of NFRs for a particular system;

Task 2: Identifying NFRs as NFR softgoals and decomposing them into a finer level;

Task 3: Identifying the possible design alternatives for meeting NFRs in the target system as operationalizing softgoals;

Task 4: Dealing with ambiguities, tradeoffs, priorities, and interdependencies among NFRs and operationalizations (which are the solutions to implement the NFRs);

Task 5: Selecting operationalizations;

Task 6: Supporting decisions with a design rationale;

Task 7: Evaluating the impact of operationalization selection decisions on NFR satisfaction.

A cornerstone of this framework is the concept of the "softgoal," which is used to represent the NFR. A softgoal is a goal that has no clear-cut definition or criteria to determine whether or not it has been satisfied. The framework speaks of softgoals being "satisficed" rather than satisfied, to underscore their ad hoc nature, both with respect to their definition and to their satisfaction. The term "satisfice" was coined by Herbert Simon. Satisficing is a decision-making strategy that attempts to meet criteria for adequacy rather than to identify an optimal solution.

The operation of the framework can be visualized in terms of the incremental and interactive construction, elaboration, analysis, and revision of a softgoal interdependency graph (SIG). NFRs softgoals are depicted by a cloud in the SIG, while operationalizations are depicted with clouds with thicker borders. Figure 5.3 illustrates an example produced by using the NFRs framework for a smart home system. The example shows two appointed soft goals: good performance of responding to critical events and secure access to the system. High-level softgoals are refined into more specific subgoals or operationalizations. In each refinement, the offspring can contribute fully or

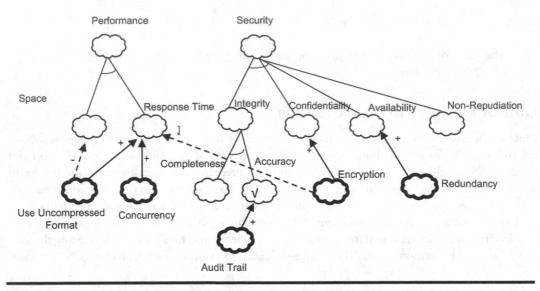

Figure 5.3 Softgoal interdependency graph for performance and security in a smart home system.

partially, and positively or negatively, toward satisficing the parent. In Figure 5.3, both *space* and *response time* should be satisficed for the performance to be satisficed. The AND contribution is represented by a single arc, and the OR by double arcs. Other types of contributions are MAKE (++), HELP (+), HURT (-), and BREAK (--).

While making choices in pursuit of a particular softgoal, it is very likely that other softgoals may be affected in this decision-making process. This is shown with interdependencies among the softgoals (the dashed lines in the figure). For example, *implementing Encryption* has a negative contribution with respect to *Response Time*.

Incorporating NFRs with UML Models

In Moreira et al. (2002), Park et al. (2004), Araujo et al. (2002), and many others, early integration of NFRs is accomplished by extending UML models to integrate NFRs to the functional behavior.

Supakkul et al. (2004) for example propose a use case and goal-driven approach to integrate FRs and NFRs. They use the UML use case model to capture the functionality of the system, and they also use the NFR framework (Chung et al. 2000) to represent NFRs. They propose to associate the NFRs with four use case model elements: *actor, use case, actor-use case association*, and the *system boundary*.

They name these associations "Actor Association Point," "Use Case Association Point," "Actor-Use Case Association (AU-A) Point," and "System Boundary Association Point," respectively. Having such an extension to the UML use case model, NFRs can be integrated at the requirements analysis level with FRs and can provide a better understanding of the requirements model. Figure 5.4 shows the proposed NFR association points in the UML use case model.

In Figure 5.4:

- Cloud "A" represents the NFRs related to an *actor* of a use case model. These NFRs are related to the *actor* by "Actor Association Point." For example, associating scalability NFR to customer actor would indicate that the system must handle a potentially large number of users accessing system functionality represented by use cases available to the actor.

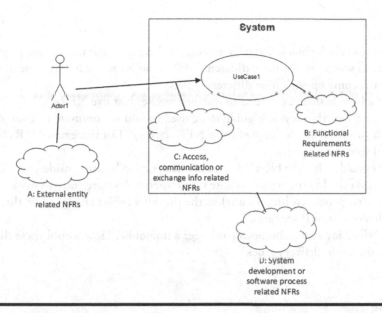

Figure 5.4   NFR association points in a use case diagram (with adaptation).

- Cloud "B" represents the NFRs related to the *use case* of the use case model. These NFRs are related to the *use case* by "Use Case Association Point." For example, associating fast response time NFR to withdraw fund use case of an automated teller machine (ATM) system would indicate that the system must complete the functionality described by the withdraw fund use case within an acceptable duration.
- Cloud "C" represents the NFRs related to the *actor-use case association* of the use case model. These NFRs are related to this association by "Actor-Use Case Association (AU-A) Point." For example, associating security NFR to an AU-A between customer and withdraw fund use case would indicate that withdraw fund must be secured, which also precisely implies that user interface to other AU-A is not required to be secured.
- Finally, cloud "D" represents the NFRs related to the *system boundary* of the use case model. These NFRs are related to this boundary by "System Boundary Association Point." For example, associating portability NFR to the system boundary would intuitively specify that the NFR is global and that the system must be operational in multiple platforms, which globally affects every part of the system. These four NFRs association points are the authors' proposed extensions to the UML use case model.

The book *Non-functional Requirements: Modeling and Assessment* (Kassab 2009) provided also a summary of the other related work to incorporate NFRs against all types of UML 2.0 diagrams.

Exercises

1. For a common cinder block (a type of brick), list three different use cases/scenarios for it. For each of these scenarios, list three different NFRs—no NFR can be the same for any scenario so you must come up with nine different NFRs.
2. Consider a bicycle that you intend to use for exercise. List five NFRs.
    a. Consider that the bicycle is going to be used by you to commute to your job site, about 5 miles away. Which (if any) of these NFRs change? List three new NFRs for this operational scenario.
    b. Now consider that the bicycle is actually designed to be used inside your home and connected via the Internet to an exercise system/network where you can work out with other connected people. Which (if any) of the previous NFRs change? List three new NFRs for this operational scenario.
3. List five NFRs for an autonomous passenger automobile. How would these differ (if at all) for an autonomous delivery truck?

References

Antón, A. (1997). *Goal Identification and Refinement in the Specification of Information Systems*, Ph.D. Thesis, Georgia Institute of Technology, Georgia, USA.

Araujo, J., Moreira, A., Brito, I., & Rashid, A. (2002). Aspect-oriented requirements with UML. *Workshop on Aspect-Oriented Modeling with UML (Held with UML 2002*, Dresden, Germany.

Babcock, C. (1985, September). New Jersey Motorists in Software Jam. *Computerworld*, 30: 1–6.

Barbacci, M. R., Robert E., Anthony J. L., Judith A. S., & Charles B. W. (2003). *Quality attribute workshops (qaws)*. CARNEGIE-MELLON UNIV PITTSBURGH PA SOFTWARE ENGINEERING INST.

Bass, L., Clements, P., & Kazman, R. (2021). *Software Architecture in Practice, 4th edition*. Addison-Wesley Professional, Boston, MA, USA.

Boehm, B., & In, H. (1996). Identifying quality-requirement conflicts. *IEEE Software*, IEEE Computer Society Press, pp. 25–35. Washington, D.C., United States.

Boehm, B. W., Brown, J. R., & Lipow, M. (1976). Quantitative evaluation of software quality. In *Proceeding of the 2nd International Conference on Software Engineering* (pp. 592–605). IEEE Computer Society, San Francisco, CA, Long Branch, CA.

Breitman, K. K., Leite J. C. S. P., & Finkelstein, A. (1999). The world's stage: A survey on requirements engineering using a real-life case study. *Journal of the Brazilian Computer Society*, 1(6): 13–37.

Chung, L., Nixon, B. A., Yu, E., & Mylopoulos, J. (2000). *Nonfunctional Requirements in Software Engineering*. Springer, Boston, MA.

Davis, A. (1993). *Software Requirements: Objects, Functions and States*. Prentice Hall, Hoboken, New Jersey, United States.

Fenton, N. E., & Pfleeger, S. L. (1997). *Software Metrics: A Rigorous and Practical Approach*. International Thomson Computer Press.

Finkelstein, A., & Dowell, J. (1996). A comedy of errors: The London Ambulance Service case study. *Proceedings of the 8th International Workshop Software Specifications and Design* (pp. 2–5), Schloss Velen, Germany.

Gilb, T. (2005). *Competitive Engineering: A Handbook for Systems Engineering, Requirements Engineering, and Software Engineering Using Planguage*. Elsevier.

Grady, R. B. (1992). *Practical Software Metrics for Project Management and Process Improvement*. Prentice-Hall, Hoboken, New Jersey, United States.

ISO 25000 STANDARDS. https://iso25000.com/index.php/en/iso-25000-standards

Jacobson, I., Booch, G., & Rumbaugh, J. (1999) *The Unified Software Development Process*. Addison Wesley, Reading, MA.

Kassab, M. (2009). *Non-Functional Requirements: Modeling and Assessment*. VDM Verlag, Germany.

Kassab, M., Daneva, M., & Ormandjieva, O. (2007, August). Scope management of non-functional requirements. In *33rd EUROMICRO Conference on Software Engineering and Advanced Applications (EUROMICRO 2007)* (pp. 409–417). IEEE, Lubeck, Germany.

Kassab, M., DeFranco, J., & Laplante, P. (2020). A systematic literature review on Internet of things in education: Benefits and challenges. *Journal of Computer Assisted Learning*, 36(2): 115–127.

Kassab, M., DeFranco, J., & Voas, J. (2018). Smarter education. *IT Professional*, 20(5): 20–24.

Keller, S. E., Kahn, L. G., & Panara, R. B. (1990). Specifying software quality requirements with metrics. In R. H. Thayer & M. Dorfman (Eds.), *System and Software Requirements Engineering* (pp. 145–163). IEEE Computer Society Press, Washington, DC.

Kotonya, G., & Sommerville, I. (1998). *Requirements Engineering: Processes and Techniques*. John Wiley & Sons, Hoboken, New Jersey, United States.

Laplante, P. A., Kassab, M., Laplante, N. L., & Voas, J. M. (2017). Building caring healthcare systems in the Internet of Things. *IEEE Systems Journal*, 12(3): 3030–3037.

Leffingwell, D., & Widrig, D. (2003). *Managing Software Requirements: A Unified Approach*. The Addison-Wesley Object Technology Series, Boston, Massachusetts, United States.

Lyu, M. R. (1996). *Handbook of Software Reliability Engineering*. McGraw-Hill, New York, New York, United States.

Moreira, A., Araujo, J., & Brito I. (2002). Crosscutting quality attributes for requirements engineering. In *14th International Conference on Software Engineering and Knowledge Engineering 2002* (pp. 167–174). Ischia, Italy.

Mylopoulos, J., Chung, L., & Nixon, B. (1992). Representing and using nonfunctional requirements: A process oriented approach. *IEEE Transactions in Software Engineering*, 18(6): 483–497.

Ncube, C. (2000). *A Requirements Engineering Method for COTS-Based Systems Development*, Ph.D. Thesis, City University, London.

O'Neil, F. (2015, July). Target data breach: Applying user-centered design principles to data breach notifications. In *Proceedings of the 33rd Annual International Conference on the Design of Communication* (pp. 1–8). Limerick, Ireland.

Park, D., Kang, S., & Lee, J. (2004). Design phase analysis of software performance using aspect oriented programming. In *5th Aspect-Oriented Modeling Workshop in Conjunction with UML 2004*. Lisbon, Portugal.

Robertson, S., & Robertson, J. (1999). *Mastering the Requirements Process*. Addison-Wesley Professional, Boston, Massachusetts, United States.

Seffah, A., Desmarais, M., & Metzger, M. (2005). *Human-Centered Software Engineering*. Springer, Dordrecht, The Netherlands.

Siemens Warns of Possible Hearing Damage in Some Cell Phones. (2004). http://www.consumeraffairs.com/news04/siemens_mobile.html, last visited on 04 August 2009.

Standard Glossary of Software Engineering Terminology. (1990). *IEEE Standard 610.12-1990*. DOI: 10.1109/IEEESTD.1990.101064.

Supakkul, S., & Chung, L. (2004). Integrating FRs and NFRs: A use case and goal driven approach. In *Proceedings of the 2nd International Conference on Software Engineering Research, Management and Applications (SERA)* (pp. 30–37). Los Angeles, CA.

Weber, M., & Wesbrot, J. (2003). Requirements engineering in automotive development: Experiences and challenges. *IEEE Software*, 20(1): 16–24.

Wiegers, K. (2003). *Software Requirements*, 2nd edition. Microsoft Press, Washington, USA.

Wikipedia: *Non-Functional Requirements*. http://en.wikipedia.org/wiki/Non-functional_requirements (visited May 2021).

Wikipedia: Requirements Analysis. http://en.wikipedia.org/wiki/Requirements_analysis (visited 05 July 2007).

Chapter 6

Requirements Validations and Verifications

What Is Requirements Risk Management?

Poor requirements engineering practices have led to some spectacular failures. Bahill and Henderson (2005) analyzed several high-profile projects to determine if their failure was due to poor requirements development, requirements verification, or system validation. Their findings confirmed that some of the most notorious system failures were due to poor requirements engineering. For example, the HMS Titanic (1912), the IBM PCjr (1983), and the Mars Climate Orbiter (1999) suffered from poor requirements design. The Apollo 13 (1970), Space Shuttle Challenger (1986), and Space Shuttle Columbia (2002) projects failed because of insufficient requirements verification. More recently, the rollout of the U.S. Healthcare.gov healthcare exchange site (2013) experienced numerous and very conspicuous problems due to a rushed implementation and failure to conduct proper requirements elicitations. More catastrophically, crashes of the Boeing 737 Max aircraft were eventually attributed to poor requirements design (Table 6.1).

Requirements can be inadequate in many ways including:

- Inaccurate or incomplete stakeholder identification
- Insufficient requirements validation
- Insufficient requirements verification
- Incomplete requirements
- Incorrect requirements
- Incorrectly ranked requirements
- Inconsistent requirements (Laplante 2010)

Requirements risk management involves the proactive analysis, identification, monitoring, and mitigation of any factors that can threaten the integrity of the requirements engineering process. Requirements risk factors can be divided into two types: technical and management. Technical risk factors pertaining to the elicitation, analysis, agreement, and representation processes have already been discussed in Chapters 3 and 4. In Chapter 7, we discuss the use of formal methods in

Table 6.1 A History of Project Failures Due to Poor Requirements Practices

System Name	Year	Requirements Process Failure
HMS Titanic	1912	Poor requirements design
Apollo-13	1970	Insufficient requirements verification
IBM PCjr	1983	Poor requirements design
Space Shuttle Challenger	1986	Insufficient requirements verification
Mars Climate Orbiter	1999	Poor requirements design
Space Shuttle Columbia	2002	Insufficient requirements verification
HealthCare.gov	2013	Poor requirements design
Boeing 737 Max	2019	Poor requirements design

Source: Adapted from Bahill, T.A. & Henderson, S.J., *Syst. Eng.*, 8, 1–14, 2005.

improving the quality of requirements representation through mathematically rigorous approaches. Requirements management risk factors tend toward issues of expectation management and interpersonal relationships, and these are discussed in Chapter 10.

Share your Opinion: Do you have a real-world story of a project failure due to poor requirements practices?
https://phil.laplante.io/requirements/failures.php

In this chapter, we wish to focus on the mitigation of requirements risk through the analysis of the requirements specification document itself. That is, the validation and verification of the Software Requirements Specification (SRS) should occur early in order to avoid costly problems further downstream. There are a variety of complementary and overlapping techniques to check the quality attributes of the SRS, but the IEEE 29148 contains a set of rules that are extremely helpful in vetting the technical aspects of the SRS and, in turn, the mitigating risk later on. So in this chapter, we will look at the nature of SRS verification and validation, at various qualities of "goodness" for requirements specifications, and, finally, turn to work at NASA that can be very helpful in comprehending these qualities of goodness.

To further motivate the notion of requirements risk, consider this scenario.[1] On a road that one of the authors travels in Pennsylvania there is a sign that declares "End brake retarder prohibition." The meaning of this sign is hard to understand and is blurred by the curious use of a quadruple negative (each word in the directive has the connotation of stopping something). As it turns out, this sign is an interesting example of a "shall not" requirement as well as illustrating a poor requirements specification, namely, one that is ambiguous, vague, contradictory, incomplete, or contains a mixed level of abstraction.

What is a "brake retarder"? Briefly, it is a device used on large trucks to slow the engine down by allowing air to be exhausted out of the pistons, thus slowing the vehicle. Brake retarders are very noisy, and, as a result, many municipalities prohibit their use within city limits. By Pennsylvania law, appropriate signs must be posted with the instructions "Brake Retarders Prohibited Within Municipal Limits." The complementary sign then reads "End Brake Retarder Prohibition." The latter phrasing is, apparently, unique to the Commonwealth of Pennsylvania (O'Neil 2004). In some states, the sign pair reads "No Jake Brakes" and "End Jake Brake Prohibition," the term "Jake" being a nickname of a company that manufactures the device, Jacobs Vehicle Systems.

Let's have a little fun with the Pennsylvania form of the sign. If we substitute the synonyms:

end → stop, brake → stop, retarder → stopper, prohibition → stopping

the sign translates to:

stop stop stopper stopping

which is a quadruple negative. In theory, we should be able to replace such a quadruple negative with a null sign reading:

Ø

The original sign has the paradoxical effect, however, of causing a positive action (i.e., to allow the application of brake retarders).

Of course, this whimsical discussion ignores the fact that the word "end" is a verb, "retarder" and "prohibition" are nouns, and in this context "brake" is an adjective, rendering a logical analysis of meaning more difficult. And the sign really does have the intended effect—truck drivers know what the sign means. But it should be clear that this particular sign can be interpreted multiple ways and is, therefore, ambiguous.

The "brake retarder" sign is a poor one—but perhaps not nearly as poor as the requirements for installing one. Pennsylvania's requirements on municipalities who wish to implement a "no brake retarder" zone are:

1. Downhill grade(s) greater than 4%
2. A posted reduced speed limit for trucks due to a hazardous grade determination
3. Posted reduced gear zone(s)
4. Posted speed limits over 55 miles/hour
5. Highway exit ramps with a posted speed limit over 55 miles/hour

The crash history for the stretch of road a municipality is seeking to keep brake-retarder free *must not include:*

1. *A history of runaway truck crashes over the past 3 years*
2. *A discernible pattern of rear-end crashes over the past 3 years where the truck was the striking vehicle* (O'Neil 2004)

The italic font was added to the latter section because it highlights the fact that the requirements on brake retarder prohibitions contain an embedded "shall not" requirement.

Given all of this confusion, clearly, a better job in formulating the sign language is needed, and more work is needed in formulating the requirements for the sign. Formal methods may have been advisable in this regard. All of these activities fall under the context of requirements risk management.

Validation and Verification

Requirements validation and verification involve review, analysis, and testing to ensure that a system complies with its requirements. Compliance pertains to both functional and nonfunctional requirements. But the foregoing definition does not readily distinguish between verification and validation, and is probably too long, in any case, to be inspirational or to serve as a mission statement for the requirements engineer. Boehm (1984) suggests the following to make the distinction between verification and validation:

- Requirements validation: "am I building the right product?"
- Requirements verification: "am I building the product right?"
- In other words, validation involves fully understanding customer intent and verification involves satisfying the customer intent.

There are great benefits of implementing a requirements verification and validation program. These include:

- Early detection and correction of system anomalies
- Enhanced management insight into the process and product risk
- Support for life cycle processes to ensure conformance to program performance and budget
- Early assessment of software and system performance
- Ability to obtain objective evidence of software and system conformance to support process
- Improved system development and maintenance processes
- Improved and integrated systems analysis model (IEEE Std 1012 2012)

Requirements validation involves checking that the system provides all of the functions that best support the customer's needs and that it does not provide the functions that the customer does not need or want. Additionally, validation should ensure that there are no requirements conflicts and that satisfaction of the requirements can actually be demonstrated. Finally, there is some element of sanity check in terms of time and budget—a requirement may be able to be literally met, but the cost and time needed to meet it may be unacceptable or even impossible.

Requirements verification (testing) involves checking the satisfaction of a number of desirable properties of the requirements (e.g., IEEE 29148 rules). Usually, we do both validation and verification (V&V) simultaneously, and often the techniques used for one or the other are the same.

Bahill and Henderson (2005) offer an informal verification/validation demonstration for a set of requirements for an electric water heater controller.

- If 70° <temperature <100°, then the system shall output 3,000 W.
- If 100° <temperature <130°, then the system shall output 2,000 W.
- If 120° <temperature <150°, then the system shall output 1,000 W.
- If 150° <temperature, then the system shall output 0 W.

We notice that this set of requirements is incomplete because the behavior for temperature <0° is not defined. The requirements are also inconsistent—for example, what happens when temperature = 125°?

The requirements are also incorrect because the temperatures given are not specified as being in degree Fahrenheit or degree Celsius. Clearly, this set of requirements could have benefited from effective V&V processes.

Techniques for Requirements V&V

V&V techniques may include some of the requirements elicitation techniques that were discussed in Chapter 3. For example, group reviews/inspections, focus groups, prototyping, viewpoint resolution, or task analysis (through user stories and use cases) can be used to simplify, combine, or eliminate requirements. In addition, we can use comparative product evaluations to uncover missing or unreasonable requirements and task analysis to uncover and simplify requirements. Systematic manual analysis of the requirements, test-case generation (for testability and completeness), using an executable model of the system to check requirements, and automated consistency analysis may also be used. Wideband Delphi or the Analytical Hierarchy Process (see Chapter 10) can be used for requirements reconciliation, harmonization, and negotiation. Finally, certain formal methods such as model checking and consistency analysis can be used for V&V, and these will be discussed in Chapter 7.

Walkthroughs

Walkthroughs or peer/team reviews are an informal methodology to detect errors and improve requirements quality. A typical walkthrough scenario involves requirements engineers, a supervisor, and peers participating in a semi-structured meeting to review the requirements before release. The goal is to identify ambiguities and inconsistencies and to determine if the requirements can be tested. The IEEE 29148 rules can be used as a framework or checklist for the walkthrough.

Inspections

Inspections are a method of requirement quality control that can be informal (ad hoc) or highly structured. More formal inspection processes, such as those described by Fagan (1986), provide close procedural control and repeatability. Fagan inspections define the following:

- What can be inspected
- When the code can be inspected
- Who can inspect the code
- What preparation is needed for the inspection
- How the inspection is to be conducted
- The data to be collected
- The follow-up activities

Inspection can be used in many kinds of requirements engineering settings but is especially appropriate for safety-critical systems.

The inspection team is typically comprised of about several people, who are selected from among the following candidates:

- Author(s) of the requirements being inspected.
- The reader who reads the requirements to the team during the inspection meeting.

- One or more inspectors who find errors, omissions, and inconsistencies. Appropriate inspectors of a requirements document include the project manager and a representative of the actual users.
- Representatives of anyone who has to do work based on the document bring a vital perspective. These include architects, designers, system test engineers, documentation writers, and support representatives.
- The moderator who chairs the inspection meeting and notes discovered errors.
- Scribe taking notes on the inspection process results.
- Representatives from other stakeholder groups such as program sponsors or primary users of the system.

The inspection process stages include:ss

The Preparation Stage: During preparation, inspectors examine the requirements document to understand it and to find possible defects. Inspectors use checklists of typical requirements defects to help focus their attention.

Inspection Meeting: During the inspection meeting, the reader describes their interpretation of each requirement in their own words. Such paraphrasing allows the other participants to compare their understanding with that of someone other than the author. Differences in interpretation can reveal omissions and surface assumptions. During the meeting, the record keeper documents the change requests, and the meeting moderator ensures the review is on track.

Rework: In this stage, a list of change requests is assembled, and a record of the minutes meeting is documented. The requirements engineer then works on the change requests to fix the found issues.

Goal-Based Requirements Analysis

Stakeholders tend to express their requirements in terms of operations and actions rather than goals. A risk is posed when goals evolve as stakeholders change their minds and refine and operationalize goals into behavioral requirements. To reduce this risk, stakeholder goals need to be evolved until they can be structured as requirements.

Goal evolution is facilitated through goal elaboration and refinement. Useful techniques for goal elaboration include identifying goal obstacles, analyzing scenarios and constraints, and operationalizing goals. Goal refinement occurs when synonymous goals are reconciled, when goals are merged into a subgoal categorization, when constraints are identified, and when goals are operationalized (Hetzel 1988). Regarding operationalization, we will refer to goals-based analysis when discussing metrics generation.

The 2020 survey on requirements engineering state of practice (Kassab and Laplante survey 2022) gives some clues to the prevalence of the aforementioned V&V techniques in industry (Figure 6.1).

For example, the survey results showed that team review techniques were used widely (about 38% reporting), and checklists were used by 23% of those reporting.

Stay Updated: For the up-to-date data from the RE state of practice survey.
https://phil.laplante.io/requirements/updates/survey.php

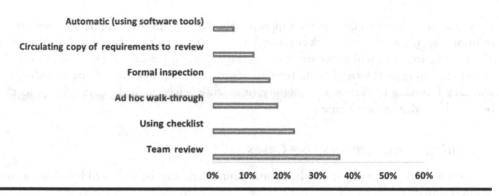

Figure 6.1   Requirements inspection techniques used (Kassab and Laplante 2022).

Other techniques reported included ad hoc walkthroughs (20%) and circulating copy (where an electronic markup copy of the SRS is shared by the team) (12%). Automated review of the SRS document, which was only noted by 4% of respondents, will be discussed shortly.

Requirements Understanding

In an early and influential book on software testing, Bill Hetzel proposed several paradigms for requirements verification and validation. To set the stage for this V&V, Hetzel addressed the problem of requirements understanding through the following analogy.

Imagine you are having a conversation with a customer in which he says that he would like for you to develop some kind of health management system, which, among other things, ensures that patients are eating a "well-balanced meal." You readily agree to this requirement. Many weeks later as you begin thinking about a system design, you reconsider the requirement to provide a well-balanced meal. What does that mean? In one interpretation, it could mean adding up everything consumed; were minimum nutritional guidelines in terms of calories, protein, vitamins, and so forth met? Another interpretation is that the patient ate exactly three, similar sized meals. Yet another interpretation is that the meals were well balanced in the sense that each food item weighed the same amount (Hetzel 1988).

Aside from the more ridiculous interpretations of "well-balanced meal," the above example illustrates a problem. The phrase "well-balanced meal" may have no language equivalent in French, Hindi, Mandarin, or any other language. So "well-balanced meal" could create a problem for some nonnative English speakers. What other colloquialisms do we use in our specifications and then ship out for offshore development? Clearly, there are various problems that can arise from language and cultural differences.

One solution to the requirements understanding problem is offered by Hetzel. He suggests that for correct problem definition it is best to specify the test for accepting a solution along with the statement of the requirement. For example, if we propose a requirement for a "well-balanced meal," we also offer a test for satisfying the requirement—that the meal must fall within the specified ranges of protein, fat, salt, sugar, and caloric content.

When the statement and test are listed together, most problems associated with misunderstanding requirements disappear. In particular, we want to derive requirements-based test situations and use them as a test of requirement understanding and validation.

For example, when a requirement is found to be incomplete, we can use the test case to focus on missing information. That is, design the test case to ask the question, "What should the system do in this case when this input is not supplied?" Similarly, when a requirement is found to be fuzzy or imprecise, use a test case to ask the question, "Is this the result I should have in this situation?" The specific instance will focus attention on the imprecise answer or result and ensure that it is examined carefully (Hetzel 1988). Then an appropriate requirement can be introduced to deal with any lingering imprecision, ambiguity, or missing behavior. Today, Hetzel's approach would be called test-driven development.

Validating Requirements Use Cases

When use cases comprise part of the requirements, these can be validated by asking a simple set of questions:

- Are there any additional actors that are not represented?
- Are there any activities that are not represented?
- Are each actor's goals being met?
- Are there events in the use case that do not address these goals?
- Can the use case be simplified?

Other related questions can also be generated.

Prototyping

Prototypes are useful in V&V when very little is understood about the requirements or when it is necessary to gain some experience with the working model in order to discover requirements. The principle behind using working prototypes is the recognition that requirements change with experience and prototyping yields that experience.

In software systems, incremental and evolutionary development approaches are essentially based on a series of non-throwaway prototypes. The difference between the two approaches is, essentially, that in incremental development the functionality of each release is planned, whereas, in evolutionary development, subsequent releases are not planned out. In both incremental and evolutionary development, lessons learned from prior releases inform the functionality of future releases' incremental and evolutionary development. In essence, early versions are prototypes used for future requirements discovery.

One point mentioned in Chapter 3, which should be reemphasized, is that prototypes are not always effective in uncovering nonfunctional requirements. This problem arises especially in novel systems and when certain requirements can only be found through an analysis of prevailing standards, applicable laws, customs, and so forth.

Tools for V&V

In the last two decades, there have been many attempts to build tools to support the formulation, documentation, and verification of natural language requirements. There are numerous tools that use natural language processing (NLP) to mitigate the problems and increase the quality of natural language requirements (see Table 6.2).

For example, Femmer et al. (2014) built a lightweight tool to detect "requirements smells," that is, various forms of phrases and word combinations that could be considered ambiguous, vague, incomplete, or problematic. Another tool, quality analyzer for requirements specification

Table 6.2 Tools That Focus on Finding Requirements Defects and Deviations

Tool Name	Year	Aim	Input	Automation
Circe	2006	Quality of requirements	Requirements document	Semi-automated
QuARS	2004	Quality of requirements	Requirements document	Automated
CRF Tool	2012	Uncertainty	Requirements document	Automated
Text2Test	–	Quality of use cases	Use cases	Unknown
AQUSA	2015	Quality of user stories	User stories	Automated
MaramaAI	2011	Quality of requirements	Requirements document	Not automated
SREE	2013	Ambiguity	Requirements document	Not available
Dowser	2008	Ambiguity	Requirements document	Not automated
RQA	2011	Quality of requirements	Requirements document	Automated
UIMA	2009	Use case model	Use case description	Not automated
Qualicen	2014	Quality of requirements	Requirements document	Automated
EARS	2010	Requirements Template Violations	Requirements document	Automated
RETA	2013	Requirements Template Violations	Requirements document	Automated
Planguage	2005	Quality of requirements	Requirements document	Automated

(QuARS), uses automated lexical and syntactic analyses to "identify requirements that are defective because of language usage" (Lami 2005). Another natural language analysis tool, the NASA automated requirements measurement (ARM), is discussed in detail later in the chapter.

Other automated analysis tools require that the SRS conforms to standard templates and boilerplates. For example, the easy approach to requirements engineering (EARS) tool can find template violations. Similarly, the requirements template analyzer (RETA) can find template violations and also identify certain problematic requirements based on keywords (Arora et al. 2015).

Restricting the language that can be used in the specification document is another way to facilitate automated requirements V&V. One such tool, planning language (Planguage), uses a programming language like syntax to structure requirements and then analyze them (Gilb 2005).

Gervasi and Nuseibeh (2002) proposed a lightweight validation of natural language requirements (Circe) that can detect violations of quality characteristics in a more exact way by building logical models of the requirements specifications. Their approach, however, assumes that the specifications are written in certain patterns. This expectation is often not the case in the industry.

Share your Opinion: From your professional experience, which validation tools for software requirements specification documents are your favorite, and why? https://phil.laplante.io/requirements/validation.php

Other automated tools for requirements V&V are designed to be used with formal methods notations (see Chapter 7).

Requirements V&V Matrices

A requirements validation matrix is an artifact that associates high-level requirements with certain system attributes for the purposes of trade-off analysis and confirmation of requirement intent. Appropriate system attributes can include business need, safety, requirement volatility, and other factors. A matrix that connects system requirements to business and stakeholder requirements is an example of a validation matrix. Creating this validation matrix is important because failure to do so may mean missing an important connection between a feature and cost means risking unplanned expenses. These missing links may cause a ripple effect and create slowdowns in product launches, weaken stakeholder confidence, and adversely affect the overall project budget. Feature cost could also be included in the validation matrix, but this issue is discussed in Chapter 11.

One format for the requirements validation matrix for several requirements from the baggage handling system is shown in Table 6.3.

While the requirements validation matrix is used early in the project life cycle, a requirement verification matrix is used later. The requirements verification matrix associates requirements with test cases that verify that the requirement has been met. Such a matrix facilitates a review of requirements and the tests and provides an easy mechanism to track the status of test-case design and implementation. An excerpt from a requirements verification matrix for the smart home system SRS documents is shown in Table 6.4.

Here the requirements forming the SRS are listed verbatim in the left column, the tests that verify those requirements are listed in the middle column, and the test case status is in the right column. The status field can contain "passed," "failed," "not run," "omitted," or any variation of these keywords. Additional columns can be added to the matrix to indicate when the test was run, who conducted the test, where the test was run, the status of testing equipment used, and for comments and other relevant information.

The requirements verification matrix is easily made part of the master test plan and can be updated throughout the project to give a record of all requirements testing.

Table 6.3 Requirements Validation Matrix

Requirement Number	Safety Impact (high=10)	Volatility (high=10)	Business Need (high=10)
3.1	10	2	10
3.2	2	1	5
…	…	…	…
3.210	3	6	7
3.211	5	10	1

Table 6.4 A Sample Requirements Verification Matrix for the Smart Home System SRS in the Appendix

Requirement	Test Cases	Status
9.13.1 System shall provide wireless support for driving any number of wall-mounted monitors for picture display.	T-1711	Passed
	T-1712	Passed
	T-1715	Passed
9.13.2 System shall provide web-based interface for authenticated users to publish new photos for display on wall monitors.	T-1711	Passed
	T-1715	Failed
	T-1811	Passed
9.13.3 System shall allow users to configure which pictures get displayed.	T-1712	Passed
	T-1715	Passed
	T-1811	Passed
	T-1812	Passed
	T-1819	Not run
9.13.4 System shall allow users to configure which remote users can submit pictures to which wall monitor.	T-1712	Passed
	T-1715	Passed
	T-1716	Passed
	T-1812	Passed

Standards for V&V

There are various international standards for the processes and documentation involved in V&V of systems and software. Many of these have been sponsored or cosponsored by the Institute for Electrical and Electronics Engineers (IEEE).

Whatever requirements V&V techniques are used, a software requirements V&V plan should always be written to accompany any major or critical software application.

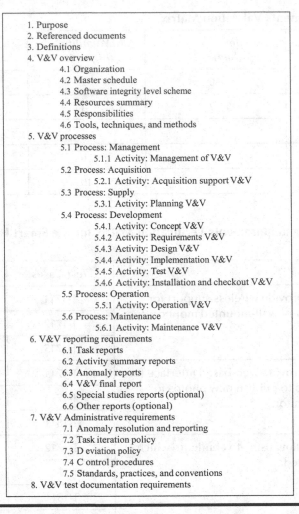

1. Purpose
2. Referenced documents
3. Definitions
4. V&V overview
    4.1 Organization
    4.2 Master schedule
    4.3 Software integrity level scheme
    4.4 Resources summary
    4.5 Responsibilities
    4.6 Tools, techniques, and methods
5. V&V processes
    5.1 Process: Management
        5.1.1 Activity: Management of V&V
    5.2 Process: Acquisition
        5.2.1 Activity: Acquisition support V&V
    5.3 Process: Supply
        5.3.1 Activity: Planning V&V
    5.4 Process: Development
        5.4.1 Activity: Concept V&V
        5.4.2 Activity: Requirements V&V
        5.4.3 Activity: Design V&V
        5.4.4 Activity: Implementation V&V
        5.4.5 Activity: Test V&V
        5.4.6 Activity: Installation and checkout V&V
    5.5 Process: Operation
        5.5.1 Activity: Operation V&V
    5.6 Process: Maintenance
        5.6.1 Activity: Maintenance V&V
6. V&V reporting requirements
    6.1 Task reports
    6.2 Activity summary reports
    6.3 Anomaly reports
    6.4 V&V final report
    6.5 Special studies reports (optional)
    6.6 Other reports (optional)
7. V&V Administrative requirements
    7.1 Anomaly resolution and reporting
    7.2 Task iteration policy
    7.3 D eviation policy
    7.4 C ontrol procedures
    7.5 Standards, practices, and conventions
8. V&V test documentation requirements

Figure 6.2 Recommended V&V plan table of contents. (From *IEEE Std 1012-2012, IEEE Standard for Software Verification and Validation*, Institute for Electrical and Electronics Engineers, Piscataway, NJ, 2004. With permission.)

The IEEE Standard 1012–2012, IEEE Standard for Software Verification and Validation, provides some guidelines to help prepare V&V plans. Figure 6.2 shows the recommended V&V plan outline (IEEE Std 1012-2012).

This template can be customized for most situations.

ISO/IECIEEE Standard 29148

ISO/IEC/IEEE 29148 (we refer to it as IEEE 29148 for short) is a very important standard for requirements engineers. The IEEE 29148 "describes recommended approaches for the specification of software requirements." The standard attempts to help:

1. Software customers to accurately describe what they wish to obtain
2. Software suppliers to understand exactly what the customer wants

3. Individuals to accomplish the following goals:
   a. Develop a standard software requirements specification (SRS) outline for their own organizations
   b. Define the format and content of their specific SRSs
   c. Develop additional local supporting items, such as an SRS quality checklist or an SRS writer's handbook (ISO/IEC/IEEE Std 29148 2011)

But from a risk mitigation standpoint, we are most interested in the qualities of goodness for the requirements document that are described. The IEEE 29148 specifies qualities for individual requirements and then for the set of requirements taken as a whole. First, the mandatory qualities for an individual requirement are that it must be:

■ Singular
■ Feasible
■ Unambiguous
■ Complete
■ Consistent
■ Verifiable
■ Traceable

Then, taken collectively, the set of all requirements in an SRS document should exhibit the following properties:

■ Complete
■ Consistent
■ Bounded
■ Affordable

The first two properties are the same as for a single requirement as they are for a group of requirements and they will be discussed shortly. Next, a set of requirements is bounded if it does not exceed the scope of the system intent. Finally, affordable requirements are those that can be satisfied within budgetary constraints. Chapter 11 covers the topic of value engineering of requirements, which deals with the issue of affordability.

We now look at each of the qualities for individual requirements in some detail.

Singularity

The IEEE 29148 mandates that a requirement specifies a single behavior and has no conjunctions. For example, consider this requirement for the wet-well control system described in Appendix B.

3.1.1 The pump control unit shall start the submersible pump motors to prevent the wet well from running over and stop the pump motors before the wet well runs dry.

This requirement could be separated into two:

3.1.1.1 The pump control unit shall start the submersible pump motors to prevent the wet well from running over.
3.1.1.2 The pump control unit shall stop the pump motors before the wet well runs dry.

The problem with the conjunctions "and" and "or" is that they can introduce some ambiguity. For example, is the "or" an inclusive "or" or an exclusive one. If the requirement contains two parts separated by the word "and," this could imply a sequence of events or it could mean that the events can occur in parallel. In the example for the wet-well control system given above, this ambiguity does not arise, but it could in other instances. It is better to leave out conjunctions when possible.

Feasibility

A requirement is feasible if it can be satisfied with current technology and cost constraints, that is, it is not a ridiculous requirement. As of examples of infeasible requirements:

■ The system shall consume no energy.
■ The length of the beam shall be less than 0 feet.

Various techniques can be used to assess feasibility including reviews, inspections, and competitive analysis. A value engineering analysis (see Chapter 11) also needs to be conducted to ensure feasibility.

Ambiguity

We define ambiguity by complementation—a requirement is unambiguous if it can have only one interpretation.

Here is an example of an ambiguous requirement:

If the valve is open and the drain is open, then it should be closed.

It is not clear whether "it" refers to the "valve" or the "drain." Conjunctions can cause ambiguity and that is why they should be avoided.

Here is another example of a specification document in which ambiguous behavior was described. In a certain automobile (belonging to one of the authors), an indicator light is displayed (in the shape of an engine) when certain exceptional conditions occur. These conditions include poor fuel quality, fuel cap not tightened properly, and other fuel-related faults. According to the user's manual, if the cause of the problem is relatively minor, such as the fuel cap not tightened, the system will reset the light upon:

Removing the exceptional condition followed by three consecutive error-free cold starts. A cold start is defined as a start-up that has not been preceded by another engine start-up in the last eight hours, followed by several minutes of either highway or city driving.

Can you see the problem in this functional definition? Aside from its confusing wording, the requirement does not make sense. If you wait 8 hours from the previous start-up, then start the engine to drive somewhere, you have to wait at least 8 hours to start up and drive back to your origin. If you have any warm start before three consecutive cold starts, the sequence has to begin again. Is the only possible way to satisfy this condition to drive somewhere, wait there for 8 hours and then drive back to the origin (three times in a row)? Or, drive around for a while, return to the origin, wait 8 hours, then do it again (two times more)? This sequence of events is very hard to follow, and in fact, after 1 month of trying, the first author (Laplante) could not get the light to reset without disconnecting and reconnecting the battery.

Another reason why ambiguity is so dangerous is that, in an ambiguous requirements specification, literal requirements satisfaction may be achieved but not customer satisfaction. "I know that is what I said I wanted, but now that I see it in action, I realized that I really meant something else" is an unfortunate refrain. Or consider this fictitious quote—"oh, you meant 'that' lever; I thought you meant the other one." We would never want this scenario to be played out late in the system's life cycle.

Some of the techniques that could be used to resolve the ambiguity of a requirement include formal reviews, viewpoint resolution, and formal modeling of the specification.

Completeness

Boehm (1984) postulates that a requirement is complete "to the extent that all of its parts are present and each part is fully developed." This definition is corroborated by Menzel et al. (2010): "[completeness is] the sense of whether the specification contains all requirements and whether all requirements contained in the specification are completely specified." According to the IEEE 29148, a single requirement is complete if it is "measurable and sufficiently describes the capability and characteristics to meet the stakeholder's need."

We will use the following definition for completeness of a set of requirements: an SRS document is complete if there is no missing functionality, that is, all appropriate desirable and undesirable behaviors are specified. Recall from Figure 1.2 that there is usually a mismatch between desired behaviors and specified behaviors— there is always some unspecified behavior, as well as undesirable behavior, that finds its way into the system that should be explicitly prohibited. Either case can lead to literal requirements satisfaction but not customer satisfaction.

Completeness is a difficult quality to prove for a set of requirements. If a single requirement contains "to be defined" (TBD) or some variation of this phrase, then clearly the requirement is incomplete. And if the requirement is missing measurable indicators of satisfaction, then that is a problem of incompleteness. But if those indicators are absent, how do you know when something is missing in a requirement specification? Use the following techniques to look for missing requirements:

- Check if you have received requirements from all your product's user classes.
- Make sure that the source and usage of all data items are stated.
- Check whether requirements are present in all pertinent functional categories for the system. These categories might include reporting, managing, and customizing users, edit operations, security, printing, and transaction logging, or whatever functional areas your system typically includes.
- Check whether NFRs, such as quality attributes and design and implementation constraints, have been specified.
- Ensure that the requirements conform to all applicable regulations.
- Build tables of similar requirements that fit a pattern to avoid duplications or oversights.
- Represent requirements information in multiple ways. Models such as UML provide a high-level view and can help in catching missing use cases and requirements.

Consistency

Since requirements may be added to the project at different times and come from different sources, they are vulnerable to become inconsistent. Figure 6.3 is a taken picture from corner streets in downtown Montreal, Canada. It shows multiple signs added a year to year to explain the parking rules on that corner. As these signs were added over time, they collectively began contradicting one other. Instead, this forest of signs could be simplified. Everyone should be able to figure out

Figure 6.3 Multiple signs can create contradicting parking rules. (Photo taken from downtown Montreal, Canada, M. Kassab.)

within a few seconds, "Can I park here now, or a little bit later, or for how long?" Can you reduce this hodgepodge of signs into a single, consistent one?

In the context of the SRS document, the consistency can take two forms:

- **Internal Consistency**: that is, the satisfaction of one requirement does not preclude the satisfaction of another.
- **External Consistency**: that is, the SRS is in agreement with all other applicable documents and standards.

Here is a simple example of illustrating an internal inconsistency. The following are requirements for some kind of pneumatic control system, possibly relating to the baggage handling system.

3.1 If the lever is in position 1, then valve 1 is opened.
3.2 If the lever is in position 1, then valve 1 is closed.

Clearly, requirements 3.1 and 3.2 are inconsistent and therefore invalid—when the lever is in position 1, the valve cannot be both open and closed at the same time. More complex consistency checking will be discussed in Chapter 7.

When either internal or external inconsistency is present in the SRS, it can lead to difficulties in meeting requirements, delays, and frustration downstream. Internal and external consistency can be checked through reviews, viewpoint resolution, various formal methods, and prototyping.

Verifiability

An SRS is verifiable if satisfaction of each requirement can be established using measurement or some other unambiguous means. This quality is important because a requirement that cannot be shown to be met has not been met. When requirements cannot be measured, they cannot be met and disputes will follow.

As we discussed in Chapter 5, NFRs are the most vulnerable to be written in an unverifiable format. For example, consider the following usability requirement:

The system shall be reliable

Clearly, the satisfaction of this requirement can't be demonstrated as it is subjective and what could be reliable for one user may not be reliable enough for another. One way to improve the above requirement is by associating metrics to quantify how the acceptance of "reliability" will be measured. We will discuss the Importance of Measurement in Requirements V&V shortly.

Overall, verifiability can be explored through various reviews, through test-case design (design-driven development), and through viewpoint resolution.

Traceability

Traceability, defined as "the ability to describe and follow the life of a requirement in both a forwards and backwards direction" from inception throughout the entire system's life cycle, provides useful support mechanisms for managing requirement changes during the ongoing change process (Gotel 1995). Moreover, the extent to which traceability is exploited is viewed as an indicator of system quality and process maturity and is mandated by many standards. In Aizenbud-Reshef et al. (2006), the specific challenges faced in state-of-the-art traceability practice are described in more detail.

In general, an SRS is traceable if each requirement is clearly identifiable, and all linkages to other requirements (e.g., dependencies, subordinate, and predominant) are clearly noted. In some cases, the relationship is clear in the hierarchy of the numbering system. Traceability matrices, to be discussed shortly, are also designed for this purpose. In other cases, the linkages are made by requirements tools.

Traceability is an essential quality for effective communications about requirements, facilitating easy modification, and even for legal considerations. For example, in the case of a dispute, it is helpful to show that responsible linking of related requirements was done.

In practice, many organizations either focus their traceability efforts on functional requirements or else fail entirely to implement an effective traceability process. Tracing NFRs has, on the whole, been neglected. As we discussed in Chapter 5, this is mainly because NFRs tend to become scattered among multiple modules when they are mapped from the one-dimensional requirements domain to the n-dimensional solution space. Furthermore, NFRs can often interact in the sense that attempts to achieve one NFR can help or hinder the achievement of other NFRs at certain functionality. Such interaction creates an extensive network of interdependencies and trade-offs between NFRs which is not easy to trace. Although prior work on tracing NFRs has been rather limited, a number of traceability approaches have in fact been developed to support related activities while incorporating NFRs in software engineering processes.

In Chung et al. (1995), the authors adopt the NFR framework we explained in Chapter 5 to show how a historical record of the treatment of NFRs during the development process can also serve to systematically support the evolution of the software system. The authors treat changes in terms of (i) adding or modifying NFRs, or changing their relative importance, and (ii) changing design decisions or design rationale. While this study has provided some support for extensions to the NFR framework, particularly in representing changes to goal achievement strengths, the impact of changes to functional models on nonfunctional models, and vice-versa, has not been discussed.

In Cleland-Huang (2005), the author proposed an approach named Goal Centric Traceability, a holistic traceability environment that provides systems analysts with the means to manage the impact of functional change on NFRs. Nevertheless, the impact of changes to an NFR on other NFRs and the functional model is not solved with this solution.

In Kassab et al. (2008) a traceability metamodel for change management of NFRs is presented to tackle the traceability problem for NFRs.

Traceability can be measured using network-like analyses. For example, we could count the efferent (inward) and afferent (outward) coupling as indicated by the key phrases "uses," "is referenced by," "references," "is used by," and so on.

Group reviews and inspections and automated tools can also be used to check for traceability between requirements to/from tests. We will discuss traceability further in Chapter 9.

Ranking

The IEEE 29148 standard does not mandate requirements ranking, but the authors think the ranking of requirements is essential. A requirements set is ranked if the items are prioritized for importance and/or stability. "Importance" is a relative term, and its meaning needs to be resolved on a case-by-case basis. Stability means the likelihood that the requirement would change. For example, a hospital information system will always have doctors, nurses, and patients (but governing legislation will change). The ranking could use a numerical scale (positive integers or real numbers) and a simple rating system (e.g., mandatory, desirable, and optional), or could be ranked by mission criticality.

For example, NASA uses a four-level ranking system. Level 1 requirements are mission-level requirements that are very high level and that very rarely change. Level 2 requirements are "high level" with minimal change. Level 3 requirements are those requirements that can be derived from level 2 requirements, that is, each level 2 requirement traces to one or more level 3 requirements. Contracts usually bid at this level of detail. Finally, level 4 are detailed requirements and are typically used to design and code the system (Wilson et al. 1997).

The ranking is an extremely important quality of an SRS document. Suppose in the course of system design, two requirements cannot be met simultaneously. It becomes easier to decide which requirement to relax based on its ranking. In addition to being useful for trade-off engineering, ranking can be used for cost estimation and negotiation, and for dispute resolution. Ranking validation is easy enough through reviews and viewpoint resolution (to agree upon the rankings).

Example Validation of Requirements

Consider requirements 5.1.1–5.1.18 for the smart home SRS found in Appendix A. Now we list the IEEE 29148 criteria and evaluate the requirements along these criteria.

Singular

There is at least one possible violation. Consider the following requirement:

5.2.1 Smart home shall have at least one multipurpose detector for detecting smoke and carbon monoxide on each floor.

It contains a conjunction and could be separated into two identical requirements except for one detecting smoke, and the other detecting carbon monoxide. But the requirement also states that the detector is "multipurpose" so it is expected that more than one thing is to be detected.

There are other similar instances throughout the specification, but none seem to introduce ambiguity.

Feasible

The requirements are feasible as they appear to be consistent with a common understanding of how television recording works in a home environment. But it would be hard to be certain if this set is correct without further analysis using techniques such as focus groups and experts.

Unambiguous

There are quite a few unclear or ambiguous phrases in these requirements that are revealed by asking "how do I test this?" These problems will need to be resolved. For example, 8.1.1 says that the system shall record "any show on television." There are many problems with this statement: any show on United States television or around the world? What about pay-per-view, and so on? Requirement 8.1.4 says the system shall "make storage" for recorded shows "expandable." What does making storage mean (allocation)? How much expansion should be allowed? Requirement 8.1.6 talks about searching—but based on what? Program name, the time it was recorded, subject? Requirement 8.1.10 describes selecting "quality for recording." What does "quality" mean? Requirement 8.1.12 mentions storing "X episodes." Is "X" a natural number, an integer, a complex number, a quaternion? There are other ambiguities here too.

Complete

The individual requirements could bear some improvement. For example, for smoke and carbon monoxide detection, "dangerous levels" are mentioned. These should be stated as measurable targets derived from some standard or regulation.

As a set, while there are no TBD requirements in the document, we cannot really say if this set of requirements is complete without using such validation techniques as QFD or focus groups.

Consistent

Without conducting a formal consistency analysis, we cannot be sure. For this segment of requirements, however, it probably wouldn't be too hard since the "shall" statements are relatively straightforward. So, although there would be a large number of logical variables, there would not be many compound propositions to evaluate. You can use automatic consistency checkers to validate this segment of requirements, or you could actually use a spreadsheet program to do the job. However, in this situation, an informal reading is probably adequate. An informal reading of the requirements does not reveal any obvious inconsistencies.

Verifiable

Because of the aforementioned ambiguities, these requirements are not fully verifiable.

Traceable

The requirements are numbered in a way that appears to be consistent. There are no internal references or usages of one requirement to another in this set.

Ranked

We know from reading the opening paragraphs of Section 8 of the SRS that this set of requirements is ranked as "medium" in importance. Ranking a large cluster of requirements as having the same priority does not particularly help; however, when the need arises to sacrifice some requirements in order to meet budget and/or schedule, it would be better if each major requirement were individually ranked.

The Importance of Measurement in Requirements V&V

Imagine an argument involving the question of which boxer was better: Muhammad Ali or Joe Louis? All kinds of strategies have been tried to settle this argument. For example, you can try to argue superiority based on particular fighting characteristics. Some have modeled these characteristics to create computer-simulated fights. Others have looked at the fighting characteristics and results against similar opponents. You could even compare training techniques and the quality of their managers, trainers, and promoters. But there is no conclusive evidence that one fighter was better than the other or that they were evenly matched because there are no direct metrics available to make such a judgment.

Now consider an argument involving who is the best long jumper of all time. The answer should be easy—Mike Powell, who set the record at 8.95 m (with no wind at his back) in 1991. No one has ever jumped farther, and he broke Bob Beamon's record, which stood for 23 years. You can try to argue that Beamon was better in terms of other characteristics—competitive spirit, tenacity, resiliency, sportsmanship, etc., but those are all unmeasurable.

Now imagine an argument with a customer involving whether or not a requirement that "the software shall be easy to use" was met or not. You contend that the requirement was met because "everyone says that the software is easy to use." The customer disagrees because she feels that "the software is too hard to use." But you can't settle the argument because you have no metrics. It is rather disappointing that as software engineers we are no better off than two boxing fans arguing from barstools (Laplante et al. 2007).

So which qualities or "ilities" can be measured? There are many possibilities.

- Accuracy
- Completeness
- Consistency
- Correctness
- Efficiency

- Expandability
- Interoperability
- Maintainability
- Manageability
- Portability
- Readability
- Reusability
- Reliability
- Safety
- Security
- Survivability
- Testability
- Understandability
- Usability

This is not an exhaustive list. The ilities generate a set of nonfunctional requirements, and for each of these requirements, there needs to be an associated metric to determine if the requirement has been met. Associating a metric to the itility should be considered with diligence to ensure that the overall meaning of the requirement is what is really intended. Consider for example the following requirement statement:

The system shall not lose track of too much baggage.

Obviously, this requirement is flawed. One may argue that a better reworked version would be:

The system shall be able to know the location of 99.99% of all baggage identified by users.

Now, consider what will happen if the system knows the location of 99.999% of the baggage identified? This means the above requirement is not satisfied. The requirement should have used the term "at least" to make it more inclusive. Also to be able to verify the above requirement, we need to measure it against certain duration. So, the requirement should state that "The system shall be able to know the location of at least 99.99% of all baggage identified by users within each X time period (or maybe for each flight)"; otherwise, how would we be able to verify it?

Goal/Question/Metric Analysis

We previously mentioned the importance of uncovering stakeholder goals in Chapter 2 and the use of goal-based analysis for requirements V&V in Chapter 3. But how can we generate the metrics that we need? The goal/question/metric (GQM) paradigm is an analysis technique that helps in the selection of an appropriate metric.

To use the technique, you have to follow three simple rules. First, state the goals of the measurement, that is, "what is the organization trying to achieve?" Next, derive from each goal the questions that must be answered to determine if the goals are being met. Finally, decide what must be measured in order to be able to answer the questions (Basili et al. 1994).

Here is an example of using GQM to define metrics that are "useful." Suppose that the stated goal for the system is "The system shall be easy to use." Clearly "easy to use" is impossible to objectively measure. So how do we approach its measurement? We can do it by creating questions that help describe what "easy to use" means. For example, one question that fits this description is "How many expert, intermediate, and novice users use the system?" The rationale for this question is that an easy-to-use system should be used by everyone. Now we need to know an appropriate metric to answer this question. Here is one way to obtain that metric— provide the system in an open lab for a period of time and measure the number and percentage of each user type who uses the system during that time. If a disproportionate number of users are expert, for example, then it may be concluded that the system is not easy to use. If an equal proportion of expert, intermediate, and novice users use the system, then it might be that the system is "easy to use."

Consider another question that addresses the goal of "easy to use": How long does it take a new user to master features 1–25 with only 8 hours of training? The rationale is that certain minimum features needed to use the system adequately ought not to take too long to train. An associated metric for this question then is obtained by taking a random sample of novice users, giving them the same 8 hours of training, and then testing the students to see if they can use features 1–25 to some minimum standard.

Following such a process to drive questions and associated metrics from goals is a good path to deriving measurable requirements and at the same time helping to refine and improve the quality of the requirements themselves.

VIGNETTE 6.1   Validation and Verification

Consider the following requirement for an autonomous passenger car:

3.1.1 THE VEHICLE SHOULD NOT BE CAPABLE OF FORWARD MOVEMENT UNLESS A PERSON IS IN THE DRIVERS' SEAT FOR AT LEAST –10 SECONDS

This requirement contains four errors, one grammatical, one conceptual, one logical, and one mathematical. The grammatical error is that the word "drivers" should be replaced with "driver's" since there is only one driver's seat. The logical error arises from the fact that not only should forward movement be prohibited without an operator, but all movement—forward, backward, and any other that the vehicle might be capable of. The conceptual (validation) error is that the word "should" needs to be replaced with "shall" since it is a safety risk if the vehicle is capable of movement without an operator in a position to override the autonomous system if necessary, and therefore, this requirement is not optional. The mathematical error is that –10 seconds was intended to be 10 seconds. An improved version of this requirement is:

3.1.2 THE VEHICLE SHALL NOT BE CAPABLE OF MOVEMENT UNLESS A PERSON IS IN THE DRIVERS' SEAT FOR AT LEAST 10 SECONDS

Clearly, we want to have a set of techniques that can identify all of these types of errors. This is why one form of verification and/or validation is generally insufficient, particularly for critical systems.

NASA Requirements Testing

One would think that the American space agency NASA is a place where rigorous requirements engineering is conducted. This is a correct assumption. Given that NASA is engaged in the engineering of very high profile, high-cost, and, most importantly, life-critical systems, the techniques used and developed here are state of the art. Table 6.5 contains an excerpt from NASA Procedural Requirements for requirements engineering (which is now obsolete but still useful for illustrative purposes) (NASA 2008).

NASA is heavily invested in the use of formal methods for requirements verification and validation, and a number of techniques and tools for this purpose have been developed.

Note that verification matrices are specifically mentioned as helping to accomplish the software requirements engineering goals. Requirements management, which will be discussed in Chapter 10, is specifically mentioned in the directive.

Table 6.5   Excerpt from NASA Procedural Requirements for Requirements Engineering

3.1.1 Requirements Development
3.1.1.1 The project shall document the software requirements. [SWE-049]
Note: The requirements for the content of each Software Requirement Specification and Data Dictionary are defined in Chapter 6.
3.1.1.2 The project shall identify, develop, document, approve, and maintain software requirements based on analysis of customer and other stakeholder requirements and the operational concepts. [SWE-050]
3.1.1.3 The project shall perform software requirements analysis based on flowed-down and derived requirements from the top-level systems engineering requirements and the hardware specifications and design. [SWE-051]
Note: This analysis is for safety criticality, correctness, consistency, clarity, completeness, traceability, feasibility, verifiability, and maintainability. This includes the allocation of functional and performance requirements to functions and subfunctions.
3.1.1.4 The project shall perform, document, and maintain bidirectional traceability between the software requirement and the higher-level requirement. [SWE-052]
Note: The project should identify any orphaned or widowed requirements (no parent or no child) associated with reused software.
3.1.2 Requirements Management
3.1.2.1 The project shall collect and manage changes to the software requirements. [SWE-053]
Note: The project should analyze and document changes to requirements for cost, technical, and schedule impacts.
3.1.2.2 The project shall identify inconsistencies between requirements, project plans, and software products and initiate corrective actions. [SWE-054]

Source: Adapted from NASA (National Aeronautics and Space Administration). (2008). Assurance Process for Complex Electronics. Updated 28 January 2008.

Note: A verification matrix supports the accomplishment of this requirement.

NASA ARM Tool

Quick Access to: NASA ARM Tool
https://arm.laplante.io/

The NASA ARM tool is a good example of a natural language analyzer that had moderate success in government and industrial use. ARM was developed at NASA's Software Assurance Technology Center at Goddard Space Flight Center in Greenbelt, MD, in the late 1990s. NASA ceased supporting ARM and a successor tool, e-Smart, sometime around 2009. Still, the tool is worth studying because it highlights several key points about the careful use of language and well-structured requirements specifications. In addition, one of the author's students reconstructed ARM for classroom and experimental use (Carlson and Laplante 2014). The tool can be accessed at http://arm.laplante.io/.

Here is how ARM works. The tool conducts an analysis of the text of the SRS document and reports certain metrics. The metrics are divided into two categories: micro- and macro-level metrics. Micro-level indicators count the occurrences of specific keyword types. Macro-level indicators are coarse-grained metrics of the SRS documentation.

Micro-level indicators include:

- Imperatives
- Continuances
- Directives
- Options
- Weak phrases

Macro-level indicators include:

- Size of requirements
- Text structure
- Specification depth
- Readability

These micro- and macro-level indicators will be described in some detail.

In addition, various ratios can be computed using macro- and micro-level indicators. No particular thresholds for the metrics are given (research is still being conducted in this regard). However, at the end of this section, summary metrics for 56 NASA projects are given for comparison.

A description of the metrics and some excerpts from the ARM report for creating the Smart Home SRS document are found in the appendix. The definitions are derived from those reported by the tool and described by the authors of the tool in a related report (Hammer et al. 1998).

Imperatives

The first metric, imperatives, is a micro-indicator that counts the words and phrases that command that something must be provided. Imperatives include "shall," "will," "must," and others as described in Table 4.2.

A more precise specification will have a high number of "shall" or "must" imperatives relative to other imperative types. Note that the word "should" is not recommended for use in an SRS. From both a logical and a legal point of view, "should" places too much discretion in the hands of system designers.

The counts of imperatives found in the Smart Home SRS document are shown in Table 6.6. An excerpt of the imperatives captured from the ARM output is shown in Figure 6.4.

Table 6.6 ARM Counts of Imperatives Found in the Smart Home SRS Document

Imperatives	Occurrences
Shall	308
Must	0
is required to	0
are applicable	0
are to	1
responsible for	0
will	51
should	7
Total	367

shall # 1: In Line No. 169, ParNo. 3.1.1, @ Depth 3
3.1.1 System SHALL operate on a system capable of multi-processing.

will # 51: In Line No. 624, ParNo. 9.12.3, @ Depth 3
9.12.3 System SHALL allow users to record greeting message that WILL be played after user defined number of rings.

should # 7: I Line No. 557, ParNo. 9.8, @ Depth 2
In the future this SHOULD be extended such that any commands can be programmed to control any device or system interfaced by the SH.

be able to # 1: In Line No. 324, ParNo. 7.3, @ Depth 2
Occupants and users of the SH's system should BE ABLE TO monitor the home from anywhere they wish.

normal # 1: In Line No. 490, ParNo. 9.3.14, @ Depth 3
9.3.14 Hot tub cover shall close with button press or if no activity / motion is detected for some time range, and water displacement levels are NORMAL (no one in the tub).

provide for # 1: In Line No. 104, ParNo. 2., @ Depth 2
The summation and harmonization of all the six categories of the SH will PROVIDE FOR a truly rewarding living experience for the occupants of the SH.

easy to # 1: In Line No. 193, ParNo. 4.1.2, @ Depth 3
4.1.2 System shall be EASY TO use.

can # 1: In Line No. 124, ParNo. 2., @ Depth 2
Ensuring the existing structure CAN support the improvements.

may # 1: In Line No. 180, ParNo. 3.1.9, @ Depth 3
3.1.9 System MAY contain separate SAN device for storage flexibility.

Figure 6.4 Excerpt of ARM output for Smart Home requirements specification document.

Continuances

Continuances are phrases that follow an imperative and precede the definition of lower-level requirement specifications. Continuances indicate that requirements have been organized and structured. Examples of and counts of continuances found in the Smart Home SRS document are shown in Table 6.7. The symbol ":" is treated as a continuance when it follows an imperative and precedes a requirement definition. These characteristics contribute to the ease with which the requirement specification document can be changed. Too many continuances, however, indicate multiple, complex requirements that may not be adequately reflected in resource and schedule estimates.

Directives

The micro-indicator "directives" count those words or phrases that indicate that the document contains examples or other illustrative information. Directives point to information that makes the specified requirements more understandable. Typical directives and their counts found in the Smart Home SRS document are shown in Table 6.8. Generally, the higher the number of total directives, the more precisely the requirements are defined.

Options

Options are those words that give the developer latitude in satisfying the specifications. At the same time, options give less control to the customer. Options and their counts found in the Smart Home SRS document are shown in Table 6.9.

Weak Phrases

Weak phrases are clauses that are subject to multiple interpretations and uncertainty and therefore can lead to requirements errors. The use of phrases such as "adequate" and "as appropriate"

Table 6.7   ARM Counts for Continuances in the Smart Home SRS Document

Continuance	Occurrence
Below	0
as follows	0
following	0
Listed	0
in particular	0
support	0
and	85
:	2
Total	87

Table 6.8   Directives Found in the Smart Home SRS Document

Directive	Occurrence
e.g.	0
i.e.	14
For example	0
Figure	0
Table	0
Note	0
Total	14

Table 6.9   Options and Their Counts Found in the Smart Home SRS Document

Option Phrases	Occurrence
Can	7
May	23
Optionally	0
Total	30

indicates that what is required is either defined elsewhere or worse the requirement is open to subjective interpretation. Phrases such as "but not limited to" and "as a minimum" provide the basis for expanding requirements that have been identified or adding future requirements. The counts of weak phrases for the Smart Home SRS document are shown in Table 6.10.

The total number of weak phrases is an important metric that indicates the extent to which the specification is ambiguous and incomplete.

Incomplete

The "incomplete" micro-indicator counts words that imply that something is missing in the document, for whatever reason (e.g., future expansion and undetermined requirements). The most common incomplete notation is TBD for "to be determined."

Variations of TBD include:

- **TBD:** to be determined
- **TBS:** to be scheduled
- **TBE:** to be established, or yet to be estimated
- **TBC:** to be computed
- **TBR:** to be resolved

Table 6.10 Weak Phrases for the Smart Home SRS Document

Weak Phrase	Occurrence
Adequate	0
as appropriate	0
be able to	3
be capable of	0
capability of	0
capability to	0
Effective	0
as required	0
Normal	1
provide for	1
Timely	0
easy to	1
Total	6

■ "Not defined" and "not determined" explicitly state that a specification statement is incomplete.
■ "But not limited to" and "as a minimum" are phrases that permit modifications or additions to the specification.

Incomplete words and phrases found in the Smart Home SRS document are shown in Table 6.11.

Leaving incompleteness in the SRS document is an invitation to disaster later in the project. While it is likely that there may be a few incomplete terms in a well-written SRS due to pending requirements, the number of such words should be kept to an absolute minimum.

Subjects

Subjects are a count of unique combinations of words immediately preceding imperatives in the source file. This count is an indication of the scope of subjects addressed by the specification.

The ARM tool counted a total of 372 subjects in the Smart Home SRS document in Appendix A.

Specification Depth

Specification depth counts the number of imperatives at each level of the document and reflects the structure of the requirements. The topological structure of requirements was discussed in Chapter 4. The numbering and specification structural counts for the Smart Home SRS document as computed by the NASA ARM tool are provided in Table 6.12.

Table 6.11 Incomplete Words and Phrases Found in the Smart Home SRS Document

Incomplete Term	Occurrence
TBD	0
TBS	0
TBE	0
TBC	0
TBR	0
not defined	0
not determined	0
but not limited to	0
as a minimum	0
Total	0

Table 6.12 Numbering and Specification Structure Statistics for Smart Home SRS Document in the Appendix

Numbering Structure		Specification Structure	
Depth	Occurrence	Depth	Occurrence
1	19	1	0
2	71	2	50
3	265	3	258
4	65	4	64
5	0	5	0
6	0	6	0
7	0	7	0
8	0	8	0
9	0	9	0
Total	420	**Total**	372

These counts indicate that the SRS requirements hierarchy has a "diamond" shape in the manner of Figure 4.3c.

Readability Statistics

In Chapter 4, we discussed the importance of clarity in the SRS document. There are various ways to evaluate reading levels, but most techniques use some formulation of characters or syllables per words and words per sentence. For example, the Flesch Reading Ease index is based on the average number of syllables per word and of words per sentence. Standard writing tends to fall in the 60–70 range, but apparently a higher score increases readability.

The Flesch–Kincaid grade level index is supposed to reflect a grade-school writing level; therefore, a score of 12 means that someone with a 12th-grade education would understand the writing. But standard writing averages seventh to eighth grade, and a much higher score is not necessarily good—higher-level writing would be harder to understand. The Flesch–Kincaid grade level indicator is also based on the average number of syllables per word and on words per sentence. There are other grade level indicators as well (Wilson et al. 1997).

The NASA ARM tool does not provide the ability to count these metrics, but most versions of the popular Microsoft Word can provide at least some relevant statistics. For example, the version of Word 2007 used to prepare this manuscript will compute various words, characters, paragraphs, sentence counts, and averages. It will also compute the Flesch Reading Ease and Flesch–Kincaid grade level metrics (consult the user's manual or online help feature to determine how to compute such metrics for your word processor, if available). In any case, we used Word to calculate the statistics for the Smart Home SRS document and obtained the output shown in Figure 6.5.

The SRS document is assessed to be at a 12th-grade reading level. The low number of sentences per paragraph (1.2) is an artifact of the way the tool counted each numbered requirement as a new paragraph.

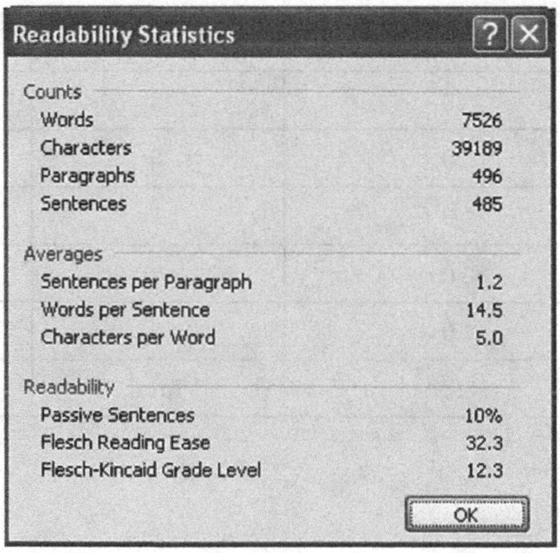

Readability Statistics	[?][X]
Counts	
Words	7526
Characters	39189
Paragraphs	496
Sentences	485
Averages	
Sentences per Paragraph	1.2
Words per Sentence	14.5
Characters per Word	5.0
Readability	
Passive Sentences	10%
Flesch Reading Ease	32.3
Flesch-Kincaid Grade Level	12.3
	[OK]

Figure 6.5 Readability statistics for Smart Home SRS document obtained from Microsoft Office Word.

Summary of NASA Metrics

To get some sense of proportion and relevance to the ARM indicators, Wilson et al. (1997) studied 56 NASA software systems ranging in size from 143 to 4,772 lines of code. They used their tool to collect statistics about these systems, which are summarized in Table 6.13.

From Table 6.12, we note that one specification was only 143 lines of text and that the longest was 28,000 lines of text. It is interesting to note from Table 6.12 that even NASA specifications have "TBDs" and often many "options." The property of feasible is not listed in the table, and modifiable, validatable, and verifiable are three additional properties (Wilson et al. 1997).

Aside from the obvious interpretations, why are these ARM indicators important? Table 6.14 gives us the intriguing answer—because there is a correlation between these indicators and the IEEE 29148 qualities.

Looking at the table, it seems that the text structure and depth are indicators of internal consistency (but not external). The numbers of directives and weak phrases are correlated (positively and negatively, respectively) with correctness. All of the micro-indicators are linked to testability (both in a positive and negative correlation—you need "just enough" directives but not too many). Finally, all quality indicators (except for readability) contribute to completeness.

NASA supported the ARM tool for several years, but by 2012 the tool was no longer available. A functionally equivalent reconstruction of the tool was developed by Carlson and Laplante (2014) and is available for classroom use at the first author's website (https://arm.laplante.io/).

VIGNETTE 6.2   NASA Assurance Process for Complex Electronics

NASA's Assurance Process for Complex Electronics Life Cycle Activities prescribes an assurance process for each phase of the project plan. In the requirements phase, the same process is used for all systems, subsystems, and safety requirements (note that safety requirements are treated separately). This assurance process comprises the risk management process.

The requirement elicitation process described is waterfall in nature and incorporates the following steps:

- Derivation of requirements from higher-level requirements
- Capture of additional requirements and constraints
- Feasibility assessment

The parallel risk analysis and process include:

- Assessment of derivation/ensuring all requirements are captured
- Evaluation of requirements
- Risk analysis

The output of this process is a set of risk assessed and managed requirements that are baselined and under configuration control (NASA).

Table 6.13 Sample Statistics from 56 NASA Requirements Specifications

	Lines of Text	Imperatives	Continuances	Directives	Weak Phrases	TBD, TBS, TBR	Option (can, may…)
Minimum	143	25	15	0	0	0	0
Median	2,265	382	183	21	37	7	27
Average	4,772	682	423	49	70	25	63
Max	28,459	3,896	118[a]	224	4[a]	32	130
Std Dev	759	156	99	12	21	20	39
Level 3 Specs	1,011	588	577	10	242	1	5
Level 4 Specs	1,432	917	289	9	393	2	2

Source: Wilson, W.M., et al., Automated analysis of requirement specifications, in *Proceedings of the 19th International Conference on Software Engineering*, pp. 161–171, 1997.

[a] These two figures are clearly wrong—how can the maximum numbers be less than the averages? However, all instances of this table in the literature show the same error, and the authors have been unable to obtain the correct ones.

Table 6.14 Cross Reference of NASA Indicators to the IEEE 29148 Qualities

Categories of Quality Indicators	Indicators of Quality Attributes — Quality Attributes										
	Complete	Consistent	Correct	Modifiable	Ranked	Testable	Traceable	Unambiguous	Understandable	Validatable	Verifiable
Imperatives	X			X		X	X	X	X	X	X
Continuances	X			X	X	X	X	X	X	X	X
Directives	X		X			X		X	X	X	X
Options	X		X			X		X	X	X	
Weak phrases	X					X	X	X	X	X	X
Size	X					X		X	X		X
Text structure		X		X	X		X		X		X
Specification depth	X	X		X			X		X		X
Readability								X	X	X	X

Source: Adopted from Wilson, W.M., et al., Automated analysis of requirement specifications, in *Proceedings of the 19th International Conference on Software Engineering*, pp. 161–171, 1997.

Exercises

6.1   What can be some pitfalls to consider when ranking requirements?

6.2   Describe two different ways to identify ambiguity in an SRS.

6.3   Which of the IEEE Standard qualities for individual requirements seem most important? Can you rank these?

6.4   Conduct an informal assessment of the IEEE 29148 qualities for the SRS given in Appendix B.

6.5   For an available SRS document, conduct an informal assessment of its IEEE 29148 qualities.

6.6   Should implementation risk be discussed with customers?

6.7   What are the advantages and risks of having requirements engineering conducted (or assisted) by an outside firm or consultants?

6.8   Create a traceability matrix for the SRS in
        6.1.1   Appendix A
        6.1.2   Appendix B

6.9   Calculate the requirements per test and tests per requirements metrics for the data shown in Table 6.3. Do you see any inconsistencies?

6.10   Consider the requirements in the SRS of Appendix A.
        6.10.1 Which of these could be improved through the use of visual formalisms such as various UML diagrams?
        6.10.2 Select three of these and create the visual formalism.

6.11   The requirements in Appendix B are not ranked. Using reasonable assumptions, rank these requirements (use a requirements ranking matrix for brevity).

6.12   Why is the NASA ARM tool, and others like it, useful in addition to objective techniques of SRS risk mitigation?

6.13   Run the NASA ARM tool (found at arm.laplante.io) on an available SRS document. What can you infer from the results?

6.14   Convert the collection of signs in Figure 6.3 into a single, consistent sign. If signs are not consistent then prove it (you can use Google translate to convert the French into English).

*6.15   Reconstruct the ARM tool in your favorite programming language. Note that the generation of the statistics is relatively simple—the challenge is in parsing the SRS document (Carlson and Laplante 2014). Place restrictions on the input to the tool as needed.

Supplemental Materials from the authors on requirements validations and verifications
https://phil.laplante.io/requirements/requirementsVV.php

Note

1 A variation of this story first appeared in Voas and Laplante (2010).

References

Aizenbud-Reshef, N., Nolan, B. T., Rubin, J., & Shaham-Gafni, Y. (2006). Model traceability. *IBM System Journal*, 45(3): 515–526.

Arora, C., Sabetzadeh, M., Briand, L., & Zimmer, F. (2015). Automated checking of conformance to requirements templates using natural language processing. *IEEE Transactions on Software Engineering*, 41(10): 944–996.

Bahill, T. A., & Henderson, S. J. (2005). Requirements development, verification, and validation exhibited in famous failures. *Systems Engineering*, 8(1): 1–14.

Basili, V., Caldiera, G., & Rombach, H. D. (1994). Goal question metric approach. In J. J. Marciniak (Ed.), *Encyclopedia of Software Engineering* (pp. 528–532). Wiley, New York.

Boehm, B. W. (1984). Verifying and validating software requirements and design specifications. *IEEE Software*, 1(1): 75–88.

Carlson, N., & Laplante, P. (2014). The NASA automated requirements measurement tool: A reconstruction. *Innovations in Systems and Software Engineering*, 10(2): 77–91.

Chung, L., Nixon, B. A., & Yu, E. (1995). Using non-functional requirements to systematically support change. *Proceedings of the Second IEEE International Symposium on Requirements Engineering* (pp. 132–139). York.

Cleland-Huang, J. (2005). Toward improved traceability of non-functional requirements. *Proceedings of the 3rd International Workshop on Traceability in Emerging Forms of Software Engineering* (pp. 14–19). Long Beach, CA.

Fagan, M. E. (1986). Advances in software inspections. *IEEE Transactions on Software Engineering*, 12(7): 744–751.

Femmer, H., Fernández, D. M., Juergens, E., Klose, M., Zimmer, I., & Zimmer, J. (2014, June). Rapid requirements checks with requirements smells: Two case studies. In *Proceedings of the 1st International Workshop on Rapid Continuous Software Engineering* (pp. 10–19). ACM, Hyderabad, India.

Gervasi, V., & Nuseibeh, B. (2002). Lightweight validation of natural language requirements. *Software: Practice and Experience*, 32(2): 113–133.

Gilb, T. (2005). *Competitive Engineering: A Handbook for Systems Engineering, Requirements Engineering, and Software Engineering Using Planguage*. Butterworth-Heinemann, Oxford.

Gotel, O. C. Z. (1995). *Contribution Structures for Requirements Traceability*, Doctoral dissertation, University of London, London, UK..

Hammer, T. F., Huffman, L. L., Rosenberg, L. H., Wilson, W., & Hyatt, L. (1998). Doing requirements right the first time. *CROSSTALK The Journal of Defense Software Engineering*, 1: 20–25.

Hetzel, B. (1988). *The Complete Guide to Software Testing*, 2nd edition. QED Information Sciences, Inc., London.

IEEE Std 1012-2012, IEEE Standard for Software Verification and Validation. (2004). Institute for Electrical and Electronics Engineers, Piscataway, NJ.

ISO/IEC/IEEE Std 29148, Systems and Software Engineering—Life Cycle Processes—Requirements Engineering. (2011). IEEE, Piscataway, NJ.

Kassab, M., & Laplante, P. (2022). The current and evolving landscape of requirements engineering state of practice. *IEEE Software*. DOI: 10.1109/MS.2022.3147692.

Kassab, M., Ormandjieva, O., & Daneva, M. (2008). A traceability metamodel for change management of non-functional requirements. In *2008 Sixth International Conference on Software Engineering Research, Management and Applications* (pp. 245–254). IEEE, Prague, Czech Republic.

Lami, G. (2005). *QuARS: A Tool for Analyzing Requirement (CMU/SEI-2005-TR-014)*. Software Engineering Institute, Carnegie Mellon University. http://resources.sei.cmu.edu/library/asset-view.cfm?AssetID=7681 (accessed 7 January 2017).

Laplante, P. A. (2010). Stakeholder analysis for smart grid systems. In *IEEE Reliability Society Annual Technical Report*. http://rs.ieee.org/images/files/Publications/2010/2010-02.pdf (accessed June 2017).

Laplante, P. A., Agresti, W. W., & Djavanshir, G. R. (2007). Guest editor's introduction, special section on IT quality enhancement and process improvement. *IT Professional*, 9: 10–11.

Menzel, I., Mueller, M., Gross, A., & Doerr, J. (2010). An experimental comparison regarding the completeness of functional requirements specifications. In *2010 18th IEEE International Requirements Engineering Conference (RE)* (pp. 15–24). IEEE, Sydney, NSW.

NASA (National Aeronautics and Space Administration). (2008). *Assurance Process for Complex Electronics*. Updated 28 January 2008. https://www.hq.nasa.gov/office/codeq/software/ComplexElectronics/1_requirements.htm (accessed January 2016).

O'Neil, R. (2004). *Pennsylvania's "No Jake Braking" Signs*. OLR Research Report #004-R-0515. http://www.cga.ct.gov/2004/rpt/2004-R-0515.htm (accessed March 2013).

Voas, J., & Laplante, P. (2010). End brake retarder prohibitions: Defining "shall not" requirements effectively. *Computer*, 12(3): 46–53.

Wilson, W. M., Rosenberg, L. H., & Hyatt, L. E. (1997). Automated analysis of requirement specifications. In *Proceedings of the 19th International Conference on Software Engineering* (pp. 161–171). Boston, Massachusetts, USA.

Chapter 7

Formal Methods

Motivation

The title of Truss's book on punctuation, *Eats Shoots and Leaves* (Truss 2003), could refer to either

- A panda, if the punctuation is as published, or
- A criminal who refuses to pay his restaurant bill if a comma is added after the word "eats."[1]

Clearly, the title of the book is not that of a system or software specification, but this anecdote illustrates that simple punctuation differences can convey a dramatically different message or intent. As has been stated by many, with respect to specifications, "syntax is destiny."[2]

Complex systems have tremendous sensitivity to errors in a requirements specification, design document, or computer code—even a single erroneous character can have severe consequences. In fact, in 1962, a missing hyphen character in a Fortran code statement led to the loss of the Mariner 1 spacecraft, the first American probe to Venus (NASA 2012).

Aside from punctuation, there are a number of problems with conventional software specifications built using only natural language and informal diagrams. These problems include ambiguities, where unclear language or diagrams leave too much open to interpretation; vagueness, or insufficient detail; contradictions, that is, two or more requirements that cannot be simultaneously satisfied; incompleteness or any other kind of missing information; and mixed levels of abstraction where very detailed design elements appear alongside high-level system requirements. To illustrate, consider the following hypothetical requirement for a missile launching system.

> 5.1.1 If the LAUNCH-MISSILE signal is set to TRUE and the ABORT-MISSILE signal is set to TRUE then do not launch the missile, unless the ABORT-MISSILE signal is set to FALSE and the ABORT-MISSILE-OVERRIDE is also set to FALSE, in which case the missile is not to be launched.

This requirement is written in a confusing manner, and the complexity of the logic makes it difficult to know just exactly what the user intends. And if the user's intent is wrongly depicted in the language of the requirements, then the wrong system will be built.

DOI: 10.1201/9781003129509-7

It is because we need precision beyond that which can be offered by natural languages that we frequently need to reach for more powerful tools, which can only be offered by mathematics.

There is another important benefit of using formal methods in requirements engineering and elsewhere. Formal methods represent a universal language with very strict rules. With modern engineering teams consisting of individuals with different primary speaking languages, a universal language of expression can provide a mechanism for the clear communication of ideas. More research on this proposition must be conducted, however.

What Are Formal Methods?

Formal methods involve mathematical techniques. To be precise, we define formal methods as follows:

A software specification and production method based on a precise mathematical syntax and semantics that comprises:

- A collection of mathematical notations addressing the specification, design, and development phases of software production, which can serve as a framework or adjunct for human engineering, and design skills and experience.
- A well-founded logical inference system in which formal verification proofs and proofs of other properties can be formulated.
- A methodological framework within which software may be developed from the specification in a formally verifiable manner. Formal methods can be operational, denotational, or dual (hybrid) (Laplante 2005).

It is clear from the definition that formal methods have a rigorous, mathematical basis.

Formal methods differ from informal techniques, such as natural language, and informal diagrams like flowcharts. The latter cannot be completely transliterated into a rigorous mathematical notation. Of course, all requirements specifications will have informal elements, and there is nothing wrong with this fact. However, there are apt to be elements of the system specification that will benefit from formalization.

Techniques that defy classification as either formal or informal because they have elements of both are considered to be semiformal. For example, UML (the current version 2.5) and its close relative SySML can be classified as semiformal since while all of its meta-models have precise mathematical equivalents, these need to be augmented with semantics in order to enable their use for formal analysis of specifications. For example, Graves and Bijan (2011) showed how formal semantics built on SySML could be used to test aerospace systems.

In any case, what advantage is there in applying a layer of potentially complex mathematics to the already complicated problem of behavioral description? The answer is given by Forster, a giant of formal methods:

> One of the great insights of twentieth-century logic was that, in order to understand how formulae can bear the meanings they bear, we must first strip them of all those meanings so we can see the symbols as themselves ... [T]hen we can ascribe meanings to them in a systematic way ... That makes it possible to prove theorems

about what sort of meanings can be borne by languages built out of those symbols. (Forster 2003)

Formal methods are especially useful in requirements specification because they can lead to unambiguous, concise, correct, complete, and consistent specifications. But as we have seen, achieving these qualities is difficult, and even with formalization correctness and completeness, it can be elusive. A thorough discussion of these issues along with a very detailed example can be found in Bowen et al. (2010).

Formal Methods Classification

There are several formal methods classes. The first class, model-based, methods provide an explicit definition of state and operations that transform the state. Typical model-based formal methods include Z, B, and the Vienna development method. The next class, algebraic methods, provides an implicit definition of operations without defining the state. Algebraic methods include Larch, PLUSS, and OBJ. Process algebras provide explicit models of concurrent processes—representing behavior with constraints on allowable communication between processes. Among these are communicating sequential process (CSP) and calculus of communicating systems (CCS). Logic-based formal methods use logic to describe the properties of systems. Temporal and interval logics fall into this category. Finally, net-based formal methods offer implicit concurrent models in terms of data flow through a network, including conditions under which data can flow from one node to another. Petri nets are a widely used net-based formal method.

It should be noted that formal methods are different from mathematically based specifications. That is, specifications for many types of systems contain some formality in the mathematical expression of the underlying algorithms. Typical systems include process control, avionics, medical, and so on. Such use of mathematics does not constitute the use of formal methods, though such situations may lend themselves to formal methods.

A Little History

Formal methods have been in use by software engineers for quite some time. The Backus–Naur (or Backus-Normal) Form (BNF) is a mathematical specification language originally used to describe the Algol programming language in 1959. Since then a number of formal methods have evolved. These include:

- The Vienna development method (VDM), 1971
- CSP, 1978
- Z (pronounced "zed"), late 1970s
- Pi-calculus, early 1990s
- B, late 1990s

Finite state machines, Petri nets, and statecharts, or other techniques regularly used by systems and software engineers can be used formally. Other formal methods derive from general mathematical frameworks, such as category theory (CT), and a number of domain-specific and general languages have been developed over the years, often with specialized compilers or other toolsets.

Formal methods are used throughout Europe, particularly in the UK, but not as widely in the United States. However, important adopters of formal methods include NASA, IBM, Lockheed, HP, and AT&T.

Using Formal Methods

Formal methods are used primarily for systems specification and verification. Users of UML 2.5 could rightly be said to be employing formal methods, but only in the specification sense. The languages B, VDM, Z, Larch, CSP, and statecharts, and various temporal logics are typically used for systems specification. And while the clarity and precision of formal systems specification are strong endorsements for these, it is through verification methods that formal methods really show their power. Formal methods can be used for two kinds of verification: theorem proving (for program proving) and model checking (for requirements validation). We will focus on the former.

Formal methods can be used in any setting, but they are generally used for safety-critical systems, for COTS validation/verification (e.g., by contract), for high financial risk systems (e.g., in banking and securities), and anywhere that high-integrity software systems are needed. For example, IBM used a theorem prover (ACL2) in the AMD x86 processor development process. NASA applied formal methods in several projects including the next-generation air transportation system and sponsors an annual conference on the subject. Alstom and Siemens used B-method to develop safety automatisms for the various subways installed throughout the world.

Fabian et al. (2010) studied several different formal methods—including an augmented version of UML—and the goal-oriented formal methods—KAOS, i* and Tropos (see Chapter 2)—to show how they could be used in creating and validating security requirements.

Formal methods activities include writing a specification using a formal notation, validating the specification, and then inspecting it with domain experts. Further, one can perform automated analysis using theorem (program) proving or refine the specification to an implementation that provides semantics-preserving transformations to code. You can also verify that the implementation matches the specification (testing).

Another important use of formal methods is in modeling and analyzing requirements interaction. Requirements interactions can create conflicts between requirements due to differing stakeholder viewpoints. Formal methods such as finite state machines, temporal logic, CSP, Z, and many others have been used in an attempt to resolve this vexing problem. An analysis of the challenges of identifying requirements interactions and research directions can be found in Shehata and Eberlein (2010).

Examples

To illustrate the use of formal methods in requirements engineering, we present a number of examples using several different techniques. Our purpose is not to present anyone's formal method and expect the reader to master it. Rather, we wish to show a sampling of how various formal methods can be used to strengthen requirements engineering practice.

Formalization of Train Station in B

In their most straightforward use, formal methods can express system intent. In this use of formal methods, we exploit the conciseness and precision of mathematics to avoid the shortcomings of natural languages and informal diagrams. B is a model-based formal language that is considered an "executable specification" (collection of abstract machines) that can be translated to C++ or Ada source code. In this case, we use the modeling language B to provide a specification for a train station, similar to Paris's Line 14 (Meteor) and New York City's L Line (Canarsie). The model used is based on an example found in Lano (1996). Those familiar with Z will see a great deal of similarity in the structure and notation to B.[3]

The model begins as shown in Figure 7.1. The specification starts with the name of the machine and any included machines (in this case, **TrackSection**, which is not shown here). Then a list of variables is given and a set of program **INVARIANT**s—conditions that must hold throughout the execution of the program. Here **N** represents the natural numbers {0, 1, 2, 3, ...}. We continue with a partial description of the behavioral intent of this model.

Notice that the three variables of interest are the **platforms**, **max_trains**, and **trains_in_station**. The **INVARIANT** sets constraints on the number of trains in the station (must be a nonnegative number) and the maximum number of trains in the system (must be a nonnegative number and there can be no more than that number in the station). It continues with platforms as a sequence of train sections, in the range of the number of sections, which acts as an upper bound on the maximum number of trains (since only one train can occupy each section of track).

The next part of the B specification is the initialization section (Figure 7.2).

This specification initializes the platform sequence to null and the maximum number of trains and trains in a station to zero.

Now we introduce a set of operations on the train station dealing with arrivals, departures, openings and closings, and resetting the maximum number of trains. For example, the arrival

```
                    TRAIN STATION

MACHINE Station
INCLUDES TrackSection


VARIABLES
     platforms, max_trains, trains_in_station

INVARIANT
     trains_in_station ∈ N ^
     max_trains ∈ N ^ trains_in_station ≤ max_trains ^
     size(platforms ∈ card(ran(platforms)) ^
     max_trains ≤ size(platforms)
```

Figure 7.1   Setup specification for train station.

```
INITIALIZATION
     platforms, max_trains, trains_in_station := [], 0, 0
```

Figure 7.2   Initialization specification for train station.

```
OPERATIONS

    train_arrives(ts) ≅

        PRE trains_in_station < max_trains ∧ ts  ran(platforms) ∧

            tstate(ts) = free

        THEN

            arrival(ts) ||

            trains_in_station := trains_in_station + 1

        END;
```

Figure 7.3 Train arrival behavior for station.

```
train_departs(ts) ≅

    PRE ts  ran(platforms) ∧

        tstates(ts) = blocked

    THEN

        departure(ts) ||

        trains_in_station := trains_in_station - 1

    END;
```

Figure 7.4 Train departure behavior for station.

specification is shown in Figure 7.3. A precondition (**PRE**) before train **ts** can arrive is that there is room in the station for the train and that a platform is available. If so, mark the train as arriving and increment the count of trains in the station.

The next operation involves train departure (Figure 7.4). The precondition is that the train is in the range of the sequence of platforms (on a platform) and that the train is not moving (blocked). If so, then train **ts** is marked as having departed, and the number of trains in that station is decreased by one.

The next operation is the opening platform **ts** (Figure 7.5). The precondition is that the platform is in the range of the sequence of platforms for that station and that the current platform is closed. If so, then platform **ts** is marked as being opened.

The closing of a platform operation is similar (Figure 7.6). The precondition is that the platform is in the range of the sequence of platforms for that station and that the current platform is open. If so, then platform **ts** is marked as being closed.

Finally, an operation is needed to set the maximum number of trains, **mt,** that can be at a given station (Figure 7.7). The preconditions are that **mt** must be a natural number and cannot exceed the maximum number of platforms and that the maximum number of trains for a station cannot be less than the number of trains already there.

The point behind this is that the mathematical specification is far more precise than the natural language "translation" of it. In addition, the mathematical representation of the behavior is less verbose than the natural language version. Finally, the representation in B enables correctness-proofs, via standard predicate calculus. And as stated before, the B specification can actually be converted to C++ or Ada code. For example, the safety-critical portion of New York City Transit's L (subway)

```
open_platform(ts) ≜

    PRE ts  ∈ran(platforms)

        tstate(ts) = closed


    THEN

        open(ts)

    END;
```

Figure 7.5   Platform opening behavior for station.

```
close_platform (ts) ≅

    PRE ts ran(platforms)

        tsstate(ts) = free

    THEN

        close(ts)

    END;
```

Figure 7.6   Platform closing behavior for station.

```
set_max_trains(mt) ≅

    PRE mt N ∧ mt ≤ size(platforms) ∧

        trains_in_station ≤ mt

    THEN

        max_trains := mt

    END
```

Figure 7.7   Behavior for setting the maximum number of trains for a train station.

line, whose communication-based train control (CBTC) system has been in revenue service for several years, was specified in B (over a decade ago). The L line's B-based specifications were auto-translated into more than 25,000 proof obligations and (their) proofs, over 95% of which were proved by machine, with the remaining 5% proven using semiautomatic (human-guided) proof using a proof engine. The proven B specifications were then translated automatically into Ada code (Hacken 2007).

Formalization of Space Shuttle Flight Software Using MurΦ

The next example of formalization uses the prototype verification system (PVS) language to model and prove various aspects of the space shuttle flight software system (Crow and Di Vita 1998). PVS is a state-driven language, meaning it implements a finite state machine to drive behavior.

A finite state machine model, M, consists of a set of allowable states, S; a set of inputs, I; a set of outputs, O; and a state transition function, described by Equation 7.1.

$$M : I \times S \to [O \times S]$$

(7.1)

```
pf_result: TYPE = [# output: pf_outputs, state: pf_state #]

principal_function ( pf_inputs, pf_state,
                             pf_I_loads, pf_K_loads,
                             pf_constants ) : pf_result =

    (# output := <output expression>,
        state := <next-state expression> #)
```

Figure 7.8 Initialization code in PVS for space shuttle functionality. (From Crow, J., & Di Vita, B., *ACM Trans. Softw. Eng. Methodol.*, 7: 296–332, 1998.)

The initial and terminal states need to be defined as well.

Part of the implementation code in PVS is shown in Figure 7.8. Without explicating the code, it is interesting to point out its expressiveness in modeling the finite state machine.

Here the **principal_function** defines an interface incorporating the inputs (**pf_inputs**) and states (**pf_state**) needed to drive the behavior. The output is an output expression and the next state in the machine.

The behavior specification continues as an infinite sequence of state transitions (functional transformations) (Figure 7.9, top part) and as a set of functions associated with each state—in this case, on the state associated with firing small control jet rockets (vernier rockets). Even without fully understanding the syntax of PVS, because it is similar to C, C++, or Java, we hope you can see how this behavior of the system can be clearly described.

In addition to clarity and economy, another advantage of using PVS to describe the behavior is that an interactive proof checker and Stanford University's MurΦ (pronounced "Murphy") can be used for theorem proving (specification verification).

Formalization of an Energy Management System Using Category Theory [4]

Category theory (CT) permits functional notation to be used to model abstract objects and is therefore useful in formal specification of those objects (Awodey 2005).

This formalization facilitates the good design and can assist in user interface design, and permits users of the system to build on their domain knowledge as they learn the constraints that affect design decisions. These design decisions then become evident as the rules are applied from requirements, to design, to user interface functionality. Finally, for ease of maintenance and troubleshooting for the designers and programmers, the use of CT can show where an error might be contained while allowing for the group of functions to be analyzed and troubleshooting on an input-by-input basis. The following is a brief introduction to CT for software specification.

Suppose we have classes of abstract objects, A, B, and C. Further, suppose there are functions f and g such that $f: A \rightarrow B$ and $g: B \rightarrow C$. A category is composed of the objects of the category and the morphisms of the category. A basic functional relationship can be seen in Figure 7.10 where $g \circ f$ is the composition of f and g.

CT is widely used to model structures and their transformations over time (represented by finite or infinite sequences of functional compositions). For example, in Figure 7.11, the sequence of composed relationships between f and g would be given by f, fg, fgf, and so on.

```
finite_sequences[n: posnat, t: TYPE]: THEORY
BEGIN

   finite_seq: TYPE = [below[n] -> t]

END finite_sequences
requirements: THEORY
BEGIN
IMPORTING finite_sequences
. . .

   jet: TYPE
   rot_jet: TYPE FROM jet

   jet_bound: posint
   jet_count: [rot_jet -> below[jet_bound]]
   jet_count_ax: AXIOM injective?(jet_count)

   finite_number_of_jets: LEMMA . . .

   vernier?: pred[rot_jet]
   vernier_jet: TYPE = (vernier?)
   there_is_a_vernier: AXIOM (EXISTS j: vernier?(j)

   primary_rot?(j): bool = NOT vernier?(j)
   primary_rot_jet: TYPE = (primary_rot?)
   there_is_a_primary: AXIOM
      (EXISTS j: primary_rot?(j))

   downfiring?: pred[vernier_jet]

. . .
END requirements
```

Figure 7.9 Behavior of vernier rockets in the space shuttle system. (From Crow, J., & Di Vita, B., *ACM Trans. Softw. Eng. Methodol.*, 7, 1998.)

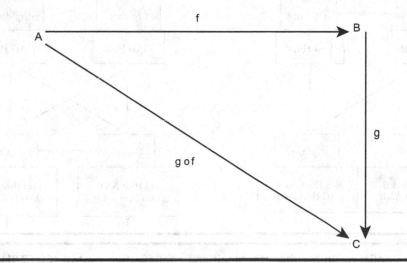

Figure 7.10 A simple functional relationship.

Figure 7.11   An infinite category.

These sequences also represent categories. For example, we just saw the space shuttle behavioral specification represented as an infinite sequence of state transitions using the PVS notation.

The composition of a system is also appropriate to model the modularity of a larger system and its interconnections. In comparison to a truth table, which is a great formalizing and testing tool, CT can associate related items and then decompose each into something that can be examined on a lower level. For example, business logic can be placed in a category either intuitively or architecturally.

Example: An Energy Management System

The example we describe is a partial formalization of a power company's energy management system (EMS). Any power generation utility has to deal with a complex, real-time system involving a high level of automation, fault tolerance, and redundancy.

In an EMS, an open access gateway (OAG) serves as a communication liaison for many communication protocols between computer systems and remote terminal units (RTUs). The OAG is also used to communicate with the plant digital interface (PDI) via intercompany communications protocol (ICCP) and to other entities via the interconnection diagram shown in Figure 7.12.

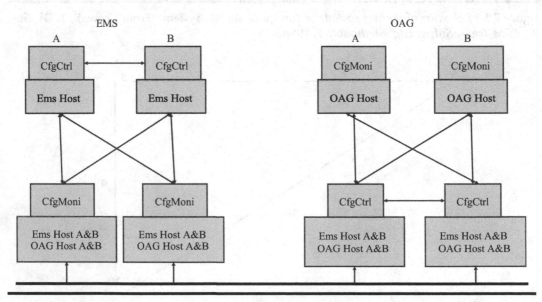

Figure 7.12   Configuration monitor (CfgMoni) and configuration control (CfgCTRL) EMS and OAG relationship.

A1—EMSA CFGCTRL	A2—OAGA CFGCTRL
B1—EMSB CFGCTRL	B2—OAGB CFGCTRL
C1—EMSA CFGMONI	C2—OAGA CFGMONI
D1—EMSB CFGMONI	D2—OAGB CFGMONI
f1—EMS CFGCTRL	f2—OAG CFGCTRL
g1—EMS CFMONI	g2—OAGEMS CFMONI
j(g)—EMS CFGMONI to OAG CFGCTRL	h(g)—OAG CFGCTRL to EMS CFGMONI

Figure 7.13 Data dictionary and category reference.

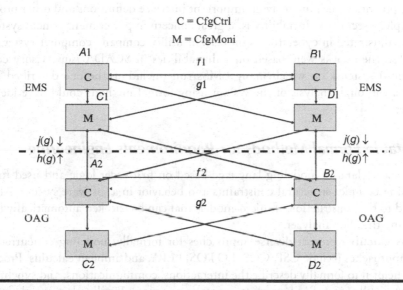

Figure 7.14 Partial formalization of Figure 7.12 using relabeled components.

To formalize Figure 7.12 using CT, we first translate the entities using short-hand symbols shown in Figure 7.13.

The resultant relabeled system is shown in Figure 7.14. Functional notations have been added to Figure 7.14, but these would need to be defined further.

Let's discuss a small portion of the formalization in Figure 7.14. We begin with EMS A and EMS B. If A1 = EMS A and B1 = EMS B, then suppose $f1$ is a bidirectional mapping between EMS A and EMS B that describes communications between CfgCtrl and CfgMoni. Taking one side of the mapping, suppose we have $f1: A1 \rightarrow B1$. Next, we assert that $f1$ is defined on all $A1$ and all the values of $f1 \in B1$. That is, *range* $(f1) \subseteq B1$.

Continuing, suppose there is another bidirectional mapping, $g1$ between EMS A and EMS B. Defining one side of g1: $A1 \rightarrow D1$ we have an associated composed function $g1 \circ f1 : A1 \rightarrow C1 \cap D1$, that is, $(g1 \circ f1)(a1) = g1(f1(a1))$ and $a1 \in A1$. The composed function is associative, as follows: if there is another function $j: C1 \rightarrow A2$ and forming $j \circ g1$ and $g1 \circ f1$, then we can compare $j \circ g1$ $\circ f1$ and $j \circ (g1 \circ 1)$. As indicated in Figure 7.14, it becomes apparent that these two functions are always identical.

At this point, it should be noted that the set of all inputs and outputs from $A1 \cap B1 \subset C1 \cup D1$ where $C1$ and $D1$ are CfgMoni on both the EMS A and B servers. The set of inputs and outputs from $C1 \cap D1 \subset A1 \cup B1$ where $A1$ and $B1$ are the CfgCtrl on the EMS A and B servers. CfgCtrl communicates with itself on both the EMS and OAG servers. The function $j(g)$: $C1 \rightarrow A2 \cup B2$ shows the configuration from the EMS CfgMoni to OAG CfgCtrl. OAG also communicates with CfgCtrl on its home server, that is, there is a function $g1(g) : C1 \rightarrow A1$. This feature is needed because CfgCtrl communicates with each redundant server so that controls can be issued while CfgMoni sends and receives status to CfgCtrl for up-to-date status on the server that it is monitoring.

To relate this to the EMS and OAG configurations, the set $A1$ is EMS CfgCtrl A, the set $B1$ is EMS CfgCtrl B, and $j(g) : (C1 \cup D1) \rightarrow (A2 \cup B2)$ shows that CfgMoni from either EMS A or B maps to CfgCtrl on the OAG A or B. We could continue with this formalization, but you should realize by now that a high level of attention to detail and rigor is necessary to create this kind of specification, potentially helping to avoid important interface definitions and other problems.

For example, a security vulnerability is of great concern in power management systems, a fact that has been illustrated in cyberattacks on certain utility company computer systems, causing power outages. The attacks were based on vulnerabilities in SCADA (supervisory control and data acquisition) systems, on which many EMSs run (including the one described here). It is possible that a thorough analysis of the system using formal methods could have identified the vulnerabilities.

Other Notable Formal Methods for Requirements Engineering

- **Alloy** is a declarative modeling language based on first-order logic and used for formally expressing complex structural constraints and behavior in a software system. Alloy can be targeted to the construction of micro-models that can be checked automatically for correctness using the alloy analyzer.
- **Process calculi** represent diverse approaches for formally modeling concurrent systems. These approaches include CSP, CCS, LOTOS, PEPA, and ambient calculus. Process calculi can be helpful to formally describe the interactions, communications, and synchronizations between a collection of independent processes or agents. Finally, they can define the algebraic laws for the process operators, which allow process expressions to be changed using equational reasoning.
- **FORM-L** results from the MODRIO project (Model-DRIven physical systems Operation), which improved MODELICA (an object-oriented notation for modeling the behavior of physical systems). FORM-L addresses the early stage of system development, with a level of detail sufficient to support some early validation through model-checking using the Stimulus tool.
- **A Petri net** is a mathematical modeling language used to describe distributed systems. It offers a graphical notation for stepwise processes that include representing iterations and concurrent executions. Because of their formality, Petri nets are often used to augment UML diagrams.

Requirements Validation

We already discussed requirements validation in Chapter 6 using informal techniques, but formal techniques are particularly appropriate for testing the requirements to ensure, for example, that the requirements are consistent. By consistent we mean that the satisfaction of one requirement does not preclude the satisfaction of any other.

Consistency Checking Using Truth Tables

One way to formally prove consistency is with a truth table. In this case, we rewrite each requirement as a Boolean proposition. Then we show that there is some combination of values for the Boolean variables of this collection of requirements for which all propositions (requirements) are true.

To illustrate, consider the following requirements from part of the pet store POS system:

1.1 If the system software is in debug mode, then the users are not permitted to access the database.
1.2 If the users have database access, then they can save new records.
1.3 If the users cannot save new records, then the system is not in debug mode.

We convert the requirements into a set of Boolean propositions. Let p, q, and r be Boolean variables and let

p: be the statement "the system software is in debug mode"
q: be the statement "the users can access the database"
r: be the statement "the users can save new records"

Clearly, the specifications are equivalent to the following propositions:

1.1 $p \Rightarrow \neg q$
1.2 $q \Rightarrow r$
1.3 $\neg r \Rightarrow \neg p$

Now we construct a truth table (Table 7.1) and determine if any one of these rows has all "T"s in the columns corresponding to propositions 1.1, 1.2, and 1.3, meaning each of the requirements is satisfied. If we can find such a row, then the requirements are consistent.

We see that there are three rows where all three propositions representing the requirements are true for all combinations of Boolean values. Therefore, this set of requirements is consistent.

Table 7.1   Truth Table Proof of Consistency for a Collection of Requirements

p	q	r	$\neg p$	$\neg q$	$\neg r$	$p \Rightarrow \neg q$	$q \Rightarrow r$	$\neg r \Rightarrow \neg p$
T	T	T	F	F	F	F	T	T
T	T	F	F	F	T	F	F	F
T	F	T	F	T	F	T	T	T
T	F	F	F	T	T	T	T	F
F	T	T	T	F	F	T	T	T
F	T	F	T	F	T	T	F	T
F	F	T	T	T	F	T	T	T
F	F	F	T	T	T	T	T	T

Notice that requirement 1.3 likely does not make logical sense. If the users cannot save new records, then it is likely it was intended to say that the system is in debug mode (as opposed to not in debug mode). We deliberately left this logically questionable requirement in place to make a point—this system of requirements is consistent but not necessarily correct (in terms of what the customer intended). This consistency checking process will not uncover problems of intent (validity)—these must be uncovered through other means such as requirements reviews.

Consistency Checking by Inspection

Sometimes a set of requirements can be inspected for consistency without using the truth table. Here is an example. Consider these three requirements for the baggage handling system:

3.4.1 If the red emergency button is pressed, then the system shall be turned off immediately.
3.7.2 Display lighting for the system shall use LED technology.
3.11.4 Audible alarms shall have an output not to exceed 65 decibels.

Do you need a truth table to prove consistency for these requirements? No, because it should be obvious that none of these statements are logically connected— they can all be independently true or false. Therefore, there are no consistency problems. This alternate analysis approach is especially helpful if there is a large set of requirements to be considered—necessitating a very large and cumbersome truth table. You can omit any requirements that have no logical connection to any other from the analysis, thus reducing the truth table size.

Consistency Checking Limitations

Consistency checking using Boolean satisfiability is a powerful tool. However, although the process can be automated, the problem space grows large very quickly. For n logical variables in the requirements set, the problem is $O(2^n)$. This kind of problem quickly becomes intractable even for supercomputers—it is a well-known NP-complete problem, the Boolean satisfiability problem. There are some special cases in which the Boolean satisfiability problem can be structured so that it can be solved efficiently using computer programs. Even so, we would likely use this technique for requirements verification only for very critical logic situations, for example, the launch/no-launch decision logic for a weapon, dosage administration logic for some kind of medical device, or shut-down logic for a nuclear power plant.

VIGNETTE 7.1   Formal Methods for Safety-Critical Systems: Proving an Unsafe Condition Is Impossible

Clearly, it would be computationally difficult to prove consistency for a large set of requirements due to combinatorial state explosion. A more practical and tractable use for formal methods in requirements is to show that an unsafe or forbidden state in the system cannot be reached. It's easy to imagine unsafe states in all kinds of systems, but especially in critical ones such as nuclear power plants, self-driving vehicles, avionics, medical devices, and space systems. But even consumer appliances have to be designed for safety.

Here is a simple example proving that a microwave oven, such as the kind found in most homes, is safe with respect to one particular danger. In a microwave oven if the door is open, the system should not emit radiation as this represents a burn danger to anyone nearby. Here's a set of requirements for the oven:

If the microwave generator is on, the oven light shall be on

1.2 If the oven light is off, the microwave generator is off
1.3 If the door is open, the microwave generator shall be off

We need to show that these requirements preclude the danger. First, we assign Boolean variables, P, Q, and R, to the conditions defined in the requirements:

P="the microwave generator is on"
Q="the oven light is on"
R=" the door is open"

We can rewrite the requirements as:

1.1: $P \Longrightarrow Q$
1.2: $\neg Q \Longrightarrow \neg P$
1.3: $R \Longrightarrow \neg P$

Now we want to show that the following condition is precluded in this set of requirements:

The door is open and the microwave generator is on
In terms of the Boolean variables, this can be represented as

$$X: R \wedge P$$

We could do a full consistency analysis taking the three requirements together with the forbidden condition, showing that the set is inconsistent. But all we really have to do is show that any one of the requirements precludes the forbidden condition (i.e., they are a contradiction). In this case, requirement 1.3 and X are a contradiction. We show this with a simple truth table:

R	P	1.3: $R \Longrightarrow \neg P$	$X: R \wedge P$
T	T	F	T
T	F	T	F
F	T	T	F
F	F	T	F

From the table, it is clear that anytime X is true, 1.3 cannot be. There are automated tools to do this kind of proof, but for a small set of simple requirements, a manual approach is acceptable and not tedious.

Theorem Proving

Theorem proving techniques can be used to demonstrate that specifications are correct. That is, axioms of system behavior can be used to derive proof that a system (or program) will behave in a given way. Remember, a specification and program are both the same thing—a model of execution. Therefore, program proving techniques can be used for appropriately formalized specifications. These techniques, however, require mathematical discipline.

Program Correctness

A system is correct if it produces the correct output for every possible input, and if it terminates. Therefore, system verification consists of two steps:

- Show that, for every input, the correct output is produced (this is called partial correctness).
- Show that the system always terminates.

To deal with this situation, we need to define assertions. An assertion is some relation that is true at that instant in the execution of the program. An assertion preceding a statement in a program is called a precondition of the statement. An assertion following a statement is called a postcondition. Certain programming languages, such as Eiffel and Java, incorporate run-time assertion checking. Assertions are also used for testing fault tolerance (through fault injection).

Hoare Logic

Hoare logic features straightforward mathematical notation and proof techniques. It has been used to represent and validate privacy requirements testing suites in various critical applications such as banking (Agrafiotis et al. 2011).

Much of the following follows the presentation found in Gries and Gries (2005). Suppose we view a requirements specification as a sequence of logical statements specifying system behavior. In 1969, C.A.R. (Tony) Hoare introduced the notation (called the Hoare triple) (Hoare 1969):

$$P\{S\}Q$$

where P and Q are assertions (pre- and postconditions, respectively) and S is a system behavioral segment. Then, the precondition, statement, postcondition triple has the following meaning: Execution[5] of the statement begun with the precondition true is guaranteed to terminate, and when it terminates, the postcondition will be true.

The intent is that P is true before the execution of S, then after S executes, Q is true. Note that Q is false if S does not terminate.

Hoare's logic system can be used to show that the system segment (specification) is correct under the given assertions. For example, suppose we have the following:

$$// \{x = 5\} \text{an assertion} \left(\text{precondition}\right)$$

$$z = x + 2;$$

$$// \{z = 7\} \text{an assertion} \left(\text{postcondition}\right)$$

Proof: Suppose that {x=5} is true, then z=5 + 2 =7 after the system begins execution. So the system is partially correct. That the system terminates is obvious, as there are no further statements to be executed. Hence, correctness is proved.

Hoare added an inference rule for two rules in sequence. We write this as:

$$P\{S_1\}Q$$
$$\underline{Q\{S_2\}R}$$
$$P\{S_1;S_2\}R$$

This inference rule means that if the postcondition (Q) of the first system segment (S_1) is the antecedent (precondition) of the next segment, then the postcondition of the latter segment (R) is the consequent of the concatenated segments. The horizontal line means "therefore" or, literally, "as a result."

To illustrate, we show that the following specification is correct under the given assertions:

$$//\{x = 1 \wedge y = 2\}\text{an assertion}\,(\text{precondition})$$

$$x = x + 1;$$

$$z = x + y;$$

$$//\{z = 4\}\text{an assertion}\,(\text{postcondition})$$

Proof: Suppose that {x=1} is true, then the system executes the first instruction, x=2. Next, suppose that {y=2} is true. Then after the execution of the second statement z=2+2=4. Hence, z=4 after the execution of the last statement and the final assertion is true, so the system is partially correct. That the system terminates is obvious, as there are no further statements to be executed.

An inference rule is needed to handle conditionals. This rule shows that

$$(P \wedge \text{condition})\{S\}Q$$
$$\underline{(P \wedge \neg\,\text{condition}) \Rightarrow Q}$$
$$\therefore P\{\text{if condition then } S\}Q$$

To illustrate, we show that the specification segment is correct under the given assertions:

$$//\{\text{Any}\}\text{an assertion}\,(\text{precondition})\,\text{that is always true}$$

$$\text{if } x > y \text{ then}$$

$$y = x;$$

$$//\{x \leq y\}\text{ an assertion }(\text{postcondition})$$

Proof: If, upon entry to the segment, $x \leq y$, then the if statement fails and no instructions are executed and clearly the final assertion must be true. On the other hand, if upon entry, $x > y$, then the statement $y=x$ is executed. Subsequently, $y=x$ and the final assertion $x \leq y$ must be true. That the system terminates is obvious, as there are no further statements to be executed.

Another inference rule handles if-then-else situations, that is,

$$(P \wedge \text{condition})\{S_1\}Q$$

$$\frac{(P \wedge \neg \text{condition})\{S_2\}Q}{\therefore P\{\text{if condition then } S_1 \text{ else } S_2\}Q}$$

To illustrate, we show that the specification segment is correct under the given assertions:

```
//{Any} an assertion (precondition) that is always true
if x < 0 then
abs = -x; else
abs = x;
//{abs = |x|} an assertion (postcondition)
```

Proof: If $x < 0$, then $abs = -x \Rightarrow abs$ is assigned the positive x. If $x \geq 0$ then abs is also assigned the positive value of x. Therefore, $abs = |x|$. That the system terminates is obvious, as there are no further statements to be executed.

Finally, we need a rule to handle "while" statements:

$$\frac{(P \wedge \text{condition})\{S\}P}{\therefore (P \wedge \text{while condition})\{S\}(\neg \text{condition} \wedge P)}$$

We will illustrate this rule through an example shortly.

So far the specification snippets (one might say they are really "code" snippets) that have been proven represent very simplistic but low-level, detailed behavior unlikely to be found in a requirements specification. Or is this the case? It is not so hard to imagine that for critical behavior, it might be necessary to provide specifications at this level of detail. For example, the behavior required for the dosing logic for an insulin pump (or machine that delivers controlled radiation therapy) might require such logic. The decision to launch a missile, control the life support system in the space station, or shut down a nuclear plant might be based on simple, but critical logic. This logic, incidentally, could have been implemented entirely in hardware.

Even our running examples might have some critical logic that needs formal proof. For example, the baggage counting logic for the baggage handling system or certain inventory control logic might require the following behavior:

```
//{sum = 0 ∧ count = n ≥ 0}
while count > 0
{
sum = sum + count;
count = count--;
}
//{sum = n(x + 1)/2}
```

And so it would be necessary to show that the specification segment is correct under the given assertions. To prove it, we use induction.

Basis

Suppose that *sum* = *0* and *n* = *0* as given by the assertion. Now upon testing of the loop guard, the value is false and the system terminates. At this point, the value of *sum* is zero, which satisfies the postcondition. That the system terminates was just shown.

Induction Hypothesis

The program is correct for the value of *count* = *n*. That is, it produces a value of *sum* = $n(n+1)/2$ and the system terminates.

Induction Step

Suppose that the value of *sum* = 0 and *count* = $n + 1$ in the precondition.

Upon entry into the loop, we assign the value of *sum* = $n+1$, and then set *count* = *n*. Now, from the induction hypothesis, we know that for *count* = *n*, *sum* = $n(n+1)/2$. Therefore, upon continued execution of the system, *sum* will result in $(n+1) + n(n+1)/2$, which is just $(n+1)(n+1+1)/2$ and the system is partially correct.

Since by the induction hypothesis, by the nth iteration of the loop *count* has been decremented n times (since the loop exited when *count* was initialized to n). In the induction step, *count* was initialized to $n+1$, so by the nth iteration it has a value of 1. So after the $n+1$st iteration it will be decremented to 0, and hence the system exits from the loop and terminates.

Does this approach really prove correctness? It does, but if you don't believe it, try using Hoare logic to prove an incorrect specification segment to be correct, for example:

```
//{sum = 0 ∧ count = n ≥ 0}
while count < n
{
sum = sum + count;
count = count++;
}
//{sum = n(n + 1)/2}
```

This specification is incorrect because n is left out of *sum*, and therefore, no proof of correctness can be obtained.

It is easy to show that a *for* loop uses a similar proof technique to the *while* loop. In fact, by Böhm and Jacopini (1966), we know we can construct verification proofs just from the first two inference rules. For recursive procedures, we use a similar, inductive, proof technique applying the induction to the n+1st iteration in the induction step, and using strong induction if necessary.

Model Checking

Given a formal specification, a model checker can automatically verify that certain properties are theorems of the specification. Model-checking has traditionally been used in checking hardware designs (e.g., through the use of logic diagrams), but it has also been used with software. Model checking's usefulness in checking software specifications has always been problematic due to the combinatorial state explosion and because variables can take on non-Boolean values. Nevertheless,

Edmund M. Clarke, E. Allen Emerson, and Joseph Sifakis won the 2007 A. M. Turing Award for their body of work making model checking more practical, particularly in the design of chips. A great deal of interesting and important research continues in model checking and practical implementations exist.

VIGNETTE 7.2   Model-Based Systems Engineering

Model-based systems engineering (MBSE) is "the formalized application of modeling to support system requirements, design, analysis, verification and validation activities beginning in the conceptual design phase and continuing throughout development and later life cycle phases" (INCOSE). MBSE makes systems engineering more rigorous, precise, and repeatable through the use of multiple, interconnected systems models. These include models for

- requirements,
- hardware and software architectures,
- design elements (including hardware drawings and software design documents),
- embodiments of those designs (including circuit diagrams and software code),
- test cases from unit tests through acceptance tests (including both manual and automated ones),
- and maintenance documentation and information.

These artifacts use natural languages, graphical notations, formal syntax, and semantics (where appropriate), and are stored in a shared system information base under strict configuration control. MBSE is well known to manage complexity, provide quality and productivity improvements, and lower risk through its use of rigor and precision and by fostering better communications between the development team and customers (Ramos). The International Council on Systems Engineering INCOSE promotes MBSE as a best practice, and it is used widely in many application domains.

In order to fully embrace MBSE and realize its benefits, investments need to be made in tools for modeling, tracking, and maintaining the models and for strict configuration management. Additional investments, such as training, may also be needed to allow engineers to faithfully build these models and to maintain them under configuration management.

Integrated Tools

Any realistic formal methods application requires tool support. There are several integrated toolsets that combine formal specification tools with model checkers, consistency checkers, and code generators.

For example, consider a requirements-focused formal method called software cost reduction (SCR), which was first introduced in the 1970s. SCR uses a tabular notation (inherently formal) to represent the system behavior in a way that is understandable to stakeholders. Then a set of formal tools can be used to check the consistency, completeness, and correctness of the specification. In addition, the toolset includes a model checker, a front-end to PVS, an invariant generator, a property checker, and a test case generator.

Many companies including Lockheed Martin have been using SCR for many years. Notable successes include certifying that a security critical module of an embedded software device enforced data separation and specifying the requirements of three safety-critical software modules for NASA systems (Heitmeyer 2007).

Which Formal Approach to Use?

The matter of assessing the quality of requirements has received significant attention, and we cover this in Chapter 6. But how do we assess the suitability of the formal methods to be used for a software/system project. Bruel et al. (2019) recommend using nine criteria to assess the suitability of requirements approaches:

■ **Criterion 1:** System vs. Environment refers to the ability of the approach to support the distinction between the environment (or "domain") in which the system operates, including constraints it imposes; and the system (or "machine") which the project will build. Does the approach cover both or only the system part?
■ **Criterion 2:** Audience addresses the level of expertise in formal methods expected of people who will use the requirements.
■ **Criterion 3:** Level of abstraction addresses the level of detail which the requirements may or must cover.
■ **Criterion 4:** Associated method assesses whether the approach includes a comprehensive methodology to guide the requirements process.
■ **Criterion 5:** Traceability support assesses how the keeps track of one- or two-way relations between requirement elements and their counterparts in design, code, and other project artifacts.
■ **Criterion 6:** Nonfunctional requirements support addresses whether the approach covers only the description of functional properties of the system or extends to non-functional properties such as performance and security.
■ **Criterion 7:** Semantic definition assesses whether there exists a precise definition of the approach.
■ **Criterion 8:** Tool support covers the availability of tools to support the approach.
■ **Criterion 9:** Verifiability assesses whether an approach supports the possibility of formally verifying properties of the requirements.

Objections, Myths, and Limitations

There are a number of "standard" objections to formal methods, for example, that they are too hard to use or expensive, are hard to socialize, or require too much training. The rebuttals to these objections are straightforward but situational. For example, how much is too much training? Certain organizations spend millions of dollars to attain certain capability maturity model (CMM) levels but balk at investing significantly less money in training for the use of formal methods. In any case, be wary of out-of-hand dismissal of formal methods based on naïve points. But if you needed an implantable insulin pump or defibrillator what kind of requirements verification and validation would you want? You probably would want every kind of verification and validation available including formal techniques. The same goes for a nuclear plant near your home, a radiation dosing machine that you needed, the safety and restraint system for your automobile, and so on. Even in relatively mundane home devices, such as a smart clothes washer, high fidelity requirements are required, and these can be achieved only with a combination of formal and informal techniques. Formal methods are not for every situation, but they really ought to be considered in mission-critical situations.

Objections and Myths

Some objections to using formal methods include the fact that they can be error-prone (just as mathematical theorem proving or computer programming are) and that sometimes they are unrealistic for some systems. These objections are valid, sometimes. But formal methods are not intended to be used in isolation, nor do they take the place of testing. Formal methods are complementary to other quality assurance approaches. Some of the other "standard" objections to formal methods are based on myth. Papers by Hall, Bowen, and Hinchey help capture and rebut these misconceptions. The first set of myths is as follows (Hall 1990):

1. *Myth: Formal methods can guarantee that software is perfect.* Truth: Nothing can guarantee that software will be perfect. Formal methods are one of many techniques that improve software qualities, particularly reliability.
2. *Myth: Formal methods are all about program proving.* Truth: We have shown that formal methods involve more than just program proving and involve expressing requirements with precision and economy, requirements validation, and model checking.
3. *Myth: Formal methods are only useful for safety-critical systems.* Truth: Formal methods are useful anywhere that high-quality software is desired.
4. *Myth: Formal methods require highly trained mathematicians.* Truth: While mathematical training is very helpful in mastering the nuances of expression, using formal methods is really about learning one or another formal language and about learning how to use language precisely. Requirements engineering must be based on the precise use of language, formal or otherwise.
5. *Myth: Formal methods increase the cost of development.* Truth: While there are costs associated with implementing a formal methods program, there are associated benefits. Those benefits include a reduction in necessary rework downstream in the software lifecycle—at a more expensive stage in the project to make corrections.
6. *Myth: Formal methods are unacceptable to users.* Truth: Users will accept formal methods if they understand the benefits they impart.
7. *Myth: Formal methods are not used on real, large-scale systems.* Truth: Formal methods are used in very large systems, including high-profile projects at NASA.

Five years after Hall's myths, Jon Bowen and Mike Hinchey gave us *Seven More Myths of Formal Methods* (Bowen and Hinchey 1995a):

1. *Myth: Formal methods delay the development process.* Truth: If properly incorporated into the software development lifecycle, the use of formal methods will not delay, but actually accelerate the development process.
2. *Myth: Formal methods lack tools.* Truth: We have already seen that there are a number of tools that support formal methods—editing tools, translation tools, compilers, model checkers, and so forth. The available toolset depends on the formal method used.

3. Myth: Formal methods replace traditional engineering design methods. Truth: Formal methods enhance and enrich the traditional design.

4. *Myth: Formal methods only apply to software.* Truth: Many formal methods were developed for hardware systems, for example, the use of Petri nets in digital design.

5. *Myth: Formal methods are unnecessary.* Truth: This statement is true only if high-integrity systems are unnecessary.

6. *Myth: Formal methods are not supported.* Truth: Many organizations use, evolve, and promote formal methods. Furthermore, see Myth 2.

7. *Myth: Formal methods people always use formal methods.* Truth: "Formal methods people" use whatever tools are needed. Often that means using formal methods, but it also means using informal and "traditional" techniques too.

Parnas (2010) raised another compelling objection to traditional formal methods noting that

> Funding agencies often require that larger research-funded projects include some cooperation with industrial organizations and demonstrate the practicality of an approach on 'real' examples. When authors report such efforts, they state that they are successful. Paradoxically, such success stories reveal the failure of the industry to adopt formal methods as standard procedures; if using these methods was routine, papers describing successful use would not be published.

It is difficult to dispute his analysis—the use of formal methods in real industrial-strength systems is still relatively rare. Parnas' suggested antidote, however, is to invent a new set of formal methods that are more likely to be adapted by mainstream engineers. To date, no such antidote has been found.

Limitations of Formal Methods

No method can guarantee absolute correctness, safety, and so on, and formal methods have some limitations. For example, they do not yet offer good ways to reason about alternative designs or architectures. Formal specifications must be converted to a design, then a conventional implementation language. This translation process is subject to all the potential pitfalls of any programming effort.

For example, consider the following specification segment for an accelerometer processing function in an avionics system. This operation is to be initiated by an interrupt to read the accumulated acceleration pulses collected every 10 ms. From these, velocity and position (with respect to one axis) can be determined.

3.2.2 Compute accumulated acceleration every 10 ms as follows:

$$\text{Precondition: } x = x_{prev}$$

$$x \Leftarrow x + \Delta x;$$

$$\text{Postcondition: } x = x_{prev} + \Delta x$$

Let's use a modified Hoare logic approach to show that the specification is correct under the given assertions. Suppose that $x = x_{prev}$ is true, then after the addition and assignment operations are performed, $x = x_{prev} + \Delta x$. So the specification fragment is correct. There is no need to show "termination" since there are no more operational requirements in this fragment. So this specification fragment is verified. But what if x is a 16-bit number and Δx is a 64-bit number? Then there is the possibility of an overflow condition in x. These are design details and errors therein cannot be identified through verification or validation of the requirements specification.

Jean-Raymond Abrial, the father of both Z and B, and the intellectual leader behind the formal verification of the Paris Meteor railway line previously described suggested the distinction between requirements and design verification in the context of the crash of the European Union's Ariane 5 in 1996 (Abrial 2009). In this case, an unmanned rocket crashed 30 seconds into flight due to a loss of guidance and attitude information. The lost information was caused by an incorrectly handled data conversion from a 64-bit floating-point to 16-bit signed integer value, a problem that was not caught in requirements verification but was more aptly due to an implementation error or testing inadequacy (Nuseibeh 1997).

In other words, a formal proof like the one just demonstrated would not likely have prevented the Ariane 5 disaster. But we shouldn't conclude from the Ariane 5 case that formal methods are inadequate. We can conclude, however, that formal methods used for verification or validation in one phase of systems engineering do not eliminate the need for other forms of verification and validation such as system testing.

Combining Formal and Informal Methods

Bowen and Hinchey's Advice

Some final advice on the use of formal methods is adapted from "Ten Commandments of Formal Methods" by Bowen and Hinchey (1995b).

1. Thou shalt choose the appropriate notation.
   Comment: Whether you use B, Z, VDM, PVS, or whatever, pick the formal notation that is most appropriate for the situation. We do not advocate one formal language or another. What is appropriate for a situation will depend largely on what the organization is most comfortable using and the supporting tools available. An organization is not typically fluent in multiple formal languages, however, and it is not recommended that a company changes notation from one situation to the next. Training in the use of a formal language has to be considered if there are not enough engineers capable of using the tool of choice.
2. Thou shalt formalize, but not over formalize.
   Comment: Everything in moderation (including moderation).
3. Thou shalt estimate costs.
   Comment: Modern project management practice requires cost estimation at every step of the way.
4. Thou shalt have a formal method guru on call.
   Comment: It is always recommended to have access to an expert in formal methods. That expert may be an in-house specialist or a consultant.
5. Thou shalt not abandon thy traditional development methods.
   Comment: It has been continuously noted that formal methods are augmentative in nature.
6. Thou shalt document sufficiently.
   Comment: Documentation is critical in all systems engineering activities.

7. Thou shalt not compromise thy quality standards.

   Comment: Any objection here?
8. Thou shalt not be dogmatic.

   Comment: Dogmatism is not the way to promote the benefits of formal methods. We have tried to be balanced in our approach to formal methods usage.
9. Thou shalt test, test, and test again.

   Comment: Formal methods do not replace testing, they are supplemental to testing. An appropriate, lifecycle testing program is essential to modern systems engineering.
10. Thou shalt reuse.

    Comment: Take advantage of the power of formal methods, and one of the greatest of these is confidence in reuse. A software module that has been formally validated (as well as tested in the traditional sense) will be a good candidate for reuse because it is trusted.

Finally, don't try to try to push or overuse formal methods. Overhyping their value where formal methods are not as useful can lead to resistance and failure (Bowen et al. 2010).

VIGNETTE 7.3   Recent Success Using Formal Methods

There have been a few, publicized successes using formal methods for requirements engineering (there may be other, unpublished successes). The first example involves Quark, a formally verified web browser that is compatible with Gmail, Facebook, and Amazon, among other well-used social media and commerce sites. The code was written in Python, C/C++, and Occam, and with the code representing a requirements specification. Because a web browser involves a great deal of code, the approach involved forcing

> all components to communicate through a small, lightweight shim which ensures the components are restricted to only exhibit allowed behaviors. Formal shim verification only requires one to reason about the shim, thus eliminating the tremendously expensive or infeasible task of verifying large, complex components in a proof assistant.

An automated proof assistant, Coq, was used to validate the shim components. The verified browser was shown to preserve cookie integrity and confidentiality, and address bar integrity and correctness. The performance of Quark was determined to be as good as typical browsers (Jang et al. 2012).

A second example involves the formal verification of embedded software for several Airbus systems. Here the formal analysis was conducted directly on the binary representation, using the C source code as the requirements. Using a proof assistant tool, properties such as worst-case execution time and maximum stack usage of executables were formally verified. This approach has replaced some of the testing activities for critical software embedded on certain military and commercial aircraft at Airbus. Others in the aerospace industry are also using formal methods to ensure the integrity of critical functions. For example, Dassault Aviation has used formal verification techniques to replace integration robustness testing, that is, the ability to continue to operate under abnormal conditions (Moy et al. 2013).

Share Your Opinion: Formal methods for requirements engineering—too much trouble, or worth the effort?
https://phil.laplante.io/requirements/formal.php

Exercises

7.1   Are customers more likely to feel confident if formal methods are explained to them and then used?

7.2   Where in the software development process lifecycle do formal methods provide the most benefit?

7.3   Rewrite the train station specification in another formal language, such as Z or VDM.

7.4   Conduct a consistency check for the requirements found in Section 8.2 of the Smart Home SRS (video entry).

7.5   Conduct a consistency check for the requirements found in Section 8.3 of the Smart Home SRS (video playback).

7.6   Determine if the following requirements for a wet-well control system, such as the one given in Appendix B, are consistent:

7.6.1   If the pump is on, then the valve shall be open.

7.6.2   If the valve is closed, then the water level is ≤ 10 m.

7.6.3   If the water level is >10 m, then the pump is on.

7.7   Consider the following set of requirements for an insulin pump system:

7.7.1   If the insulin dose button is pressed, then the insulin dose is administered.

7.7.2   If the insulin dose is administered, then the dose indicator light is off.

7.7.3   If the insulin dose button is pressed and the dose indicator light is on, then the insulin dose is administered.

7.7.4   If the insulin dose is administered, then the dose indicator light is on. Determine using a truth table if these requirements are consistent or not.

7.8   Which requirements (groups) in the requirements in the SRS of Appendix A would benefit from formalization (be specific as to formalization type)?

7.9   Why is it easier to show that a set of requirements is inconsistent rather than prove that they are consistent?

*7.10 For the formalizations in Sections 3.2.1 and 3.2.5 in the SRS of Appendix B discuss the problems that exist. Using whatever assumptions you need, rewrite these formalizations using Z, B, or VDM or any other formal method you choose.

Notes

1 Actually, the authors heard the joke differently from members of the Royal Australian Air Force in the early 1990s. As told, the Panda is an amorous, but lazy creature and is well known for its ability to scope out a tree occupied by a panda of the opposite sex, where it "eats, roots, shoots, and leaves." In a double entendre, "rooting" refers to the act of procreation.
2 Dr. Laplante heard this quote first from Dr. George Hacken, Senior Director, Vital Systems, New York City Transit, sometime around 2007.
3 Thanks to Dr. George Hacken and Sofia Georgiadis for providing this example.
4 This section is adapted from Richards (2006).
5 "Execution" means either program execution or effective change in system behavior in the case of a nonsoftware behavioral segment.

References

Abrial, J. R. (2009). Faultless systems: Yes we can! *Computer*, 42(9): 30–36.

Agrafiotis, I., Creese, S., & Goldsmith, M. (2011, September). Developing a strategy for automated privacy testing suites. In *IFIP PrimeLife International Summer School on Privacy and Identity Management for Life* (pp. 32–44). Springer, Berlin, Heidelberg.

Awodey, S. (2005). *Category Theory*. Carnegie Mellon University, Pittsburg, PA.

Böhm, C., & Jacopini, G. (1966). Flow diagrams, Turing machines, and languages with only two formation rules. *Communications of the ACM*, 9(5): 366–371.

Bowen, J. P., & Hinchey, M. G. (1995a). Seven more myths of formal methods. *Software*, 12(4): 34–41.

Bowen, J. P., & Hinchey, M. G. (1995b). Ten commandments of formal methods. *Computer*, 28(4): 56–63.

Bowen, J. P., Hinchey, M. G., & Vassev, E. (2010). Formal requirements specification. In P. Laplante (Ed.), *Encyclopedia of Software Engineering* (pp. 321–332). Taylor & Francis, Boca Raton, FL.

Bruel, J. M., Ebersold, S., Galinier, F., Mazzara, M., Naumchev, A., & Meyer, B. (2021). The role of formalism in system requirements. *ACM Computing Surveys (CSUR)*, 54(5): 1–36.

Crow, J., & Di Vita, B. (1998). Formalizing space shuttle software requirements: Four case studies. *ACM Transactions on Software Engineering and Methodology*, 7(3): 296–332.

Fabian, B., Gürses, S., Heisel, M., Santen, T., & Schmidt, H. (2010). A comparison of security requirements engineering methods. *Requirements Engineering*, 15(1): 7–40.

Forster, T. E. (2003). *Logic, Induction and Sets*. Cambridge University Press, Cambridge.

Graves, H., & Bijan, Y. (2011). Using formal methods with SysML in aerospace design and engineering. *Annals of Mathematics and Artificial Intelligence*, 63(1): 53–102.

Gries, D., & Gries, P. (2005). *Multimedia Introduction to Programming Using Java*. Springer, New York.

Hacken, G. (2007). Private email communication to P. Laplante.

Hall, A. (1990). Seven myths of formal methods. *Software*, 7(5): 11–19.

Heitmeyer, C. (2007). Formal methods for specifying, validating, and verifying requirements. *Journal of Universal Computer Science*, 13(5): 607–618.

Hoare, C. A. R. (1969). An axiomatic basis for computer programming. *Communications of the ACM*, 12(-10): 576–583.

Jang, D., Tatlock, Z., & Lerner, S. (2012). Establishing browser security guarantees through formal shim verification. In *The 21st USENIX Security Symposium (USENIX Security 12)* (pp. 113–128). Bellevue, WA.

Lano, K. (1996). *The B Language and Method*. Springer-Verlag, London, pp. 94–98.

Laplante, P. A. (2005). *Dictionary of Computer Science, Engineering and Technology*, 2nd edition. CRC Press, Boca Raton, FL.

Moy, Y., Ledinot, E., Delseny, H., Wiels, V., & Monate, B. (2013). Testing or formal verification: Do-178c alternatives and industrial experience. *IEEE Software*, 30(3): 50–57.

NASA. (2012). National Space Science Data Center: Mariner 1. http://nssdc.gsfc.nasa.gov/nmc/masterCatalog.do?sc=MARIN1 (accessed March 2017).

Nuseibeh, B. (1997). Ariane 5: Who dunnit? *Software*, 14(3): 15–16.

Parnas, D. L. (2010). Really rethinking "formal methods." *Computer*, 43(1): 28–34.

Richards, A. (2006). *Formalization of an EMS System Using Category Theory*, Master's Thesis, Penn State University, State College, PA, under the supervision of P. Laplante.

Shehata, M., & Eberlein, A. (2010). Requirements interaction detection. In P. Laplante (Ed.), *Encyclopedia of Software Engineering* (pp. 979–998). Taylor & Francis, New York.

Truss, L. (2003). *Eats Shoots and Leaves: The Zero Tolerance Approach to Punctuation*. Profile Books, London.

Chapter 8

Requirements Specification and Agile Methodologies

Introduction to Agile Methodologies[1]

Agile methodologies comprise a family of nontraditional software development strategies that have captured the imagination of many who are leery of traditional, process-laden (plan-driven) approaches. Agile methodologies are characterized by their lack of rigid process, though this fact does not mean that agile methodologies, when correctly employed, are not rigorous nor suitable for industrial applications—they are. What is characteristically missing from agile approaches, however, are "cook-book" solutions that focus on mandatory meetings and complex documentation-prescribed development approaches.

Agile methodologies apply to software engineering. While there are elements of agile methodologies that can be applied to the engineering of systems (in particular, the human considerations), such methodologies are usually described as lightweight or lean when applied to non-software systems. This is so because agile methodologies depend on a series of rapid, non-throwaway prototypes, an approach that is not usually practical in hardware-based systems. In any case, the non-software engineer can still benefit from this chapter because agile methodologies are increasingly being employed and because the mindset of the agile software engineer includes some healthy perspectives.

In order to fully understand the nature of agile methodologies, we need to examine a document called the Agile Manifesto and the principles behind it. The Agile Manifesto was introduced by a number of leading proponents of agile methodologies to explain their philosophy (see Figure 8.1).

Signatories to the Agile Manifesto include many luminaries of modern software engineering practice such as Kent Beck, Mike Beedle, Alistair Cockburn, Ward Cunningham, Martin Fowler, Jim Highsmith, Ron Jeffries, Brian Marick, Robert Martin, Steve Mellor, and Ken Schwaber. Underlying the Agile Manifesto is a set of principles. Look at the principles below, noting the emphasis on those aspects that focus on requirements engineering, which we set in italics.

DOI: 10.1201/9781003129509-8

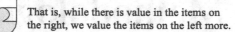

We are uncovering better ways of developing software by doing it and helping others do it.

Through this work we have come to value:
- Individuals and interactions over processes and tools
- Working software over comprehensive documentation
- Customer collaboration over contract negotiation
- Responding to change over following a plan

That is, while there is value in the items on the right, we value the items on the left more.

Figure 8.1   Manifesto for agile software development. (From Beck, K., *Extreme Programming Explained: Embrace Change*, Longman Higher Education, London, 2000.)

Principles Behind the Agile Manifesto

- At regular intervals, the team reflects on how to become more effective, then tunes and adjusts its behavior accordingly.
- Our highest priority is to satisfy the customer through early and continuous delivery of valuable software.
- *Welcome changing requirements, even late in development.* Agile processes harness change for the customer's competitive advantage.
- Deliver working software frequently, from a couple of weeks to a couple of months, with a preference for the shorter timescale.
- *Business people and developers must work together daily throughout the project.*
- Build projects around motivated individuals. Give them the environment and support they need, and trust them to get the job done.
- *The most efficient and effective method of conveying information to and within a development team is face-to-face conversation.*
- Working software is the primary measure of progress.
- Agile processes promote sustainable development. The sponsors, developers, and users should be able to maintain a constant pace indefinitely.
- Continuous attention to technical excellence and good design enhances agility.
- Simplicity—the art of maximizing the amount of work not done—is essential. "Do the simplest thing that could possibly work."
- *The best architectures, requirements, and designs emerge from self-organizing teams* (Beck 2000).

Notice how the principles acknowledge and embrace the notion that requirements change throughout the process. Also, the agile principles emphasize frequent, personal communication (this feature is beneficial in the engineering of non-software systems too). The highlighted features of requirements engineering in agile process models differ from the "traditional" waterfall and more modern models such as iterative, evolutionary, or spiral development. These other models favor a great deal of preliminary work on the requirements engineering process and the production of, often, voluminous requirements specifications documents.

Benefits of Agile Software Development

Agile software development methods are a subset of iterative methods[2] that focus on embracing change and emphasize collaboration and early product delivery while maintaining quality. Working code is considered the true artifact of the development process. Models, plans, and documentation are important and have their value but exist only to support the development of working software, in contrast with the other approaches already discussed. However, this does not mean that an agile development approach is a free-for-all. There are very clear practices and principles that agile methodologists must embrace.

- Agile methods are adaptive rather than predictive. This approach differs significantly from process models (e.g., waterfall) that emphasize planning the software in great detail over a long period of time, and for which significant changes in the software requirements specification can be problematic. Agile methods are a response to the common problem of constantly changing requirements that can bog down the more "ceremonial" up-front design approaches, which focus heavily on documentation at the start.
- While testing is testing phase is a separate phase and follows a build phase in process models, testing is completed each programming in agile development. What this implies is that users in an agile environment can frequently use the newly developed components to validate the financial value of the software. After the users know the financial value of each iteration, they can make better decisions about the software's future.
- Agile methods are also adaptive rather than predictive. This approach differs significantly from process models that emphasize planning the software in great detail over a long period of time, and for which significant changes in the software requirements specification can be problematic. Agile methods are a response to the common problem of constantly changing requirements that can bog down the more "ceremonial" up-front design approaches, which focus heavily on documentation at the start.
- Agile methods are also people-oriented rather than process-oriented. This means they explicitly make a point of trying to make development "fun." Presumably, this is because writing software requirements specifications and software design descriptions is onerous and, hence, to be minimized.

Agile methodologies include Crystal, Extreme Programming, Scrum, Kanban, Lean, Dynamic systems development method (DSDM), feature-driven development, and adaptive programming, and there are others. We will look more closely at four of the most widely used agile methodologies: XP, Scrum, Kanban, and Lean.

Extreme Programming

Extreme programming (XP) is one of the most widely used agile methodologies. XP is traditionally targeted toward smaller development teams and requires relatively few detailed artifacts. XP

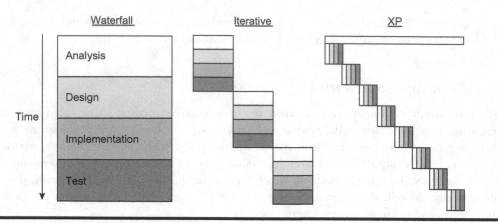

Figure 8.2 Comparison of waterfall, iterative, and XP development cycles. (From Beck, K., *Extreme Programming Explained: Embrace Change*, Longman Higher Education, London, 2000.)

takes an iterative approach to its development cycles. We can visualize the difference in process between a traditional waterfall model, iterative models, and XP (Figure 8.2). Whereas an evolutionary or iterative method may still have distinct requirements analysis, design, implementation, and testing phases similar to the waterfall method, XP treats these activities as being interrelated and continuous.

XP promotes a set of 12 core practices that help developers to respond to and embrace inevitable change. The practices can be grouped according to four practice areas:

■ Planning
■ Coding
■ Designing
■ Testing

Some of the distinctive planning features of XP include holding daily stand-up meetings, making frequent small releases, and moving people through different project roles. Coding practices include having the customer constantly available, coding the unit test cases first, and employing pair-programming (a unique coding strategy where two developers work on the same code together).

Design practices include looking for the simplest solutions first, avoiding too much planning for future growth (speculative generality), and refactoring the code (improving its structure) continuously.

Testing practices include creating new test cases whenever a bug is found and having unit tests for all code, possibly using such frameworks as XUnit.

While removal of the territorial ownership of any code unit is another feature of XP, this framework may not work in the best way possible if all team members are not collocated.

Scrum

Scrum, which is named after a particularly contentious point in a rugby match, enables self-organizing teams by encouraging verbal communication across all team members and

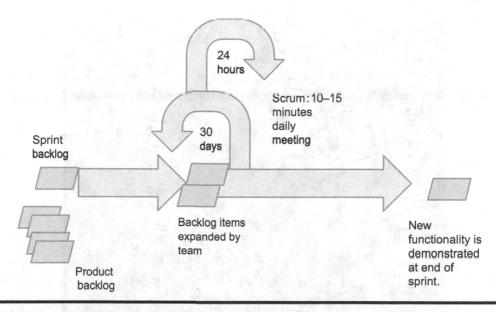

Figure 8.3   The Scrum development process from Boehm and Turner (2003). (Adapted from Schwaber, K., & Beedle, M., *Agile Software Development with SCRUM,* **Prentice Hall, Upper Saddle River, NJ, 2001.)**

across all stakeholders. A fundamental principle of Scrum is that traditional, plan-driven software development methodologies such as waterfall and iterative development focus too much on process and not enough on software. Moreover, while plan-driven development focuses on non-software artifacts (e.g., documentation) and processes (e.g., formal reviews), Scrum emphasizes the importance of producing functioning software early and often. Scrum promotes self-organization by fostering high-quality communication between all stakeholders. In this case, it is implicit that the problem cannot be fully understood or defined (it may be a wicked problem). And the focus in Scrum is on maximizing the team's ability to respond in an agile manner to emerging challenges.

Scrum features a "living" (constantly changing) backlog of prioritized work to be completed. Completion of a largely fixed set of backlog items occurs in a series of short (approximately 30 days) iterations or sprints. Each day, a brief (e.g., 15-minutes) meeting or Scrum is held in which progress is explained, upcoming work is described, and impediments are raised. A short planning session occurs at the start of each sprint to define the backlog items to be completed. A brief postmortem or heartbeat retrospective review occurs at the end of the sprint (Figure 8.3).

A Scrum Master removes obstacles or impediments to each sprint. The Scrum Master is not the leader of the team (as they are self-organizing) but acts as a productivity buffer between the team and any destabilizing influences. In some organizations, the role of the Scrum Master can cause confusion. For example, if two members of a Scrum team are not working well together, it might be expected by a senior manager that the Scrum Master fix the problem. Fixing team dysfunction is not the role of the Scrum Master. Personnel problems need to be resolved by the line managers to which the involved parties report. This scenario illustrates the need for institution-wide education about agile methodologies when such approaches are going to be employed.

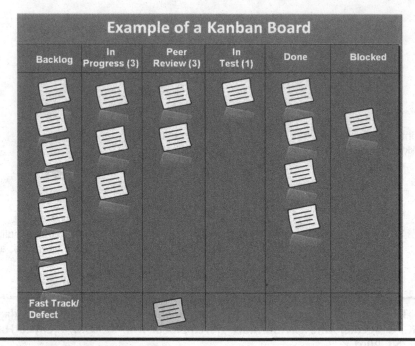

Figure 8.4   Example of Kanban Board.

Kanban

The term "Kanban" is of Japanese origin, and its meaning is linked to the concept, "just-in-time." The underlying Kanban method originated in lean manufacturing (presented next) and focuses on visualizing the progress of the work items, from start to finish (usually through a Kanban board that can be physical or digital (Figure 8.4)). Having the work progress visualized allows team members to see the state of every piece of work at any time.

Kanban teams write the work items onto cards, usually one item per card. The cards are placed on the board, which is typically divided into columns, that show each flow of the software production (e.g., In Progress, In Testing, Completed, etc.). As the development evolves, the cards contained on the board are moved to the proper column that represents its status, and when a new task comes into play, a new "card" is created. Work moves around the board as capacity permits, rather than being pushed into the process when requested.

The Kanban method requires transparency and real-time communication so that the members of a team can know exactly at what stage the development is and can see the status of the project at any time. Kanban has been effectively used to develop large software and software-intensive systems.

Lean Development

Lean Development encompasses hardware, software, and hybrid systems. "Lean Software Development" is borrowed from "Lean Manufacturing," which is a set of seven principles adapted from the Toyota Production System. The seven principles are:

■ **Eliminate Waste**: everything that does not bring effective value to the customer's project shall be deleted;

- **Build Quality In**: creating quality in development requires discipline and control of the quantity of residuals created;
- **Deliver Fast**: deliver value to the customer as soon as possible;
- **Defer Commitments**: encouraging the team not to focus too much on planning and anticipating ideas without having a prior and comprehensive understanding of the requirements of the business;
- **Optimize the Whole**: the development sequence has to be perfected enough to be able to delete errors in the code, in order to create a flow of true value;
- **Respect People**: communicating and managing conflicts are two essential points
- **Creating Knowledge**: the team is motivated to document the whole infrastructure to later retain that value.

Applying these principles in software and systems development allows the team to decrease the time needed to deliver functionalities since the principles prepare the development team in the decision-making process, hence increasing general motivation, while deleting superfluous activity, therefore saving time and money. Besides, these principles are easily scalable methodology and easily adaptable to projects of any dimension.

Requirements Engineering for Agile Methodologies

The increasing adoption of Agile practices in software development has been on the radar of the Requirements Engineering (RE) community in recent years (Wagner et al. 2018). This is mainly because "doing requirements" consumes time that could be rather spent on writing code, and business stakeholders in an agile environment often feel better if they think that the development team is "working" (that is, coding) on their problems (Orr 2004).

While many conventional RE practices are important also in agile projects, as Savolainen et al. (2010) observe, there are some major differences as well.

Paetsch et al. (2003) analyzed commonalities and differences between traditional RE approaches and agile software development. Their findings include that one of the differences is related to customers involvement. Customers are constantly involved in requirements discovery and refinement in agile methods. All systems developers should be involved in the requirements engineering activity, and each can, and should, have regular interaction with customers. In traditional approaches, the customer has less involvement once the requirements specification has been written and approved, and typically, the involvement is often not with the systems developers.

Another difference is in the timing of the requirements engineering activities. In traditional systems and software engineering, requirements are gathered, analyzed, refined, etc. at the front end of the process. In agile methods, requirements engineering is an ongoing activity; that is, requirements are refined and discovered with each system build. Even in spiral methodologies, where prototyping is used for requirements refinement, requirements engineering occurs much less so late in the development process.

A third noted difference in the agile approach to requirements engineering is much more invulnerable to changes throughout the process (remember, "embrace change") than in traditional software engineering.

The same findings were confirmed in a study by Ramesh et al. (2010) who examined how RE had been conducted in 16 organizations that were involved in agile software development.

Bose et al. (2008) suggested that while following agile practices may be effective in acquiring continuous feedback from customers, the limits of these practices on RE are not well defined. They recommended that proper requirements management should be adopted under the agile umbrella to ensure proper traceability when the requirements are likely to be changed. The verification of early requirements representations was also proposed.

Cao and Ramesh (2008) studied 16 software development organizations that were using either XP or Scrum (or both) and uncovered seven requirements engineering practices that were agile in nature.

1. Face-to-face communications (overwritten specifications)
2. Iterative requirements engineering
3. Extreme prioritization (ongoing prioritization rather than once at the start, and prioritization is based primarily on business value)
4. Constant planning
5. Prototyping
6. Test-driven development
7. Reviews and tests

Some of these practices can be found in non-agile development (e.g., test-driven development and prototyping), but these seven practices were consistent across all of the organizations studied.

As opposed to the fundamental software requirements specification, the fundamental artifact in agile methods is a stack of constantly evolving and refining requirements. These requirements are generally in the form of user stories. In any case, these requirements are generated by the customer and prioritized—the higher the level of detail in the requirement, the higher the priority. As new requirements are discovered, they are added to the stack and the stack is reshuffled in order to preserve prioritization.

There are no proscriptions on adding, changing, or removing requirements from the list at any time (which gives the customer tremendous freedom). Of course, once the system is built, or likely while is it is being built, the stack of user stories can be converted to a conventional software requirements specification for system maintenance and other conventional purposes.

Yat et al. conducted a systematic review of studies of agile software projects published between 2002 and 2012. They identified five traditional requirements engineering challenges that were either eradicated or minimized by the requirements engineering practices found in agile methodologies. These challenges were:

■ Bridging the communication gap with all stakeholders
■ Increasing customer involvement
■ Reducing the size and complexity of documentation
■ Reducing scope creep
■ Achieving requirements validation (Inayat et al. 2015a)

They noted that more research and empirical evidence were needed to strengthen and generalize these conclusions, but the results of their study still strongly point to the power of agile methods.

Ambler (2007) suggests the following best practices for requirements engineering using agile methods. Many of the practices follow directly from the principles behind the Agile Manifesto:

■ Have active stakeholder participation
■ Use inclusive (stakeholder) models

- Take a breadth-first approach
- Model "storm" details (highly volatile requirements) just in time
- Implement requirements, do not document them
- Create platform-independent requirements to a point
- Remember that smaller is better
- Question traceability
- Explain the techniques
- Adopt stakeholder terminology
- Keep it fun
- Obtain management support
- Turn stakeholders into developers
- Treat requirements like a prioritized stack
- Have frequent, personal interaction
- Make frequent delivery of software
- Express requirements as features

Ambler also suggests using such artifacts as CRCs, acceptance tests, business rule definitions, change cases, data flow diagrams, user interfaces, use cases, prototypes, features and scenarios, use cases diagrams, and user stories to model requirements (Ambler 2007). These elements can be added to the software requirements specification document along with the user stories.

For requirements elicitation, he suggests using interviews (both in-person and electronic), focus groups, JAD, legacy code analysis, ethnographic observation, domain analysis, and having the customer on-site at all times (Ambler 2007).

Finally, a systematic literature review on agile requirements engineering practices and challenges (Inayat and Daneva 2015b) identified 17 RE practices that we found to be adopted in agile software development:

- Face-to-face communication between team members and client representatives is a characteristic of agile RE.
- Customer involvement and interaction were declared the primary reasons for project success and limited failure.
- User stories are created as specifications of the customer requirements. User stories facilitate communication and better overall understanding among stakeholders.
- Iterative requirements, requirements unlike in traditional software development methods, emerge over time in agile methods.
- Requirement prioritization is part of each iteration in agile methods.
- Change management has proven to be a significant challenge for traditional approaches thus far.
- Cross-functional teams include members from different functional groups who have similar goals.
- Prototyping is perceived as a straightforward way to review requirements specifications with clients and to gain timely feedback before moving to subsequent iterations.
- Testing before coding means writing tests prior to writing functional codes for requirements.
- Requirements modeling is performed in agile software development methods, but it is different from RE models developed in traditional software development methods.
- Requirements management is performed by maintaining product backlog/feature lists and index cards.

- Review meetings and acceptance tests are the developed requirements and product backlogs that are constantly reviewed in meetings.
- Code refactoring is meant to revisit and modify developed code structure, improve on the structure, and accommodate changes.
- Shared conceptualization is a supporting concept to carry out RE activities related to gathering, clarifying, and evolving for agile methods.
- Pairing for requirements analysis is a practice that encourages the stakeholders to perform multiple roles as well.
- Retrospectives are the meetings held after the completion of an iteration.
- Continuous planning is a routine task for agile teams.

VIGNETTE 8.1   DevOps and SAFe

DevOps, a mashup of the words "development" and "operations," is "a set of practices intended to reduce the time between committing a change to a system and the change being placed into normal production, while ensuring high quality" (Bass et al. 2015). Concepts leading to DevOps far predate its origin, but it began to appear in industry around 2007/2008. DevOps practices involve four main concerns:

- Getting changes quickly into production.
- Using automated testing to find errors.
- Reducing or eliminating errors occurring during deployment.
- Finding and repairing system faults quickly

Since it is well known that the cost of fixing requirements errors late in the development process is much higher than finding them early in the process, addressing the DevOps concerns suggests that a strong RE program is needed. For example, ensuring that an effective means for capturing NFRs is in place as these are particularly difficult to address late in the process.

Scaled Agile Framework (SAFe) begin to emerge around the same time as DevOps as an agile style of development that incorporates elements of lean-agile, systems thinking, DevOps, and more. SAFe is also based on principles that predate its origin significantly. One of the major differences between SAFe, however, and other agile development methodologies is that SAFe is designed to scale up to much larger teams and projects.

Like the Agile Manifesto, SAFe is based on ten principles which include taking a systems thinking and value-based approach to managing project requirements with iterative build out of functionality in working non-throwaway prototypes (also found in DevOps). SAFe also adopts other principles from Agile development such as sustainable pace, shared ownership, and keeping participants motivated in the process (though these terms are not precisely used).

SAFe defines several requirements types that capture functionality from high level to lower level, conceptional, and NFRs. The intention of these types is to allow for a greater level of generality early in the development process with idiosyncratic requirements not being defined until later in the project. SAFe is itself adaptive; for example, it can work with many of the requirements representations and practices previously described such as User Stories, Use Cases, Scrum, XP, and Kanban.

Technical Debt

Technical debt is a metaphor referring to the eventual consequences of poor system architecture and system development within a codebase (Cunningham 1992). The debt can be thought of as work that needs to be done before a particular job can be considered complete. In agile development, the technical debt can be thought of as the amount of work that is left to be done from previous iterations (sprints), when a new iteration starts. The buildup of technical debt is the main reason for projects to miss deadlines. As with monetary debt, the cost of never paying down a technical debt may accumulate an "interest" making it harder to implement changes. The debt can be paid off by simply completing the uncompleted work.

Rubin uses the following status categories to classify technical debts:

- Known technical debt: that is made visible to the system development team.
- Happened-upon technical debt—that the development team was not familiar with its existence until it was exposed during development. For example, the team is implementing a new feature to the product and in doing so it realizes that some work was left incomplete for a while by someone who has long since departed.
- Targeted technical debt—debt that is known and has been targeted for servicing by the development team.

The technical debt can be unavoidable. The survey from (Kassab et al. 2014) shows that for the agile development projects on average 91% of these projects reported carrying some level of technical debt when moving from iteration to the next, but most of the projects (84%) reported that this level is below 25% on average. So when a new iteration or sprint starts, there was on average 25% of a work still to be completed from previous iterations or sprints. Common causes for the technical debt that are directly relevant to requirements include:

- Insufficient up-front "metaphor" definition. In this situation, design and coding tasks may start while the core requirements are still being refined. While this was may have been done to save time, often the design and the code will have to be reworked later.
- Business constraints, where the business stakeholders requesting something released sooner before all required changes are complete. Those uncompleted changes form a technical debt.
- Lack of alignment to standards. If early in the project, the regulations, standards, and guidelines are ignored, technical debt will be formed to eventually integrate what has been developed with what has been missed.
- Lack of a development process, where the business is blind to the concept of technical debt.
- Tightly coupled requirements, where the majority of requirements refer or depend on others. In this situation, the system will not be flexible enough to adapt to changes in business needs.

■ Parallel development on multiple requirements without proper coordination accrues technical debt because of the work required to merge these developments into a single base.

■ Last-minute requirements specification changes have the potential to percolate throughout a project but no time or budget to see them through with documentation.

Example Application of Agile Software Development

Consider the pet store point of sale (POS) system. Suppose that we wanted to use an agile software development methodology, specifically Scrum, to develop this system. Let's look at what the requirements engineering process might look like.

Suppose that we had a five-person development team, including the Scrum Master. For simplicity, let's assume that the only stakeholders are the pet store owner (who is really the customer, since he is paying for the system), store customers, cashiers, and accountants. We will select class representatives for the store customers, cashiers, and accountants. Collectively, let's call them the "stakeholder panel." Now consider the following activities.

A. The Scrum Master organizes a set of meetings between the development team and the customer (the store owner) to get a basic understanding of the system requirements, constraints, and ground rules. Based on these discussions, the Scrum Master organizes a set of additional elicitation activities, possibly including structured interviews, card sorting, and focus groups with the other stakeholders. A QFD study is used as a validation activity (since there are likely many other POS systems that can be compared). The goal of these activities is to produce a set of prioritized user stories (let's say around 100), including effort estimates for each. But the customer has told us that he wants a "cutting edge" system, so we know that as the system goes through development, the customer and other stakeholders are going to want new or different functionality.

B. The development team analyzes the user stories and selects a subset of these; let's say 10, for the first sprint. These 10 would be the most "user-facing" stories, leaving as much backend processing for later as possible.

C. The development team consults with the stakeholders again to get further clarity for these 10 user stories before beginning development.

D. Each day starts with a scrum. Developers use test-driven design to organically derive the system design as well as unit test cases. The Scrum Master facilitates communications between the development team and the various stakeholders so that stakeholder-facing questions can be answered quickly. After approximately 30 days, the first sprint is completed.

E. The stakeholder panel is presented with the system built during the sprint for feedback. This feedback will include changes to the current system increment, new features to be added, and possibly, features to be omitted.

F. Based on this feedback, the development team creates new user stories and adds them to the backlog, modifies other user stories as needed, and reshuffles the backlog.

G. The development team selects a new set of user stories (say 10) from the backlog for the next 30-day sprint, and process steps (C–F) are repeated until the backlog is exhausted.

Additional acceptance testing may be conducted, depending on the terms of the contract with the customer. This testing would be based on a set of criteria developed during the contract negotiation. Such acceptance testing based on a requirements specification is usual for conventional development but atypical for agile development.

There were 100 user stories originally, but even if we schedule 10 user stories per sprint (month), the total development time is likely to be longer than 10 months. This is because new user stories are being added, and old ones are being changed.

Of course, the Scrum Master is tracking the project velocity during each sprint in order to refine his project completion estimates.

When Is Agile Recommended?

The use of agile software methodologies is becoming very popular in the industry, at least as evidenced by survey data. For example, in a Forrester/Dr. Dobbs developer survey, 35% of respondents reported using some kind of agile methodology (West et al. 2010). More recently, a 2020 survey on requirements engineering state of practices (Kassab and Laplante 2022) found that 70% of software engineers surveyed used an agile methodology as Software Development Life Cycle for their project making agile the most popular SDLC (Figure 8.5). The usage of agile is almost increased by more than 24% since 2013 when a similar survey was conducted (Kassab' 2014). The 2020 survey also found that agile methods were used more frequently than waterfall style development in every industry (Figure 8.6), and it observed a shift in the "finance /banking/ insurance," "defense," and "aerospace" domains from 2013 when the waterfall was more popular than agile back then.

But the question arises, when should agile development methodologies be used?

Boehm and Turner suggest that the way to answer the question is to look at the project along a continuum of five dimensions: size (in terms of the number of personnel involved), system criticality, personnel skill level, dynamism (anticipated number of system changes over some time interval), and organizational culture (whether the organization thrives on chaos or order) (Boehm and

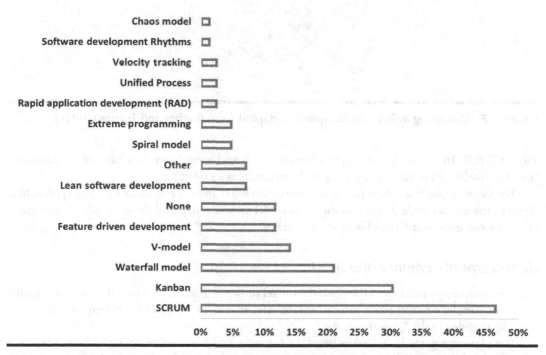

Figure 8.5   Software Development lifecycle employed (*n*=91).

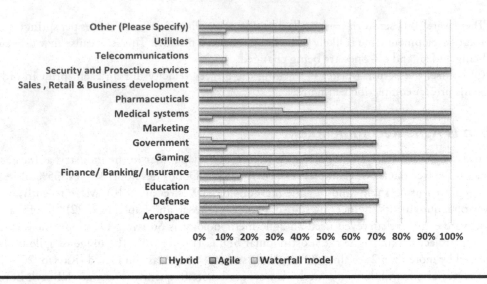

Figure 8.6   Software Development lifecycle employed across business domains.

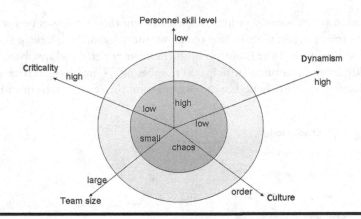

Figure 8.7   Balancing agility and discipline. (Adapted from Boehm and Turner (2003).)

Turner 2003). In Figure 8.7, as project characteristics tend away from the center of the diagram, then the likelihood of succeeding using agile methodologies decreases.

Therefore, projects assessed in the innermost circle are likely candidates for agile approaches, those in the second circle (but not in the inner circle) are marginal, and those outside of the second circle are not good candidates for agile approaches.

Requirements Engineering in XP

Requirements engineering in XP follows the model shown in Figure 8.8 where the stack of requirements in Ambler's model refers to user stories. And in XP, user stories are managed and implemented as code via the "planning game."

The planning game in XP takes two forms: release and iteration planning. Release planning takes place after an initial set of user stories has been written. This set of stories is used to develop

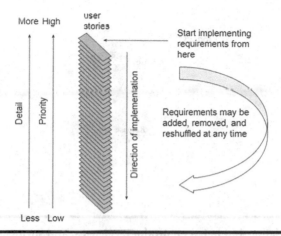

Figure 8.8 Agile requirements change management process. (Adapted from Ambler, S., http://www.agilemodeling.com/essays/agileRequirements.htm, 2007.)

the overall project plan and plan for iterations. The set is also used to decide the approximate schedule for each user story and overall project.

Iteration planning is a period of time in which a set of user stories and fixes to failed tests from previous iterations are implemented. Each iteration is 1–3 weeks in duration. Tracking the rate of implementation of user stories from previous iterations (which is called project velocity) helps to refine the development schedule.

Because requirements are constantly evolving during these processes, XP creator Kent Beck says that "in XP, requirements are a dialog, not a document" (Beck et al. 2001) although it is typical to convert the stack of user stories into a software requirements specification.

Requirements Engineering in Scrum

In Scrum, the requirements stack shown in the model of Figure 8.8 is, as in XP, the evolving backlog of user stories. And as in XP, these requirements are frozen at each iteration for development stability. In Scrum, each iteration takes about a month. To manage the changes in the stack, one person is given final authority for requirement prioritization (usually the product sponsor).

In Scrum, the requirements backlog is organized into three types: product, release, and sprint. The product backlog contains the release backlogs, and each release contains the sprint backlog. Figure 8.9 is a Venn diagram showing the containment relationship of the backlog items.

The product backlog acts as a repository for requirements targeted for release at some point. The requirements in the product backlog include low-, medium-, and high-level requirements.

The release backlog is a prioritized set of items drawn from the product backlog. The requirements in the release backlog may evolve so that they contain more details and low-level estimates.

Finally, the sprint backlog list is a set of release requirements that the team will complete (fully coded, tested, and documented) at the end of the sprint. These requirements have evolved to a very high level of detail, and hence, their priority is high.

Scrum has been adopted in several major corporations, with notable success. Some of the authors' students also use Scrum in courses. In these cases, it proves highly effective when there is little time for long requirements discovery processes.

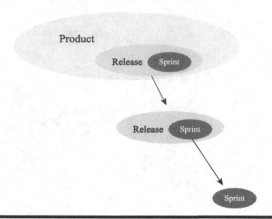

Figure 8.9 Backlog relationship between product, releases, and sprints.

Gathering User Stories

User stories are a common unit of requirements in most agile methodologies. Each user story represents a feature desired by the customer. User stories gathering goes on throughout the project because user requirements keep on evolving throughout the project. Initial user stories are usually gathered in small offsite meetings. Stories can be generated either through goal-oriented (e.g., "let's discuss how a customer makes a purchase") approaches or through interactive (stream-of-consciousness) approaches. Developing user stories is an "iterative and interactive" process. Formal requirements, use cases, and other artifacts are derived from the user stories by the software engineering team as needed.

Writing User Stories

User stories (a term coined by Beck) are written by the customer on index cards, though the process can be automated via wikis or other tools (Beck 2000).

A user story consists of the following components:

- **Title:** this is a short handle for the story. A present tense verb in active voice is desirable in the title.
- **Acceptance Test:** this is a unique identifier that will be the name of a method to test the story.
- **Priority:** this is based on the prioritization scheme adopted. Priority can be assigned based on "traditional" prioritization of importance or on the level of detail (higher priority is assigned to higher detail).
- **Story Points:** this is the estimated time to implement the user story. This aspect makes user stories helpful for effort and cost estimation.
- **Description:** this is one to three sentences describing the story.

A sample layout for these elements on an index card is shown in Table 8.1.

An example user story for a customer returning items in the pet store POS system is shown in Table 8.2.

Table 8.1   User Story Layout

Title		
Acceptance Test	*Priority*	*Story Points*
Description		

Table 8.2   User Story: Pet Store POS System

Title: Customer Returns Items		
Acceptance Test: custRetItem	*Priority: 1*	*Story Points: 2*
When a customer returns an item, its purchase should be authenticated. If the purchase was authentic, then the customer's account should be credited or the purchase amount returned. The inventory should be updated accordingly.		

Table 8.3   User Story: Baggage Handling

Title: Detect Security Threat		
Acceptance Test: detSecThrt	*Priority: 1*	*Story Points: 3*
When a scanned bag has been determined to contain an instance of a banned item, the bag shall be diverted to the security checkpoint conveyor. The security manager shall be sent an email stating that a potential threat has been detected.		

User stories should be understandable to the customers and each story should add value.

Developers do not write user stories, but users do. But stories need to be small enough that several can be completed per iteration. Stories should be independent (as much as possible); that is, a story should not refer back and forth to other stories. Finally, stories must be testablelike any requirement, if it cannot be tested, it's not a requirement. The testability of each story is considered by the development team. Table 8.3 depicts another example user story describing a security threat detection in the airport baggage handling system.

Estimating User Stories

To measure the user stories, agile teams allocate story points—arbitrary values to measure the effort required to complete that user story. These points can be allocated in many ways based on the team's preferences. In most cases, project managers define a story point complexity range as a Fibonacci series (for example, 1, 2, 3, 5, 8). Another way is to pick a small reference story and estimate the other ones regarding that using the Delphi estimation technique known as "planning poker." The stories are rated and reviewed before entering the next phase of prioritization.

Prioritizing User Stories

User stories are prioritized before iteration. The user stories are prioritized in terms of ranking (i.e., ordered as first, second, and third) and also as a group (e.g., "high priority," "low," and "medium"). The high-priority user stories are recorded in the product backlog and used as a guide to carry out the development work. Customers perform the prioritization task based on their understanding of the business value the user stories will bring using various techniques.

User Stories vs. Use Cases

It is worth noting that there is a significant difference between use cases and user stories. User stories come from the customer perspective and are simple and avoid implementation details. Use cases are more complex and may include implementation details (e.g., fabricated objects). Customers don't usually write use cases (and if they do, beware, because now the customer is engaging in "software engineering"). Finally, it's hard to say what the equivalence is for the number of use cases per user story. One user story could equal one or more than 20 use cases.

For example, for the customer return user story in Table 8.2, you can imagine that it will take many more use cases to deal with the various situations that can arise in a customer return. In agile methodologies, user stories are much preferred to use cases. Appendix D includes a further discussion of user stories in both agile and non-agile settings.

Agile Requirements Engineering vs. Requirements Engineering in Agile

We need to make a distinction between requirements engineering for agile methodologies and agile requirements engineering. Agile requirements engineering means, generally, any ad hoc requirements engineering approach purported to be more flexible than traditional requirements engineering. This definition is not to be confused with specific practices for requirements engineering in agile methodologies as we just discussed (Sillitti and Succi 2006).

A number of agile requirements engineering approaches have been introduced in the past few years, many of them not much more than sloppy requirements engineering. But some of the recent work in this area has been good too. However, for any "legitimate" agile requirements engineering methodologies that you may encounter, most of the practices can be traced to the Agile Manifesto. For example, Vlaanderen et al. (2011) showed how certain requirements engineering practices from Scrum could also be for software product management (after delivery). In particular, they identified sprint cycles, backlog administration, daily meetings, and early and frequent collaboration as necessary practices.

Story-Test-Driven Development

To illustrate an agile methodology, we describe one notable example. In this methodology called story-test-riven development (SDD), many of the usual customer-facing features of agile methodologies are incorporated along with short bursts of iterative development à la XP or Scrum. One difference in SDD, however, is that instead of the conventional user stories, customers write or review "story tests." Story tests are nontechnical descriptions of behavior along with tests that verify that the behavior is correct. The story tests help "discover a lot of missing pieces or inconsistencies in a story" (Mugridge 2008).

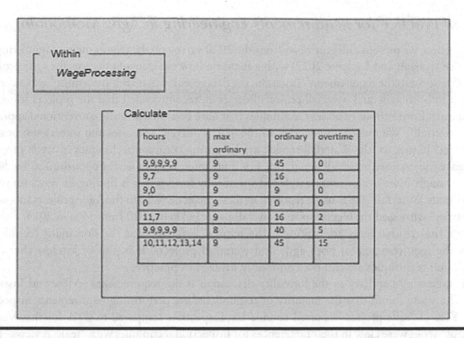

Figure 8.10   Using a story test to show how a company calculates pay for ordinary and over-time hours. (From Mugridge, R., *Software*, 25: 68–75, 2008.)

A nice feature of the story test is that it uses the Fit framework to allow users to build test cases into the stories in a tabular fashion (see Chapter 9). Hence, the users intuitively specify behavior and the tests for that behavior in the requirements.

For example, for a typical payroll system for a business (such as the pet store), a story test describing how to calculate ordinary and overtime pay for an employee based on various clocked hours is given in Figure 8.10.

The table in Figure 8.10 shows, for instance, that if an employee works 5 consecutive 9-hour workdays, then they are considered to have 45 hours in ordinary and zero-hours in over-time pay due. If a worker clocks 10, 11, 12, 13, and 14-hour days in a week, then they are entitled to 45 hours of regular and 15 hours of overtime pay. But the Fit framework is an interac-tive specification—plugging in different values into the table that are incorrect will show up as invalid.

SDD development is considered to be complementary to test-driven development (TDD) in that the former is applied to overall system development (Mugridge 2008).

State of Practice for Requirements Engineering in Agile Methodologies

In this section, we present different views from the 2020 survey on the requirements engineering state of practice (Kassab and Laplante 2022) with a focus on how the state of practices for agile projects.

With respect to the requirements elicitation, the data revealed that "Brainstorming" was a common approach in both agile and waterfall projects. Bose et al. recommended that the projects using Agile "should learn from various elicitation techniques that have been in use in the conventional approaches as the Waterfall". We indeed observe that in the "2020 Survey" use cases and interviews; originally popular techniques for the Waterfall model, are becoming common techniques in Agile projects as well, hence the increase in overall popularity. It is quite interesting to see the opposite also holds. That is, even though "user stories" were developed specifically for Agile, it is finding its way into projects that use "only Waterfall" (28% usage reported in these projects). Within the Agile projects sub-sample, "user stories" witnessed the biggest spike in overall usage (43% in 2020 from 14% in 2013).

Once the requirements are captured, the natural language was the dominant technique to express the requirements for both agile and waterfall projects. It is also of interest that formal specification techniques are still not commonly utilized in practice.

An interesting corollary to the formality discussion is the requirements review and inspection effort. The study found that the majority of respondents are performing requirements inspections. While 57% of Agile projects reported conducting inspection compared to 60% for the Waterfall, both approaches were close in their preferences for inspection techniques with "team reviews" taking first place and with automated requirements inspection using tools remaining largely unpopular.

On the requirements effort estimations, while 65% reported performing some requirements estimation activity, "Story Points" was the most popular technique reported. This is a shift from 2013 when "Expert Judgements" was the most popular back then. The majority of projects (87%) also reported consideration for Non-Functional Requirements (NFRs) during the effort estimation. This percentage was consistent among the Agile and Waterfall samples too. This was a particularly interesting finding considering the common belief that Agile approaches tend to deal less with NFRs in comparison to the Waterfall. One reason for this belief is that NFRs are not always apparent when the requirements are presented through user stories which is a common representation technique in Agile.

Overall, 49% of respondents reported that a tool was used to manage requirements. The usage of RE tools in Agile projects was 1.5 times more popular than in Waterfall projects. This was an interesting observation especially in the light of the common belief that Agile and RE tools may appear incompatible. While both involve the management of processes, their approaches differ in many ways. Agile promotes flexibility and rapid accommodation to change, whereas RE tools are based on tightly controlled processes that do not necessarily adapt well to changing business needs and requirements. The most-reported usage was for "confluence and Jira" (38%), followed by IBM Doors (18%), followed by MS office applications (13%).

Challenges for Requirements Engineering in Agile Methodologies

Of course, there are challenges in any new technology, and there are some in using agile methodologies, particularly concerning requirements engineering. Williams (2004) discusses some of these shortcomings.

For example, agile methodologies do not always deal well with nonfunctional requirements. Why should this be so? One reason is that they are not always apparent when dealing with requirements functionality only (through user stories).

Williams suggests dealing with this challenge by augmenting the user stories with appropriate nonfunctional requirements discovery techniques such as competitive analysis.

Another shortcoming is that with agile methodologies, customer interaction is strong—but mostly through prototyping. As we have seen, there are other ways to elicit nuanced requirements, and understanding stakeholder needs, for example, using various interviewing techniques would be desirable.

Furthermore, with agile methodologies, validation is strong through testing and prototyping, but verification is not so strong. Williams (2004) suggests that using formal methods could strengthen agile requirements engineering.

Rubin and Rubin (2011) identified documentation constructs that are often missed in agile software development, and they suggested ways to alleviate this problem. In particular, they noted that domain knowledge, and emergent system architecture and design, are not properly captured by traditional agile requirements practices. Their findings imply that agile requirements documents should be annotated with appropriate domain knowledge, and the corresponding agile architecture and design documents should be annotated with the final system architecture and design, respectively.

Requirements management is built into the process (e.g., XP and Scrum), but it is mostly focused on the code level. Williams suggests that requirements management could be strengthened by adding more standard configuration management practices.

Finally, Inayat et al. (2015b) summarized a set of agile requirements engineering challenges. The challenges identified by Inayat et al. were as follows:

- Minimal documentation is a vital challenge that agile methods pose to development teams.
- Customer availability is assumed and advocated by agile methods.
- Budget and schedule estimation is a challenge for organizations that follow agile methods.
- Inappropriate architecture finalized by the team in earlier stages of the project becomes inadequate in later stages with new requirements.
- Neglecting NFRs is where NRFs focus on system quality, including its internal quality, i.e. maintainability, testability and external quality usability, and security.
- Customer inability and agreement are the two main issues apart from availability.
- Contractual limitations and requirements volatility are important by not allowing changes in requirements after the signing of the contract; the changes can cause an increase in cost and sometimes failure of projects.
- Requirements change and change evaluation is an important aspect of agile methods.

These challenges echoed some of those identified by others.

Even though there are still significant challenges in using agile methodologies for software engineering, there are also significant advantages. Therefore, agile software methodologies should at least be considered for most projects. And even when agile methodologies are not considered suitable for certain software projects, or when the project is not software, many of the practices discussed for agile methodologies could be used. For example, such practices as the customer always on-site focus on product over documentation, and early development of test cases should still be incorporated in almost every project.

VIGNETTE 8.2 Affordable Care Act Website

In 2010, the United States Congress passed the Affordable Care Act requiring all citizens to obtain some form of healthcare insurance. The law was a controversial one, and adding to the controversy was a disastrous rollout of the federal government's associated website. The project was complex and involved 55 contractors. Rather than focusing on a robust development process, there was a rush to deploy the website, and, unsurprisingly, it suffered from many problems. There were many structural and security problems but usability issues such as frequent crashing, poor response times, missing or misleading content, and navigation difficulties received the most public attention (Venkatesh et al. 2014). Around 9.47 million users attempted to register during the first week of deployment but only 271,000 succeeded (Cleland-Huang 2014). Studies of the failed launch of the website cited a variety of root causes including project schedule underestimation, ill-defined scope, poor requirements specification, and inefficient risk analysis and management (Anthopoulos et al. 2016).

While many experts correctly suggested that a better traditional requirements process would have mitigated these problems, the pressure to deploy the site quickly was tremendous. It is likely, however, that an agile, dynamic requirements discovery and development process could have led to better project outcomes and public perception while permitting an earlier deployment. It is unreported that such a process was suggested or considered.

Supplemental Materials from the authors on NFRs.
https://phil.laplante.io/requirements/agile.php

Exercises

8.1 How do you fit SRS documentation into an agile framework?

8.2 Is it possible to use agile methodologies when the customer is not on site? If so, how?

8.3 Why are agile methodologies generally not suitable for hardware-based projects as opposed to software projects?

8.4 Why can it be difficult for agile methodologies to cover nonfunctional requirements?

8.5 Are there any problems in encapsulating requirements into user stories?

8.6 For the pet store POS system, generate a user story for customer purchases.

8.7 For the pet store POS system, generate use cases for various customer purchases.

8.8 For the airport baggage handling system, generate a user story for dealing with baggage that is to be diverted to another flight.

8.9 For the airport baggage handling system, generate use cases for dealing with a lost piece of baggage.

8.10 For the airport baggage handling system, generate use cases for dealing with baggage that is to be diverted to another flight.

8.11 For the following systems discuss the advantages and disadvantages of using an agile approach (for the software components):

8.11.1 Pet store POS system

8.11.2 Airline baggage handling system

8.11.3 Smart home system (Appendix A)

*8.12 There are a number of software tools, both commercial and open source, that can be used to manage the requirements activities of an agile project. Research these tools and prepare a product comparison matrix.

Notes

1 Some of this section has been excerpted from Laplante (2006) with permission.

2 Most people define agile methodologies as being incremental. But incremental development implies that the features and schedule of each delivered version are planned. In some cases, agile methodologies tend to lead to versions with feature sets and delivery dates that are almost always not as planned.

References

Ambler, S. (2007). Agile Requirements Change Management. http://www.agilemodeling.com/essays/changeManagement.htm (accessed January 2017).

Anthopoulos, L., Reddickb, C. G., & Giannakidou, I. (2016). Why e-government projects fail? An analysis of the Healthcare. Gov. *Government Information Quarterly*, 33(1): 161–173.

Bass, L., Weber, I., & Zhu, L. (2015). *DevOps: A Software Architect's Perspective*. Addison-Wesley Professional, Boston, MA, USA.

Beck, K. (2000). *Extreme Programming Explained: Embrace Change*. Longman Higher Education, London.

Beck, K., Beedle, M., van Bennekum, A., Cockburn, A., Cunningham, W., Fowler, M., Grenning, J., Highsmith, J., Hunt, A., Jeffries, R. and Kern, J. (2001). "Agile Manifesto" and "Principles Behind the Agile Manifesto." http://agilemanifesto.org/ (accessed January 2017).

Boehm, B., and Turner, R. (2003). *Balancing Agility and Discipline: A Guide to the Perplexed*. Addison-Wesley, Boston, MA.

Bose, S., Kurhekar, M., & Ghoshal, J. (2008). Agile Methodology in Requirements Engineering. SETLabs Briefings Online, pp. 13–21.

Cao, L., & Ramesh, B. (2008). Agile requirements engineering practices: An empirical study. *Software*, 25(1): 60–67.

Cleland-Huang, J. (2014). Don't fire the architect! Where were the requirements? *IEEE Software*, 31(2): 27–29.

Cunningham, W. (1992). The WyCash portfolio management system. *ACM SIGPLAN OOPS Messenger*, 4(2): 29–30.

Inayat, I., Moraes, L., Daneva, M., & Salim, S. S. (2015a, May). A reflection on agile requirements engineering: Solutions brought and challenges posed. In *Scientific Workshop Proceedings of the XP2015* (p. 6). ACM, Chicago, IL.

Inayat, I., Salim, S. S., Marczak, S., Daneva, M., & Shamshirband, S. (2015b). A systematic literature review on agile requirements engineering practices and challenges. *Computers in Human Behavior*, 51: 915–929.

Kassab, M., & Laplante, P. (2022). The current and evolving landscape of requirements engineering state of practice. *IEEE Software*. DOI: 10.1109/MS.2022.3147692.

Kassab, M., Neill, C., & Laplante, P. (2014). State of practice in requirements engineering: Contemporary data. *Innovations in Systems and Software Engineering*, 10(4): 235–241.

Laplante, P. A. (2006). *What Every Engineer Needs to Know about Software Engineering.* CRC/Taylor & Francis, Boca Raton, FL.

Mugridge, R. (2008). Managing agile project requirements with story test-driven development. *Software,* 25(1): 68–75.

Orr, K. (2004). Agile requirements: Opportunity or oxymoron? *IEEE Software,* 21(3), 71–73.

Paetsch, F., Eberlein, A., & Maurer, F. (2003, June). Requirements engineering and agile software development. In *WET ICE 2003. Proceedings. Twelfth IEEE International Workshops on Enabling Technologies: Infrastructure for Collaborative Enterprises, 2003* (pp. 308–313). IEEE, Linz, Austria.

Ramesh, B., Cao, L., & Baskerville, R. (2010). Agile requirements engineering practices and challenges: An empirical study. *Information Systems Journal,* 20(5): 449–480.

Rubin, E., & Rubin, H. (2011). Supporting agile software development through active documentation. *Requirements Engineering,* 16(2): 117–132.

Savolainen, J., Kuusela, J., & Vilavaara, A. (2010, September). Transition to agile development-rediscovery of important requirements engineering practices. In *2010 18th IEEE International Requirements Engineering Conference* (pp. 289–294). IEEE, Sydney, NSW.

Schwaber, K., & Beedle, M. (2001). *Agile Software Development with SCRUM.* Prentice Hall, Upper Saddle River, NJ.

Sillitti, A., & Succi, G. (2006). Requirements engineering for agile methods. In A. Aurum & C. Wohlin (Eds.), *Engineering and Managing Software Requirements* (pp. 309–326). Springer, New York.

Venkatesh, V., Hoehle, H., & Aljafari, R. (2014). A usability evaluation of the Obamacare website. *Government Information Quarterly,* 31(4): 669–680.

Vlaanderen, K., Jansen, S., Brinkkemper, S., & Jaspers, E. (2011). The agile requirements refinery: Applying SCRUM principles to software product management. *Information and Software Technology,* 53(1): 58–70.

Wagner, S., Méndez-Fernández, D., Kalinowski, M., & Felderer, M. (2018). Agile requirements engineering in practice: Status quo and critical problems. *CLEI Electronic Journal,* 21(1): 15.

West, D., Grant, T., Gerush, M., & D'Silva, D. (2010). *Agile Development: Mainstream Adoption Has Changed Agility.* Forrester Research, Cambridge, MA.

Williams, L. (2004). Agile Requirements Elicitation. http://agile.csc.ncsu.edu/SEMaterials/AgileRE.pdf (accessed January 2017).

Chapter 9

Tool Support for Requirements Engineering

Introduction

We have already discussed automated requirements analysis via the NASA ARM and related requirements verification and validation tools in Chapter 6. Word processors, database managers, spreadsheets, content analyzers, concept mapping programs, automated requirements checkers, and so on are also tools of interest to the requirements engineer. Table 9.1 compares the capabilities of conventional office tools used for requirements management to a feature-rich requirements management tool.

But the most celebrated requirements tools are large, commercial, or open-source packages that provide a high level of integrated functionality. The chief function of these requirements management tools is to represent and organize all "typical" requirements engineering objects such as sources for requirements, requirements, use cases, scenarios, and user stories throughout the requirements engineering life cycle. Support for user-defined entities is also usually provided.

According to Information technology—Systems and software engineering—Guide for requirements engineering tool capabilities ISO/IEC TR 24766:2009, six major capabilities exist for requirements engineering tools:

- Requirements elicitation
- Requirements analysis
- Requirements specification
- Requirements verification and validation
- Requirements management
- Other capabilities

Verification and validation features are indeed an important component of any automated requirements engineering tool. Indeed, the more sophisticated commercial requirements engineering tools provide other requirements checking, tracing, and archiving features. These are shown in Table 9.2. These features are particularly important because they provide for the accurate tracing

DOI: 10.1201/9781003129509-9

Table 9.1 Requirements Repository Metric Capabilities

	Word Processor	Spreadsheet	Relational Database	Requirements Management Tool
Document size	Yes	No	No	Not in the preformatted state
Dynamic changes over time	Possible with complex change tracking enabled	No	No	Yes
Release size	Yes	Yes	Yes	Yes
Requirement expansion profile	No	No	Yes	Yes
Requirements verification	No	No	Possible	Possible
Requirements volatility	Yes	Yes	Yes	Yes
Test coverage	No	Possible with complex equation logic	Yes	Yes
Test span	No	Possible with complex equation logic	Yes	Yes
Test types	Yes	Yes	Yes	Yes

Source: Adapted from Hammer et al. (1998).

Table 9.2 Automated Requirements Engineering Tool Features (Heindl et al. 2006)

Tool Feature	Description
Definition of workflow for requirements	A workflow (states, roles, state transitions) is configurable for requirements.
Automated generation of bidirectionality of traces	When the user creates a trace between artifact A and artifact B, it automatically establishes a backward trace from B to A.
Definition of user-specific trace types	An authorized user can define trace types and assign names.
Suspect traces	When a requirement changes, the tool automatically highlights all traces related to this requirement for checking and updating traces.
Long-term archiving functionality	All data in the tool can be archived in a format accessible without the tool if necessary.

of artifacts over time. Traceability is an important characteristic of the SRS document and bears further discussion.

Other typical functionalities for these large, commercial requirements engineering tools include may include more capabilities such as:

- Supports agile methodologies such as Scrum, Kanban, and collaborative working.
- Multiuser support and version control
- Support for traceability
- Verification and validation tools (we discussed these in Chapter 6)
- Online collaboration support
- Issue resolution management
- Configuration management
- Product management
- Project management
- Built-in support for standards templates (such as IEEE 29148)
- User-defined glossary support (Heindl et al. 2006)
- Customizable functionality through a programmable interface
- Customizable user interfaces
- Product lifecycle management
- Test management

Table 9.3 lists the top tools reported by the participants of the 2020 RE state of practices survey (Kassab and Laplante 2022).

Share your Opinion: What requirements engineering tools (if any) do you use (or does your company use) and how are they used?
https://phil.laplante.io/requirements/tools.php

Traceability Support

Requirements traceability is a subdiscipline of requirements management, and it is one of the main capabilities to look for in requirement tools. Requirements traceability is concerned with the relationships between requirements, their sources, and numerous other artifacts. Different kinds of traceability can be established:

- **Source Traceability**: This provides links requirements to stakeholders who proposed these requirements
- **Requirements-Linkage Traceability**: This provides links between dependent requirements.
- **Requirements-Design Traceability**: This provides links from the requirements to the design.
- **Requirements-Source Code Traceability**: This provides links from the requirements to the code.
- **Requirement-Test Cases Traceability**: This provides links from the requirements to the test cases.

Table 9.3 Common Commercial Tools Supporting RE Activities (Kassab and Laplante 2022)

Tool Name	Vendor	License	Capabilities
Jira	Atlassian	Commercial	Agile support, issue resolution management, project management, requirements management
Rational Doors	IBM	IBM EULA	Configurations management, requirements management, test management
Azure DevOps	Microsoft	Commercial	Agile support, issue resolution management, requirements management, test management
Jama Connect	Jama Software	Commercial	Project management, requirements management, test management
Doors Next (Jazz)	IBM	IBM EULA	Application lifecycle management, requirements management, test management
Aha!	Aha! Labs	Commercial	Product management, requirements management
Quality Center	Micro Focus	Commercial	Issue resolution management, requirements management, test management, project management
VersionOne	CollabNet	Commercial	Agile support, project management, requirements management

Traceability is especially relevant when developing safety-critical systems and therefore prescribed by safety guidelines, such as DO-178C, ISO 26262, and IEC61508. For example, the standard DO-178C, Software Considerations in Airborne Systems and Equipment Certification, which is used by federal aviation regulatory agencies in the US, Canada, and elsewhere, contains rules for artifact traceability. DO-178C requires traceability between all low-level requirements and their parent high-level requirements. Links are also mandatory between source code elements, requirements, and test cases. All software components must be linked to a requirement, that is, all elements must have a required purpose (i.e., no gold-plating). Certification activities are then conducted to ensure that these rules are followed (RTCA 2011). But for our purposes, we are only concerned with traceability artifacts found in the requirements specification document (or ancillary documents).

Within the SRS document, traceability focuses on the interrelationship between requirements and their sources, stakeholders, standards, regulations, and so on. For these purposes, traceable artifacts include:

■ Requirement
■ Stakeholder(s) who are associated with a requirement
■ Standard, regulation, or law that mandates the requirement
■ Rationale (for the requirement)
■ Keywords (for searching)

The various combinations of these and other artifacts lead to many traceability formats, a few of which will be presented shortly. Yue et al. (2011) also suggest creating traceability matrices that maintain links between requirements elements and analysis model elements. An extensive discussion of requirements tracing approaches and applications can be found in Lee et al. (2011).

One goal of traceability is to visualize the relationship between artifacts. As the number and complexity of trace links increases, techniques for traceability visualization are necessary. Common visualizations for traceability information are matrices, graphs, lists, and hyperlinks. Traceability can be realized by capturing traces either entirely manual or tool-supported. In the following discussions, we will focus on the "matrices" as a way to visualize the traceability links.

Requirements Linkage Traceability Matrix

It is not unusual for one requirement to have an explicit reference to another requirement. In the RE discipline, there is a large body of research on the topic of requirements dependencies. For example, a 2012 empirical study on requirements dependencies (Li et al. 2012) found more than 20 types of dependencies identifiable in published RE literature. Kassab et al. (2009) treated dependencies from an ontology viewpoint. Martakis and Daneva (2013) investigated handling requirements dependencies in agile projects. In general, there are different types of relationships between requirements, a non-exclusive list may include:

- Requires
- Is needed by
- Supersedes
- Is superseded by
- Is child of
- Is parent of
- Refers to
- Uses
- Refers to
- Is satisfied by
- Satisfied
- Contradicts with

Let's focus the discussion now on two relations: "Uses" and "Refers to."

Consider the following requirement:

2.1.1 The system shall provide for control of the heating system and be schedulable according to the time scheduling function (ref. requirement 3.2.2).

which depicts a "refers to" relationship. The "uses" relationship is illustrated in the following requirement:

2.1.1 The system shall provide for control of the heating system in accordance with the time-table shown in requirement 3.2.2 and the safety features described in requirement 4.1.

The main distinction between "uses" and "refers to" is that the "uses" represents a stronger, direct link.

The primary artifact of traceability is the requirements traceability matrix. The requirements traceability matrix can appear in tabular form in the SRS document, in a stand-alone traceability document, or internally represented within a tool for modification, display, and printout. There are many variations of traceability matrices depending on which artifacts are to be included.

One form of the matrix shows the interrelationship between requirements. Here, the entries in the requirements traceability matrix are defined as follows:

$R_{ij} = R$ if requirement i references requirement j (meaning "refers to" for informational purposes).
$R_{ij} = U$ if requirement i uses requirement j (meaning "depends on" directly).
R_{ij} = blank otherwise.

When a requirement both uses and references another requirement, the entry "U" supersedes that of "R." Since self-references are not included, the diagonal of the matrix always contains blank entries. We would also not expect circular referencing in the requirements so that if $R_{ij} = U$ or $R_{ij} = R$, we would expect R_{ji} to be blank. In fact, an automated verification feature in a requirements engineering tool should flag such circular references.

The format of a typical requirements traceability matrix is shown in Table 9.4. A partial traceability matrix for a hypothetical system with sample entries is shown in Table 9.5.

Since R is usually sparse, it is convenient to list only those rows and columns that are not blank. For example, for the SRS document for the smart home found in the appendix, the traceability matrix explicitly derived from the requirements is shown in Table 9.6.

A sparse traceability matrix of this type indicates a low level of explicit coupling between requirements. A low level of coupling within the requirements document is desirable—the more linked requirements, the more changes to one requirement propagate to others. In fact, explicit

Table 9.4 Format of Traceability Matrix (*R*) for Requirements Specification

Requirement ID	1	1.1	1.1.1	...	1.2	1.2.1	...	2	2.1	2.1.1	...
1											
1.1											
1.1.1											
...											
1.2											
1.2.1											
....											
2											
2.1											
2.1.1											
...											

Table 9.5 Partial Traceability Matrix for a Fictitious System Specification

Requirement ID	1	1.1	1.1.1	1.1.2	1.2	1.2.1	1.2.2	2	2.1	2.1.1	3
1									R	R	
1.1			U								
1.1.1	R								R		
1.1.2						R					
1.2										U	
1.2.1							R				
1.2.2										R	
2	U								R		
2.1				U							
2.1.1											R
3					R						

Table 9.6 Traceability Matrix for Smart Home Requirements Specification Shown in the Appendix

Requirement ID	5.11	9.11
9.1.1	R	
9.10.7		R

requirements linkages are a violation of the principle of separation of concerns, a fundamental tenet of software engineering. Therefore, link requirements only when absolutely necessary.

Generally, we would like each requirement to be verified by more than one test case. At the same time, we would like each test case to exercise more than one requirement. The "test span" metrics are used to characterize the test plan and identify insufficient or excessive testing:

■ Requirements per test
■ Tests per requirement

Research is still ongoing to determine appropriate statistics for these metrics. But at the very least, you can use these metrics to look for inconsistencies and nonuniform test coverage. Of course, there is always a trade-off between time and cost of testing vs. the comprehensiveness of testing. But testing is not the subject of this book.

Requirements Source Traceability Matrix

Yet another kind of traceability matrix links requirements to their sources. Aside from those coming directly from users, many requirements are derived from governmental regulations and from standards. Linking the requirements to these sources can be very helpful when the sources change. Table 9.7 shows the typical format for such a traceability matrix.

This kind of traceability matrix is especially useful for tracking nonfunctional requirements. Failure to trace nonfunctional requirements throughout the project life cycle can be a significant problem (Kassab et al. 2008).

Requirements Stakeholder Traceability Matrix

Another kind of traceability matrix links stakeholders to the requirements they submitted, an example of which is shown in Table 9.8.

Table 9.8 also incorporates a ranking schema, which greatly facilitates requirements negotiation and trade-off analysis.

Only three kinds of traceability matrices have been shown in the previous sections, but any variation or combination of these can be created and they can incorporate the other traceability artifacts mentioned. Commercial requirements engineering tools allow the user to customize all kinds of traceability matrices.

Extensive studies document the effectiveness, but also the difficulties of capturing traceability information. An analysis of the pre-market testing of software in medical devices at the US Food and Drug Administration (FDA) in 2013 identified significant gaps between prescribed and filed traceability information (Mäder et al. 2013). The data from the 2020 RE State of Practices Survey (Kassab and Laplante 2022) shows that only 63% of the participants reported satisfaction with

Table 9.7 Traceability Matrix Showing Requirements and Their Sources

Requirement ID	Federal Regulation #1	Federal Regulation #2	State Regulation #1	State Regulation #1	International Standard #1
3.1.1.3	X				
3.1.2.9	X	X			
3.2.1.8			X	X	
3.2.2.5			X		
3.2.2.6					X
3.3.1		X			
3.3.2		X			
3.4.1		X			
3.4.3		X			
3.4.4		X			
3.6.5.1				X	
3.6.6.4					X

Table 9.8 Traceability Matrix Showing Requirements, Their Rank, and Their Stakeholder Source

Requirement	Rank 1 (lowest importance) – 5 (highest importance)	Stakeholder Source D—Doctor, N—Nurse, A— Administrative Support Staff, P—Patient, R—Regulator
3.1.1.1	5	R
3.1.1.2	4	R
3.1.2.1	5	D,N,A
3.1.2.2	3	D,N,A

the efforts to maintain proper requirements traceability within their projects. Requirements traceability best practices include:

- Unique identifiers must be adopted for requirements and business rules.
- When tracing all requirements is simply time-prohibitive, the analyst may be selective based on effort and cost.
- A responsible party within the team must take ownership of traceability.
- An organization must adopt consistent practices in requirements management, including traceability, and the analyst must practice consistency in updates.

Requirements Management Tools

Requirements tools are extremely important in managing requirements information across the lifecycle of a project. In this section, we will review four popular tools used in the industry to manage requirements.

- **Jira** has become one of the most popular tools in the industry, not just software development teams but also by people who are working on the support desk and business teams. Jira essentially is an issue tracking tool, where "issue" is a generic term for a ticketed item that could be a task, bug, story, epic in software development projects, a simple to-do in business projects, or an incident, problem, or service request in service desk-based projects. The popularity of Jira has increased because it can also be customized to act like a help desk system, a simple test management suite, requirements management, or a project management system with end-to-end traceability for software development projects. Jira is mainly accessed using a web browser but it has a lot of integrations with other tools, and Jira comes with a RESTful API so you can interact with it programmatically.
- **Aha! tool** allows users to build and share visual roadmaps. It connects the strategic business goals into high-level requirements (features) and releases. It provides planning board capability where the team can define the features, prioritize the backlog, manage the daily work and plan upcoming releases. The tool also provides a scorecard feature to objectively rank features against the project's strategy. The tool also provides the capability to group

multiple work items that span across many releases into epics, and it automatically syncs the status of epics and features so team members can keep track of the overall effort and report on progress. Aha! is integrated with many third-party applications including JIRA, GitHub, and Azure DevOps Server.

- **Jama Connect** is another example of a requirements management tool. It provides upstream and downstream relationships traceability to understand the impact of change and coverage across the development lifecycle, real-time collaboration among team members, risk and hazard analysis, and requirements reuse and baseline management.

- **IBM's DOORS** (dynamic object-oriented requirements system)—one of the most widely used requirements management tools. DOORS is essentially an object database for requirements management. That is, each feature is represented as an object containing a feature description, feature graph, and use case diagram. The database is further organized by creating folders and projects, and a history of all module and object-level actions is maintained. There is an extension called Analyst, which integrates DOORS with another tool that allows UML models to be embedded and traced within DOORS (Hull et al. 2011). Feature attributes include mandatory, optional, single adaptor, and multiple adaptor, and the user can specify whether an attribute is required or excluded, which embodies the complementary and supporting nature of requirements. Other attributes are based on use case packages and product instances (Eriksson et al. 2005). DOORS offers linking between all objects in a project for full traceability and missing link analysis. The tool can be customized via a C-like programming language. And standard templates are available to structure requirements in compliance with ISO 12207, ISO 6592, and other standards (Volere 2013). There are many more requirements analysis and management features in the latest version of DOORS. Please consult Appendix F, DOORS NG, for more information on the current version of DOORS (as of the time of writing) called DOORS Next Generation (NG).

Tool Evaluation

There are many commercial and open-source tools, and it is important to carefully evaluate the features of any tool before adopting it in the enterprise. ISO/IEC TR 24766 (2009) Information Technology—Systems and Software Engineering—Guide for Requirements Engineering Tool Capabilities is a framework for assessing the capabilities of requirements engineering. These capabilities are organized along with the following areas: elicitation, analysis, specification, verification and validation, management, and other capabilities. Several studies of commercial tools have been conducted using the ISO/IEC TR 24766 framework (e.g., Carrillo de Gea et al. 2011, 2015; Daud et al. 2014). These studies have generally found that the tool market is rapidly changing and that tools are becoming increasingly complex and difficult to use. The complexity of the expensive commercial tools then creates opportunities for inexpensive tools to emerge but don't offer sophisticated features. Furthermore, these studies have indicated that validation functionalities such as consistency, correctness, and completeness are still lacking in most of the tools.

Sud and Arthur (2003) evaluated a number of requirements management tools using the following dimensions:

- Requirements traceability mechanism
- Requirements analysis mechanism

- Security and accessibility mechanism
- Portability and backend compatibility
- Configuration management approach
- Communication and collaboration mechanism
- Change management support
- Online publishing support
- Usability features such as word processor compatibility
- SRS documentation format

While the findings of their study are now somewhat obsolete, these evaluation dimensions can be used by engineers to compare various commercial and open-source requirements management tools prior to adoption. For example, a simple checklist could be created using these dimensions, or a Likert scale–based evaluation could be conducted.

Carrillo de Gea et al. (2021) provided an alternative list of 17 items to be used in the evaluation of RE tool capabilities:

- Organization of requirements with metadata, attributes, and reuse
- Reports, database queries, and open interface language
- Internal checks, that is, consistency, dependencies, and history
- Traceability support, that is, drag and drop (horizontal and vertical)
- Providing support for reuse
- Remote working, cloud only
- Multiple views of requirements
- Performance
- Collaboration, workflow management
- Easily adapted and integrated into business processes
- Federation and notification with ALM/PLM tools
- Export/import with standard formats
- Macros for repeated commands
- Training and learning curve effort
- Agile, CI/CD, and DevOps
- Intelligent support
- Scalability

This checklist (or an adapted version) and an appropriate consensus management approach (e.g., Wideband Delphi, AHP) can be used to select the right tool for a team or enterprise.

Open-Source Requirements Engineering Tools

There are hundreds of thousands of open-source projects many of which are full-featured requirements management tools, and you are encouraged to turn first to open-source repositories to look for tools before purchasing or trying to develop them from scratch. There are also utilities or resources for requirements engineering.

In this section, we look at two such open-source utilities that can be of use to the requirements engineer.

FreeMind

Quick access to: FreeMind
https://phil.laplante.io/requirements/freemind.php

Mind mapping tools allow the user to create concept maps, which depict the relationship between things and ideas and are useful for the requirements engineer for brainstorming, task analysis, laddering, traceability, JAD, focus groups, and many other requirements engineering activities. Concept mapping tools allow for the easy creation of concept maps, and while the foregoing discussion is based on a specific tool for illustrative purposes, any concept mapping tool can be used by the requirements engineer.

FreeMind is an open-source mind-mapping tool written in Java (FreeMind 2017). The tool has an easy-to-use graphical user interface. Concepts are mapped in a hierarchical structure based on a primary concept and subordinate ideas to that concept. Handy icons can be used to mark priorities, important criteria, or hazards for each idea. Many other icons are available to enrich the content of the mind map.

To illustrate the use of the requirements engineer's use of mindmaps, let's look at Phil's concept of the smart home control system. Starting off with the basic concept of a smart home Phil creates a parent node (represented by a bubble) as shown in Figure 9.1.

Next, Phil uses the tool to attach a feature to the system (security) and assign a priority to that feature. Although there are up to seven levels of rankings available, Phil chose a simple three-level system.

1. Mandatory
2. Optional
3. Nice to have

The resultant ranked feature and the updated mind map are shown in Figure 9.2.

Phil then adds details to the feature. For example, he wants the security system of the smart home to use and work with the existing security system in the home.

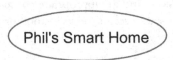

Figure 9.1   Phil's smart home system.

Figure 9.2   Adding the first feature to Phil's smart home system.

Another feature he desires is for the HVAC (heating, ventilation, and air conditioning) system to integrate with the smart home system. In addition, because Phil has delicate plants, pets, and collectibles in the house, it is important that the temperature never exceeds 100°F, so this hazard is marked with a bomb icon. The revised mind map is shown in Figure 9.3.

Phil continues with his brainstorming of features, adding music management, lawn care, and a telephone answering system. Some of the features include important details that are marked with the appropriate symbol (Figure 9.4).

As he sees the system materialize in a visual sense, Phil thinks of more features and is able to make the appropriate prioritization decisions. Phil adds to his mind map accordingly, as seen in Figure 9.5.

The mind map shown in Figure 9.5 is not complete. But this mind map can be used during elicitation discussions with requirements engineers and can be easily evolved over time.

FreeMind has a nice feature that allows you to convert the mind map into a hierarchical structure in text. Thus, if you select the entire diagram and copy it and then paste it into a Word processor, you get a nested text list. You can then add numbers to obtain a hierarchical numbered list for use in a requirements specification document. FreeMind allows you to write down ideas and

Figure 9.3 Adding a detail to the first feature to Phil's smart home system and adding a hazard.

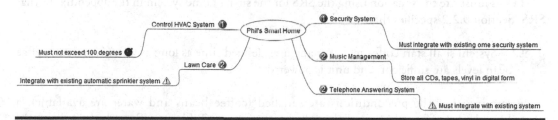

Figure 9.4 Adding more features to Phil's smart home system.

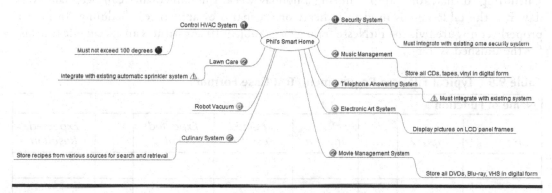

Figure 9.5 Phil's partial mind map of a smart home system.

reorganize them very quickly, and it's a great tool for getting a requirements specification document started. There is also a version called FreeMindPDA that works on many portable devices.

Of course, there are other open-source and commercial mind mapping tools that have similar functionality to FreeMind. Any of these tools can be very valuable in writing the requirements specification and in requirements elicitation.

FitNesse

Quick Access to: FitNesse
https://phil.laplante.io/requirements/fitnesse.php

FitNesse is an open-source, wiki-based software collaboration tool that provides a framework for building interactive test cases. While FitNesse was originally developed as a testing tool, our interest is in using it for integrated requirements/test specifications, for example, in an agile environment.

Before briefly describing how FitNesse works, it is appropriate to examine what a FitNesse test case looks like. The test case is, essentially, a table consisting of the name of the function (or method) to be tested, the set of input parameters and corresponding expected results parameters, and then a set of typical test cases across the rows of the table (see Table 9.9).

The FitNesse table is executable; when the actual code is developed, upon clicking a button, the results of the corresponding function will be calculated.

Let us illustrate the situation using the SRS for the smart home system in the appendix. In that SRS, Section 9.2.2 specifies that:

> 9.2.2 System shall start coffee maker at any user-defined time as long as water is present, coffee bean levels are sufficient, and unit is powered.

Assuming that the preconditions are satisfied (coffee beans and water are available), a plausible schedule for making coffee might be as given in Table 9.9. The schedule shown in Table 9.9 recognizes that the homeowner likes to sleep late on weekends. We could populate this table in many different ways. Now it is true that the functionality desired is for dynamic scheduling so that, for example, during a holiday week, the homeowner can sleep late every day. But the table here is used to indicate one scenario of appliance scheduling. In fact, if properly configured via the FitNesse framework, Table 9.10 constitutes an executable test case for the finished system.

Table 9.9 Typical FitNesse Requirement/Test Case Format

Name of Function							
Input 1	*Input 2*	*...*	*Input n*	*Expected Result 1*	*Expected Result 2*	*...*	*Expected Result m*

Table 9.10   FitNesse Requirements Specification/Test Case for Activate Coffee Pot Function in Smart Home System

Activate_Coffee_Pot		
Day	*Time*	*Output*
Monday	7:00	coffee_pot_on_signal=high
Monday	8:00	coffee_pot_on_signal=low
Tuesday	7:00	coffee_pot_on_signal=high
Tuesday	8:00	coffee_pot_on_signal=low
Wednesday	7:00	coffee_pot_on_signal=high
Wednesday	8:00	coffee_pot_on_signal=low
Thursday	7:00	coffee_pot_on_signal=high
Thursday	8:00	coffee_pot_on_signal=low
Friday	7:00	coffee_pot_on_signal=high
Friday	8:00	coffee_pot_on_signal=low
Saturday	9:00	coffee_pot_on_signal=high
Saturday	10:00	coffee_pot_on_signal=low
Sunday	10:00	coffee_pot_on_signal=high
Sunday	11:00	coffee_pot_on_signal=low

Without going into too much detail, here is roughly how FitNesse works. FitNesse is a wiki GUI built on top of the Fit framework. Fit runs the fixture code (a Java or C# class) corresponding to the test table. For example, in the top row of Table 9.10, Activate_Coffee_Pot specifies the actual class to be called. FitNesse colors the expected results cells red, green, or yellow, if the test failed, passed, or an exception was thrown, respectively (Gandhi et al. 2005).

There are other open-source tools that are similar in functionality to Fitnesse. For example, Cucumber is a tool that is widely used in Scrum environments (cucumber.io).

Requirements Engineering Tool Best Practices

Whatever requirements engineering tool(s) you use, it is appropriate to use the tool judiciously and follow certain best practices. An excellent set of such practices is offered by Cleland-Huang et al. (2007):

- **Trace for a Purpose**: That is, determine which linkages are truly important; otherwise, a large number of extraneous links will be generated.
- **Define a Suitable Trace Granularity**: For example, linkages should be placed at the appropriate package, class, or method level.
- **Support In-Place Traceability**: Provide traceability between elements as they reside in their native environments.

- **Use a Well-Defined Project Glossary**: Create the glossary during initial discovery meetings with stakeholders and use it consistently throughout the requirements engineering process.
- **Write Quality Requirements**: Make sure to follow generally accepted best practices such as IEEE 29148, which are particularly important for traceability.
- **Construct a Meaningful Hierarchy**: Experimental results show that hierarchically organized requirements are more susceptible to intelligent linking software.
- **Bridge the Intradomain Semantic Gap**: For example, avoid overloaded terminology, that is, words that mean completely different things in two different contexts.
- **Create Rich Content**: Incorporate rationales and domain knowledge in each requirement.
- Finally, be sure to use a process improvement plan for improving the requirements engineering process.

Following disciplined practices can result in better results from tool usage and a framework from which processes can be improved. Every project plan should include a description of the tools to be used and how they will be used.

Elicitation Support Technologies

We close this chapter by remarking on some technologies that can be used to support various requirements elicitation processes and techniques previously discussed. These technologies include:

- Wikis
- Mobile technologies
- Virtual environments
- Content analysis

FreeMind was already discussed as a tool for elicitation and for writing the requirements specification document.

Using Wikis for Requirements Elicitation

Wikis are a collaborative technology in which users can format and post text and images to a website. Access control is achieved through password protection and semaphore-like protection mechanisms (i.e., only one user can write to a particular page of the site at any one time).

Wikis can be used for collaboration, for example, to facilitate group work, card entry (for card sorting), template completions, surveys, and for organizing the requirements document. Moreover, wiki-based requirements can be exported directly to publishing tools and validation tools. Several researchers have successfully used wiki-based requirements techniques for various requirements activities including stakeholder collaboration (Ferreira and da Silva 2008), requirements negotiation (Yang et al. 2008), and distributed elicitation (Liang et al. 2009).

In addition, wikis can be used to build interactive documents that can help to automate test cases (recall that requirements specifications should contain acceptance criteria for all requirements).

For example, FitNesse is a free, wiki-based software collaboration GUI built on top of a Fit framework. FitNesse provides a framework for building web-based requirements documentation with embedded, interactive test cases (FitNesse 2007).

Mobile Technologies

Various mobile technologies such as cell phones and personal digital assistants can be used to capture requirements information in place. For example, while physicians are working with patients, they can transmit information about the activities they are conducting directly to the requirements engineer, without the latter having to be on site. Using mobile devices is particularly useful because they enable instantaneous recording of ideas and discoveries. Such an approach can support brainstorming, scenario generation, surveys, and many other standard requirements elicitation techniques even when the customer is not easily accessible (such as in offshore software development situations).

For example, Lutz et al. (2012) developed an application called CREWSpace that allowed users to role-play through Android-enabled devices such as smartphones and tablets and, in doing so, to interact with a representation model displayed on a shared screen. By keeping track of the role-playing states, the software was able to create CRC cards, which are a brainstorming tool used in object-oriented software systems. The CRC cards could then be incorporated into the requirements document.

A second example is the Mobile Scenario Presenter (MSP) that has been developed by the Johannes Kepler University of Linz, Austria, and the City University London, UK. The application allows for both mobile analysts and future system users to acquire requirements systematically and in situ using structured scenarios.

Arena-M (Anytime, Anyplace Requirements Negotiation Aids - Mobile) is the third example. It is a mobile application that provides support for distributed requirements negotiations.

A good discussion of the emergence of the use of mobile technologies in requirements discovery can be found in Maiden et al. (2007) and Kurtz et al. (2008).

Virtual Environments

Architectural designers have used virtual walkthroughs to elicit requirements for commercial building and home construction for many years. Virtual world environments are more complex simulations using enhanced graphics and various specialized devices such as tactile pressure sensors, force feedback peripherals, and three-dimensional display devices. Virtual world environments can be used to provide realism in testing, validating, and agreeing requirements, particularly for novel or complex application environments. Russell and Creighton (2011) suggest that using virtual environments can:

- Clarify current shortcomings
- Draw attention to potential benefits
- Show before-and-after conditions
- Personalize the impressions of various actors and roles
- Create a more shared appreciation of the affective components of valuation, and ready exploration of alternative timelines

For example, in the airline baggage handling system, a three-dimensional layout of the relevant portions of the airport can be constructed and stakeholders can use characters (avatars) to traverse the digital scene and validate requirements that have been previously communicated.

Virtual environments for complex systems can be expensive to create. But in novel and mission-critical systems, the benefits of discovering subtle requirements using virtual simulation may be well worth the cost.

Content Analysis

Content analysis is a technique used in the social sciences for structuring and finding meaning in unstructured information. That is, by analyzing writing or written artifacts of speech, things of importance to the stakeholder can be obtained. In analyzing these writings, the objective is to identify recurrent themes (Krippendorff 2004).

Content analysis can be used by requirements engineers to discover meaning within written artifacts such as transcripts from group meetings, unstructured interviews, survey data, or emails (any text or artifact that can be converted to text). These hidden meanings may reveal subtle requirements that the stakeholders could not articulate directly. In fact, these hidden requirements are often the most problematic and likely to lead to user satisfaction when the product is delivered.

Content analysis can be done manually by tagging (with colored highlighters) identical words and similar phrases that recur in various writings. For example, Figure 9.6 contains a sample content analysis of some text. The text is an excerpt from a notebook of a requirements engineer who has interviewed a customer regarding a smart home.

As we read the text, we begin noticing recurring themes, and as we do so, we highlight them with a different colored marker. For example, the word "I" is mentioned repeatedly, and for whatever reason, we decide that this noun hints at an important theme, and we highlight it in one color. We then notice that smart home or "the home" is mentioned several times—we highlight these with a different color from the previous theme. The notion of time is also mentioned frequently ("long periods of time," "time," and "periods of time") and so we highlight these with a consistent color. And so on.

Free and for-fee tools for content analysis can be used to automate this process.

I am gathering requirements for a smart home for a *customer*.

I spend long periods of time interviewing the *customer* about what *she* wants.

I spend time interacting with the *customer* as *she* goes about *her* day and ask questions like (" why are you running the dishwasher at night, why not in the morning?")

I spend long periods of time passively observing the *customer* "in action" in *her* current home to get non-verbal clues about *her* wants and desires.

I gather information from the home itself – the books on the book shelf, paintings on the wall, furniture styles, evidence of hobbies, signs of wear and tear on various appliances, etc.

Figure 9.6 Sample content analysis of notes from a smart home elicitation interview session. Different fonts are used in lieu of colors.

Artificial Intelligence

In the past few years, the thread of work on Artificial Intelligence (AI) for RE has made strides in rigorously investigating how general-purpose AI tools can be tailored best for RE tasks (Dalpiaz and Niu 2020). For example, recent research studies are investigating how human intervention in the requirement gathering processes can be reduced by using "Speech Understanding Methodology" techniques with the capability to "listen in" on a conversation and suitably collect stakeholders' declarations into a distinct vision (Spoletini and Ferrari 2017; Sharma and Pandey 2014).

Speech understanding methodology can be combined then with "Automatic Keywords Mapping," another AI technique being investigated that can enhance the requirements elicitations. Many requirements issues are related to not having the stakeholders able to depict their requirements properly, or the ignorance of the domain experts and developers to "observable" words that lead essentially to system requirements. These issues can be eliminated by automatically mapping every keyword spoken by each stakeholder. The earlier studies released a keyword mapping technique for designers so that they can recognize the keywords used by stakeholders to assist them in making ideal requirements (Sharma and Pandey 2013).

Case-based reasoning is also being investigated for requirement elicitation, which can reduce the problem of natural language understanding as well as save the time of the requirement expert. There has also been recent research on the use of machine learning algorithms to identify user preferences based on their sentiment (Li et al. 2018).

In Chapter 12, we continue our discussion on the relationship between requirements engineering and AI.

Requirements Metrics

Metrics can be calculated or collected at various stages of the requirements engineering life cycle in order to monitor and improve the requirements engineering processes and practices. The most useful requirements management tools completely automate or partially assist in the collection of metrics. Costello and Liu (1995) describe the following types of requirements metrics to be tracked by tools:

- Volatility
- Traceability
- Completeness
- Defect density
- Fault density
- Interface consistency
- Problem report and action item issues

The requirements volatility metric tracks the numbers of deletions, additions, and modifications to requirements over time. Clearly, these numbers must be tracked by tools. It is often the case that problems emerging later in the project life cycle can be traceable to those requirements that have relatively high volatility.

Traceability metrics include linkage statistics between requirements levels, such as those collected by ARM, and test coverage and span statistics. The test case coverage metric indicates that many requirements have test cases, and the test span metric indicates how many requirements a test case covers. Trace linkage metrics can also be established between requirements, architectural, and design elements. As the IEEE 29148 standard qualities dictate, a high level of internal and external traceability is desirable.

Completeness metrics relate to the number of times that placeholder terms such as TBD are used and could be related to the requirements decomposition levels as illustrated in Figure 4.3.

Defect density metrics indicate the number of requirements defects uncovered during a requirements inspection. These metrics are useful in predicting product and processes volatility and interpreting other metrics.

Requirements fault density indicates the number of requirements faults that are initially detected during test execution or posttest analysis. Requirements faults can be classified by criticality to obtain a number of different metrics. These metrics are used in determining the effectiveness of testing. Fault density metrics are used in predicting product/process volatility and quality and are essential in interpreting the meaning of other metrics (Costello and Liu 1995).

Finally, problem reports and action items are derived metrics that can be used to monitor and control the project. For example, defects per requirement, requirements churn (changes to requirements), and the requirements defect backlog. Requirements management tools can generate many of these metrics.

VIGNETTE 9.1   Tools for Agile Requirements Engineering

One of the criticisms of the requirements engineering practices in agile processes is that documentation is not always rigorously captured. Studies support the notion that agile practitioners do not always use the kinds of tools that can assist in project management and capture appropriate artifacts during systems development. For example, Kassab (2014) found that only about 50% of practitioners using agile methodologies indicated that tools were used for requirements engineering.

Certain tools, however, can play an important role in ensuring that documentation artifacts are recorded, thus enriching that agile process. For example, UML modeling tools can help to capture elements of requirements documentation. Agile teams can also use advanced communications such as web-based shared team projects and instant messaging tools to keep in touch with customers and team members when off-site. The artifacts from these tools can then be captured and converted to elements of the requirements specification.

Researchers have also found that agile projects would also benefit from using requirements traceability tools together with validation tools. Integrated acceptance testing tools such as Cucumber can also improve the quality of requirements and testing documentation (De Lucia and Qusef 2010).

Exercises

9.1   What criteria should be used in choosing an appropriate requirements engineering tool?

9.2   Are there any drawbacks to using certain tools in requirements engineering activities?

9.3   When selecting an open-source tool, what characteristics should you look for?

9.4   How can tools enable distributed, global requirements engineering activities? What are the drawbacks in this regard?

9.5   If an environment does not currently engage in solid requirements engineering practices, should tools be Introduced?

9.6   What sort of problems might you find through a traceability matrix that you might not see without one?

9.7   How is AI being proposed for knowledge acquisition and representation in requirements specifications?

9.8   Download FreeMind and use it to brainstorm a mind map for your smart home system.

9.9   Construct a FitNesse (or Cucumber) test table for the requirement described in Section 8.2 of the appendix.

9.10   Download and install the latest version of DOORS (NG at the time of this writing). Create a new project to manage the requirements for the course project you are currently completing.

*9.11 Using the Sud and Arthur dimensions for evaluating requirements management tools, conduct an assessment of five different commercial or open-source tools using publicly available information.

References

Carrillo de Gea, J. M., Nicolás, J., Alemán, J. L. F., Toval, A., Ebert, C., & Vizcaíno, A. (2011). Requirements engineering tools. *IEEE Software*, 28(4): 86–91.

Carrillo de Gea, J. M., Nicolás, J., Fernández, A. J. L., Toval, A., Ebert, C & Vizcaíno, A. (2015). Commonalities and differences between requirements engineering tools: A quantitative approach. *Computer Science and Information Systems*, 12(1): 257–288.

Carrillo de Gea, J. M., Ebert, C., Hosni, M., Vizcaíno, A., Nicolás, J., & Fernández-Alemán, J. L. (2021). Requirements engineering tools: An evaluation. *IEEE Software*, 38(3): 17–24.

Cleland-Huang, J., Settimi, R., Romanova, E., Berenbach, B., & Clark, S. (2007). Best practices for automated traceability. *IEEE Computer*, 40: 27–35.

Costello, R. J., & Liu, D. B. (1995). Metrics for requirements engineering. *Journal of Systems and Software*, 29(1): 39–63.

Dalpiaz, F., & Niu, N. (2020). Requirements engineering in the days of artificial intelligence. *IEEE Software*, 37(4): 7–10.

Daud, N., Kamalrudin, M., Sidek, S., & Ahmad, S. S. S. (2014). A review of requirements engineering tools for requirements validation. *International Journal of Software Engineering and Technology*, 1(1).

De Lucia, A., & Qusef, A. (2010). Requirements engineering in agile software development. *Journal of Emerging Technologies in Web Intelligence*, 2(3): 212–220.

Eriksson, M., Morasi, H., Borstler, J., & Borg, K. (2005). The PLUSS toolkit—Extending telelogic DOORS and IBM-rational rose to support product line in use case modeling. In *Proceedings of the 20th IEEE/ACM International Conference on Automated Software Engineering* (pp. 300–304). Long Beach, CA.

Ferreira, D., & da Silva, A. R. (2008, September). Wiki supported collaborative requirements engineering. In *Proceedings of the 4th International Symposium on Wikis*. ACM, Porto, Portugal.

FitNesse project. (2007). www.fitnesse.org (accessed January 2017).

FreeMind project. (2017). http://freemind.sourceforge.net/wiki/index.php/Main_Page (accessed February 2017).

Gandhi, P., Haugen, N. C., Hill, M., & Watt, R. (2005). Creating a living specification using fit documents. In *Agile 2005 Conference (Agile 05)* (pp. 253–258). IEEE CS Press, Denver, CO.

Hammer, T. F., Huffman, L. L., Rosenberg, L. H., Wilson, W., & Hyatt, L. (1998). Doing requirements right the first time. *CROSSTALK The Journal of Defense Software Engineering*, 20–25.

Heindl, M., Reinish, F., Biffl, S., & Egyed, A. (2006, August). Value-based selection of requirements engineering tool support. In *32nd EUROMICRO Conference on Software Engineering and Advanced Applications, SEAA' 06* (pp. 266–273), Cavtat, Croatia.

Hull, E., Jackson, K., & Dick, J. (2011). Doors: A tool to manage requirements. *Requirements Engineering*, 181–198. Springer-Verlag, London.

ISO/IEC TR 24766. (2009). *Information Technology—Systems and Software Engineering—Guide for Requirements Engineering Tool Capabilities*. International Organization for Standardization (ISO), Geneva, Switzerland.

Kassab, M. (2014). An empirical study on the requirements engineering practices for agile software development. In *2014 40th EUROMICRO Conference on Software Engineering and Advanced Applications (SEAA)*. IEEE.

Kassab, M., & Laplante, P. (2021). The current and evolving landscape of requirements engineering state of practice. *IEEE Software*. DOI Bookmark: 10.1109/MS.2022.3147692

Kassab, M., Ormandjieva, O., & Daneva, M. (2008). A traceability metamodel for change management of non-functional requirements. In *2008 Sixth International Conference on Software Engineering Research, Management and Applications* (pp. 245–254). IEEE, Prague, Czech Republic.

Kassab, M., Ormandjieva, O., & Daneva, M. (2009). An ontology based approach to non-functional requirements conceptualization. In *2009 Fourth International Conference on Software Engineering Advances* (pp. 299–308). IEEE, Porto, Portugal.

Krippendorff, K. (2004). *Content Analysis: An Introduction to Its Methodology*. Sage, Thousand Oaks, CA.

Kurtz, G., Geisser, M., Hildenbrand, T., & Kude, T. (2008). Mobile technologies in requirements engineering. In *Advances in Computer and Information Sciences and Engineering* (pp. 317–322). Springer, Dordrecht.

Lee, S.-W., Gandhi, R. A., & Park, S. (2011). Requirement tracing. In P. Laplante (Ed.), *Encyclopedia of Software Engineering* (pp. 999–1011). Taylor & Francis, Boca Raton, FL.

Li, J., Zhu, L., Jeffery, R., Liu, Y., Zhang, H., Wang, Q., & Li, M. (2012). An initial evaluation of requirements dependency types in change propagation analysis.

Li, T., Zhang, F., & Wang, D. (2018). Automatic user preferences elicitation: A data-driven approach. In *International Working Conference on Requirements Engineering: Foundation for Software Quality* (pp. 324–331). Springer, Cham, Utrecht, The Netherlands.

Liang, P., Avgeriou, P., & Clerc, V. (2009, July). Requirements reasoning for distributed requirements analysis using semantic wiki. In *Fourth IEEE International Conference on Global Software Engineering, 2009. ICGSE 2009* (pp. 388–393). IEEE, Limerick, Ireland.

Lutz, R., Schäfer, S., & Diehl, S. (2012, September). Using mobile devices for collaborative requirements engineering. In *Proceedings of the 27th IEEE/ACM International Conference on Automated Software Engineering* (pp. 298–301). ACM, Essen, Germany.

Mäder, P., Jones, P. L., Zhang, Y., & Cleland-Huang, J. (2013). Strategic traceability for safety-critical projects. *IEEE Software*, 30(3): 58–66.

Maiden, N., Omo, O., Seyff, N., Grunbacher, P., & Mitteregger, K. (2007). Determining stakeholder needs in the workplace: How mobile technologies can help. *IEEE Software*, 24: 46–52.

Martakis, A., & Daneva, M. (2013). Handling requirements dependencies in agile projects: A focus group with agile software development practitioners. In *IEEE 7th International Conference on Research Challenges in Information Science (RCIS)* (pp. 1–11). IEEE, Paris, France.

RTCA. (2011). *Software Considerations in Airborne Systems and Equipment Certification*. DO-178C. Radio Technical Commission for Aeronautics (RTCA), Washington, DC.

Russell, S., & Creighton, O. (2011, August). Virtual world tools for requirements engineering. In *2011 Fourth International Workshop on Multimedia and Enjoyable Requirements Engineering-Beyond Mere Descriptions and with More Fun and Games (MERE)* (pp. 17–20). Trento, Italy.

Sharma, S., & Pandey, S. K. (2013). Revisiting requirements elicitation techniques. *International Journal of Computer Applications*, 75(12): 35–39.

Sharma, S., & Pandey, S. K. (2014). Integrating AI techniques in SDLC: Requirements phase perspective. *International Journal of Computer Applications in Technology*, 5(3): 1362–136.

Spoletini, P., & Ferrari, A. (2017). Requirements elicitation: A look at the future through the lenses of the past. In *2017 IEEE 25th International Requirements Engineering Conference (RE)* (pp. 476–477). IEEE, Lisbon, Portugal.

Sud, R. R., & Arthur, J. D. (2003). *Requirements Management Tools: A Quantitative Assessment*. Technical Report TR-03-10. Computer Science, Virginia Tech, Blacksburg, VA.

Volere requirements resources. (2013). http://www.volere.co.uk/tools.htm (accessed February 2017).

Yang, D., Wu, D., Koolmanojwong, S., Brown, A. W., & Boehm, B. W. (2008, January). WikiWinWin: A wiki based system for collaborative requirements negotiation. In *Proceedings of the 41st Annual Hawaii International Conference on System Sciences* (pp. 24–24). IEEE, Waikoloa, Big Island, Hawaii.

Yue, T., Briand, L. C., & Labiche, Y. (2011). A systematic review of transformation approaches between user requirements and analysis models. *Requirements Engineering*, 16(2): 75–99.

Chapter 10

Requirements Management

Introduction

Requirements management involves identifying, documenting, and tracking system requirements from inception through delivery. Inherent in this definition is the understanding of the true meaning of the requirements and the management of customer (and stakeholder) expectations throughout the system's lifecycle. A solid requirements management process is the key to a successful project.

Hull et al. (2011) suggest that there are five main challenges to requirements management. The first is that very few organizations have a well-defined requirements management process, and thus, few people in those organizations have requirements management experience. The second challenge is the difficulty in distinguishing between a user or stakeholder requirements and systems. The third problem is that organizations manage requirements differently making the dissemination and transferability of experience and best practices difficult. The difficulties in progress monitoring are yet another problem. And finally, they suggest that managing changing requirements is a significant challenge. They suggest that establishing a well-defined requirements management process is key in addressing these issues.

Requirements Management Process

Most organizations do not have an explicit requirements management process in place, but this does not mean that requirements management does not occur within the organization. The requirements practices probably exist implicitly in the organization, but these practices are not usually documented. One of the first steps in improving the requirements management process in any organization is to document existing practices.

Throughout system development, even in the waterfall model, frequent customer and stakeholder interaction are essential. From the requirements engineering standpoint, interactions related to requirements changes must be managed effectively. Management of these changes includes documenting these change requests (honoring the configuration management process) and ensuring that these changes are effectively communicated to developers.

DOI: 10.1201/9781003129509-10

Along the way, there will be numerous opportunities for requirements engineers to exercise their leadership and negotiation skills to settle disputes and deal with various negative organizational forces that you have studied (including management and environmental antipatterns, negative personality types and behaviors, and cognitive dissonance).

Reinhart and Meis (2012) discuss how certain requirements management approaches adapted from production engineering can be used to improve the execution of "simultaneous engineering," that is, the parallelization of unequal activities that are generally conducted sequentially. They suggest that success factors include simulation prototyping, integration of available knowledge, frequent communication, and efficient use of tools.

Wiegers (2003) defines requirements management in terms of the following activities:

- Change control refers to the (automated) procedures used in documenting requirements changes.
- Version control is used to track the age of a document and inform users of changes.
- Status tracking defines each requirement whether it is: proposed, approved, implemented, verified, deleted, or rejected.
- Requirements tracing is a process tying user needs to requirements, and requirements to test cases and parts of the design.

By comparison, the International Council On Systems Engineering (ICOSE) explains the requirements management process in terms of the following activities:

- **Identify and Document:**
  - Capture why the end product or system needs to exist in terms of value provided to the end-user.
  - Elicit and document requirements from the client and stakeholders.
  - Gain their agreement on the written statements and on the measures of successful completion.
  - Ensure regulatory requirements are understood by all parties.

- **Analyze and Allocate:**
  - Derive explicit measurable requirements useful for the design that are directly traceable to providing the system's value within constraints (e.g., cost, regulatory, and social constraints).
  - Continue derivation and decomposition such that derived requirements can be allocated to the system and components.
  - Link requirements and their supporting bases to provide two-way traceability.
  - Demonstrate requirement validity through analysis, authority, or other demonstration of need.
  - Valid requirements are those that are necessary, clear, achievable, and verifiable.

- **Control Change:**
  - Analyze requirements changes for the full impact and approve change through a prescribed process and appropriate levels of authority.
  - Understand the impact before committing to the change. Ensure approved changes are propagated to all affected items.

- **Verify Compliance:**
  - Verify requirements are met and that the resulting system actually provides the needed value in the real environment of use.
  - Determine test conditions and acceptance criteria during design to enable test planning.

- **Provide Oversight:**
  - Manage the resulting information and the processes being implemented. The depth of requirements managed by a program needs to be set. Usually, it is sufficient to manage to the level of subcontracts or procured items.
  - Develop metrics useful to show the quality of the requirements management activities. Integrate requirements across systems to ensure a balanced solution and compatible interfaces.

These activities are part of the model-based systems engineering practice advocated by INCOSE.

Configuration Management and Control

Configuration management involves the identification, tracking, and control of important artifacts of the system. Configuration items relevant to the requirements engineer include the individual requirements, sources of requirements, stakeholders, and the requirements specification document. When items are under configuration, changes to those items can only be made by those authorized to make the changes, and all of the changes are tracked, time-stamped, and version stamped. Standards such as IEEE 29148, IEEE 12207, and ISO 9000 all require that a robust configuration management program be in place and that all requirements artifacts be placed under configuration control.

Configuration control involves requesting, evaluating, and approving requirements changes, approving releases. Placing all requirements artifacts under configuration control helps to reduce the likelihood of creating obsolete and ridiculous requirements.

Therefore, an important part of requirements management is disciplined configuration management control processes. Requirements management, configuration management, or project management tools that include the appropriate functionality are essential in enforcing the discipline necessary for successful configuration management.

Reconciling Differences

One of the most important activities in requirements management is consensus building, particularly when forming or discovering:

- Mission statements
- Goals
- Requirements
- Rankings

But achieving consensus between stakeholder groups is not easy. Tuckman's theory of team formation suggests that group formation follows a pattern of "forming, storming, norming, performing, and adjourning." The premise is that a team can dramatically change a non-cohesive group to a high-functioning one. Interested readers might want to watch Sidney Lumet's 1957 movie, *12 Angry Men*. You can actually see the jury going through the Tuckman team formation sequence, and it is also a great film (Neill et al. 2012).

Managing Divergent Agendas

Each stakeholder has a different requirements agenda. For example, business owners seek ways to get their money's worth from projects. Business partners want explicit requirements because they are like a contract. Senior management expects more financial gain from projects than can be realized. And systems and software developers like uncertainty because it gives them the freedom to innovate solutions. Project managers may use the requirements to protect them from false accusations of underperformance in the delivered product.

One way to understand why the different agendas might exist—even among persons within the same stakeholder group—is the Rashomon effect. *Rashomon* is a highly revered 1950 Japanese film directed by Akira Kurosawa. The main plot involves the recounting of the murder of a samurai from the perspective of four witnesses to that event—the samurai, his wife, a bandit, and a woodcutter—each of whom has a hidden agenda and tells a contradicting accounting of the event. Stated succinctly "your understanding of an event is influenced by many factors, such as your point of view and your interests in the outcome of the event" (Lawrence 1996). The smart requirements manager seeks to manage these agendas by asking the right questions upfront. Andriole (1998) suggests the following questions are appropriate:

1. What is the project request?
   a. Who wants it?
   b. Is it discretionary or non-discretionary?
2. What is the project's purpose?
   a. If completed, what impact will the new or enhanced system have on organizational performance?
   b. On profitability?
   c. On product development?
   d. On customer retention and customer service?
3. What are the functional requirements?
   a. What are the specific things the system should do to satisfy the purposeful requirements?
   b. How should they be ranked?
   c. What are the implementation risks?
4. What are the nonfunctional requirements, like security, usability, and interoperability?
   a. How should they be ranked
   b. What are the implementation risks?
   c. How do you trade off functional and non-functional requirements?
5. Do we understand the project well enough to prototype its functionality?
6. If the prototype is acceptable, will everyone sign off on the prioritized functionality and non-functionality to be delivered, on the initial cost and schedule estimates, on the estimates' inherent uncertainty, on the project's scope, and on the management of additional requirements?

Andriole asserts that by asking these questions upfront, hidden agendas can be uncovered and differences resolved. At the very least, important issues will be raised upfront and not much later in the process.

Consensus Building

There are numerous approaches to consensus building. Existing trade-off techniques are categorized and sorted into three categories (Berander et al. 2005): experience-based, model-based, and mathematically based.

Experience-based models rely on pure experience for supplying the needed information to perform the trade-off analysis. They are commonly used for cost and effort estimation, but rarely in literature, as they are ad hoc in nature, and the execution is up to the person performing the trade-off. These models provide quick analysis, but they are not repeatable, and they do not provide any figures presenting the trade-off.

Model-based techniques, on the other hand, rely on constructing an, for example, graphical model for illustrating and concretize the relations between trade-off entities, thus facilitating the trade-off. By applying a model-based trade-off approach, in comparison to an experience-based approach, it is possible to structure and communicate the knowledge. Dealing with trade-offs concerning quality requirements is popularly treated through the model-based technique. The NFRs framework that we explained in Chapter 5 is an example of a model-based technique.

Mathematically based trade-off techniques, on the other hand, rely on formalization for constructing and representing the trade-off, thus making it possible to feed the mathematical construct with appropriate values and receiving the best solution (either maximization or minimization or optimal with regard to certain criteria). Mathematically based trade-off techniques are widely used in management for trade-off decision support. Estimation and calculations regarding break-even, optimum production volume, and so on all use mathematical techniques. Examples of these techniques include the analytical hierarchy process (AHP) technique, Wideband Delphi technique, and reliability growth methods. Mathematically based trade-off techniques can handle large amounts of variables and come up with results that are generally more accurate than common sense. It also enables repeatable and structured analysis. If a measurement program is in place collecting metrics, and if mathematically based trade-off techniques are used as a way to estimate issues, the work can be replicated over several releases tweaking the data and choice of the technique to correspond with needs. Using the same types of techniques over an extended period of time can give an organization consistency and overview.

Analytical Hierarchical Process (AHP)

The AHP is a technique for modeling complex and multi-criteria problems and solving them using a pairwise comparison process. Based on mathematics and psychology, it was developed by Thomas L. Saaty in the 1970s and has been extensively studied and refined since then. AHP was refined through its application to a wide variety of decision areas, including transport planning, product portfolio selection, benchmarking and resource allocation, and energy rationing.

Simply described, AHP breaks down complex and unstructured problems into a hierarchy of factors. A super-factor may include subfactors. By pairwise comparison of the factors in the lowest level, we can obtain a prior order of factors under a certain decision criterion. The prior order of super-factors can be deduced from the prior order of subfactors according to the hierarchy relations.

The AHP process starts with a detailed definition of the problem; goals, all relevant factors, and alternative actions are identified. The identified elements are then structured into a hierarchy of levels where goals are put at the highest level and alternative actions are put at the lowest level.

Usually, an AHP hierarchy has at least three levels: the goal level, the criteria level, and the level of the alternative (see Figure 10.1). This hierarchy highlights relevant factors of the problem and their relationships to each other and to the system as a whole.

Once the hierarchy is built, involved stakeholders (i.e., decision-makers) judge and specify the importance of the elements of the hierarchy. To establish the importance of elements of the problem, a pairwise comparison process is used. This process starts at the top of the hierarchy by selecting an element (e.g., a goal), and then the elements of the level immediately below are compared in pairs against the selected element.

A pairwise matrix is built for each element of the problem; this matrix reflects the relative importance of elements of a given level with respect to a property of the next higher level. Saaty proposed the scale [1,3,5,..., 9] to rate the relative importance of one criterion over another (see Table 10.1).

Based on experience, a scale of 9 units is reasonable for humans to discriminate between preferences for two items. One important advantage of using the AHP technique is that it can measure the degree to which a manager's judgments are consistent. In the real world, some inconsistency is acceptable and even natural. For example, in a sporting contest, if team A usually beats team B, and if team B usually beats team C, this does not imply that team A usually beats team C. The slight inconsistency may result because of the way the teams match up overall. The point is to make sure that inconsistency remains within some reasonable limits. If it exceeds a specific limit, some revision of judgments may be required. AHP technique provides a method to compute the consistency of the pairwise comparisons.

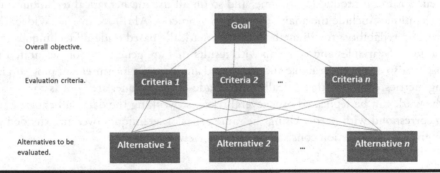

Figure 10.1   The AHP hierarchy.

Table 10.1   Pairwise Comparison Scale for AHP (Saaty 1988)

Intensity of Judgment	Numerical Rating
Extreme importance	9
Very strong importance	7
Strong importance	5
Moderate importance	3
Equal importance	1
For compromise between the above values	2, 4, 6, and 8

Some examples in using the AHP in requirements prioritizations include the integrated approach of AHP and NFRs framework (Kassab 2013; Kassab and Kilicay-Ergin 2015). In addition, Kassab (2014) has used the AHP for effort estimation at requirements time.

Wideband Delphi Technique

One of the most celebrated consensus-building techniques in systems engineering is the Wideband Delphi technique. Developed in the 1970s, but popularized in Barry Boehm in the early 1980s, Wideband Delphi is usually associated with the selection and prioritization of alternatives. These alternatives are usually posed in the form of a question:

> for the following alternative, rate your preference according to the following scale (5 = most desired, 4 = desired, 3 = ambivalent, 2 = not desired, 1 = lease desired)

The list of alternatives and associated scale is presented to a panel of experts (and in the case of requirements ranking, stakeholders) who rank these requirements silently and, sometimes, anonymously. The ranking scale can have any number of levels. The collected list of rankings is then re-presented to the group by a coordinator. Usually, there is significant disagreement in the rankings. A discussion is conducted and experts are asked to justify their differences in opinion. After discussion, the independent ranking process is repeated (Figure 10.2).

With each iteration, individual customer rankings should start to converge, and the process continues until a satisfactory level of convergence is achieved.

A more structured form of the process is:

1. The coordinator presents each expert with a specification and an estimation form.
2. The coordinator calls for a group meeting in which the experts discuss estimation issues with the coordinator and each other.
3. Experts fill out forms anonymously.
4. The coordinator prepares and distributes a summary of the estimates.

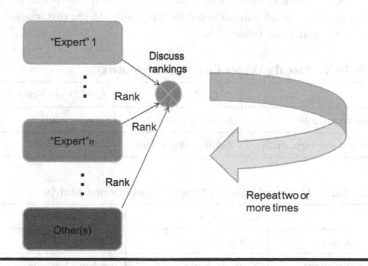

Figure 10.2   The Wideband Delphi process.

5. The coordinator calls for a group meeting, specifically focusing on having the experts discuss points where their estimates vary widely.
6. Experts fill out forms, again anonymously, and steps 4–6 are iterated for as many rounds as appropriate.

There will never be a unanimous agreement in the Wideband Delphi process, but at least everyone involved will feel that his opinion has been considered. Wideband Delphi is a kind of win-win negotiating and can be used for other types of decision-making.

Expectation Revisited: Pascal's Wager

Mathematician Blaise Pascal (1623–1662) is well known for various achievements including Pascal's Triangle (a convenient way to find binomial coefficients) and work in probability. It was his work in probability theory that led to the notion of expected value, and he used such an approach later in life when he became more interested in religion than mathematics. It was during his monastic living that he developed a theory that suggested that it was advisable to live a virtuous life, whether or not one believed in a supreme being. His approach, using expected value theory, is now called Pascal's wager, and it goes like this.

Imagine an individual is having a trial of faith and is unsure if they believe in this supreme being (let's call this being God for argument's sake) or not. Pascal suggests that it is valuable to consider the consequences of living virtuously, in the face of the eventual realization that such a God exists (or not). To see this, consider Table 10.2.

Assuming that it is equally probable that God exists or not (this is a big assumption), we see that the expected outcome (consequence) of living virtuously is half of paradise while the expected outcome of living without virtue is half of damnation. Therefore, it is in a person's best interests to live virtuously.

What does Pascal's wager have to do with expectation setting? Stakeholders will hedge their bets—sometimes withholding information or offering up inferior information because they are playing the odds involving various organizational issues. For example, does a stakeholder wish to request a feature that they believe no one else wants and for which they might be ridiculed? From a game theory standpoint, it is safer to withhold their opinion. To see this, consider the modified Pascal's wager outcome matrix in Table 10.3.

Table 10.2 Pascal's Wager Consequence Matrix

	God Exists	*God Does Not Exist*
Live virtuously	Achieve paradise	Null
Do not live virtuously	Achieve damnation	Null

Table 10.3 Modified Pascal's Wager Consequence Matrix

	Group Agrees	*Group Disagrees*
Speak out	Get praise	Get ridiculed
Remain silent	Nothing happens	Nothing happens

If the stakeholder decides to speak out about a particular feature (or in opposition to a particular feature), assuming equi-likely probability that the group will agree or disagree, the consequence matrix shows that she can expect to get some praise if the group agrees or some ridicule if the group disagrees. It is also well known that, in decision-making, individuals will tend to make decisions that avoid loss over those that have the potential for gain—most individuals are risk-averse. Furthermore, recall the possible cultural effects for stakeholders who originate from countries with high power distances and masculinity indices. The foregoing analysis also assumes that the probabilities of agreement and disagreement are the same—the expected consequences are much worse if the stakeholder believes there is a strong chance that her colleagues will disagree. Of course, later in the process, the feature might suddenly be discovered by others to be important. Now it is very late in the game, however, and adding this feature is costly. Had the stakeholder only spoken up in the beginning, in a safe environment for discussion, a great deal of cost and trouble could have been avoided.

One last comment on Pascal's wager, expectation, and risk. The first author once worked for a boss—we'll call him Bob—who greeted all new employees with a welcome lunch. At that lunch, he would declare

> I am a 'no surprises' kind of guy. You don't surprise me, and I won't surprise you. So, if there is ever anything on your mind or any problem brewing, I want you to bring that to my attention right away.

This sentiment sounded great. However, each time the author or anyone else would bring bad news to Bob, whether the messenger was responsible for the situation or not, Bob would blow his stack and berate the hapless do-gooder. After a while, potential messengers would forego bringing information to Bob. The decision was purely game-theoretic—if you had bad news and you brought it to Bob, he would yell at you. If he didn't find out about it (as often happened because no one else would tell him, either), then you escaped his rampage. If he somehow found out about the bad news, you might get yelled at—but he might yell at the first person in his sight, not you, even if you were the party responsible for the problem. So, it made sense (and it was rigorously sound via game theory) to shut up.

It was rather ironic that "no surprise Bob" was always surprised because everyone was afraid to tell him anything. The lesson here is that, if you shoot the messenger, people will begin to realize the consequences of bringing you information, and you will soon be deprived of that information. Actively seeking and inviting requirements information throughout the project lifecycle is an essential aspect of requirements management.

Global Requirements Management

Requirements engineering is one of the most collaboration-intensive activities in software development. Global and even onshore outsourcing present all kinds of challenges to the requirements engineering endeavor. These include time delays and time zone issues, the costs and stresses of physical travel to client and vendor sites when needed, and the disadvantages of virtual conferencing and telephone. Even simple email communications cannot be relied upon entirely, even though for many globally distributed projects informal emails and email distributed documents are the main form of collaboration. But email use leads to frequent context switching, information fragmentation, and the loss of nonverbal cues.

When the offshoring takes place in a country with a different native language and substantially different culture, new problems may arise in terms of work schedules, work attitudes, communication barriers, and customer-vendor expectations of how to conduct business (recall Hofstede's metrics). Moreover, offshoring introduces a new risk factor: geopolitical risk—and this risk must be understood, quantified, and somehow factored into the requirements engineering process and schedule. Finally, there are vast differences in laws, legal processes, and even the expectations of honesty in business transactions around the world. These issues are particularly relevant during the requirements elicitation phase.

When discussing the problem of global requirements engineering, Schmid (2014) suggests distinguishing between internationalization and distribution. While the latter refers to the development for a set of international customers (perhaps with a single, localized development team), the former refers to the development in a globally distributed environment, where many stakeholders are in a different location from the customer(s). While both issues often co-occur, they are different and may require slightly different strategies for handling them.

Schmid further identified eight context issues that issue that impact the requirements of a system due to the place where the software is used (internationalization):

1. **Language**: customers may use a different language than the development team.
2. **User Interface**: due to language issues and cultural issues it might be necessary to create different kinds of user interfaces.
3. **Local Standards**: the customer might use different standards like calendars or measurement systems than the developer.
4. **Laws and Regulations**: these might be fundamentally different between customer and developer, but taken as obvious, as the customer is never concerned with other rules, thus the presence of a difference might even go easily unnoticed.
5. **Cultural Differences**: due to cultural differences the expectations regarding system behavior may be profoundly different.
6. **Regional Issues**: the situation for the customer might again be regionally subdivided, so that not a single solution, but a product line or customizable system is required.
7. **Educational and Work Context-related Issues**: in different regions, different levels of educational background might be expected, which may require very different user interfaces (e.g., when operating certain machinery).
8. **Environmental Conditions**: the products might need to work in environmental conditions, fundamentally different from those the developers assume, giving rise to corresponding requirements.

The distributed nature of a development team may introduce an additional set of challenges related to:

1. **Elicitation problems** may occur due to the distributed nature of the development.
2. **Communicating the captured requirements** in a distributed software development environment.
3. **Cooperation across organizational borders** to satisfy the requirements.

Bhat et al. (2006) highlighted nine specific problems they observed or experienced when dealing with global requirements engineering. These included:

- Conflicting client–vendor goals
- Low client involvement

- Conflicting requirements engineering approaches (between client and vendor)
- Misalignment of client commitment with project goals
- Disagreements in tool selection
- Communication issues
- Disowning responsibility
- Sign-off issues
- Tools misaligned with expectation

Bhat et al. suggest that the following success factors were missing in these cases, based on an analysis of their project experiences:

- Shared goal—that is, a project metaphor
- Shared culture—in the project sense, not in the sociological sense
- Shared process
- Shared responsibility
- Trust

These suggestions are largely consistent with agile methodologies, though we have already discussed the challenges and advantages of using agile approaches to requirements engineering.

Finally, what role can tools play in the globally distributed requirements engineering process? Sinha and Sengupta (2006) suggest that software tools can play an important role in combating several of the above challenges, though there are not many appropriate tools for this purpose. Appropriate software tools must support:

- Informal collaboration
- Change management
- Promoting awareness (e.g., auto-emailing stakeholders when triggers occur)
- Managing knowledge—providing a framework for saving and associating unstructured project information

There are several commercial and even open-source solutions that claim to provide these features, but we leave the product research of these to the reader.

Antipatterns in Requirements Management[1]

Quick Access to Antipatterns Test
https://phil.laplante.io/antipatterns.php

In troubled organizations, the main obstacle to success is frequently accurate problem identification. Diagnosing organizational dysfunction is quite important in dealing with the underlying problems that will lead to requirements engineering problems.

Conversely, when problems are correctly identified, they can almost always be dealt with appropriately. But organizational inertia frequently clouds the situation or makes it easier to do the wrong thing rather than the right thing. So how can you know what the right thing is if you've got the problem wrong?

In their groundbreaking book, Brown et al. (1998) described a taxonomy of problems or anti-patterns that can occur in software architecture and design, and in the management of software projects. They also described solutions or refactorings for these situations. The benefit of providing such a taxonomy is that it assists in the rapid and correct identification of problem situations, provides a playbook for addressing the problems, and provides some relief to the beleaguered employees in these situations in that they can take consolation in the fact that they are not alone.

These antipatterns bubble up from the individual manager through organizational dysfunction and can manifest in badly stated, incomplete, incorrect, or intentionally disruptive requirements.

Our antipattern set consists of an almost even split of 28 environmental (organizational) and 21 management antipatterns. *Management* antipatterns are caused by an individual manager or management team ("the management"). These antipatterns address issues in supervisors that lack the talent or temperament to lead a group, department, or organization. *Environmental* antipatterns are caused by a prevailing culture or social model. These antipatterns are the result of misguided corporate strategy or uncontrolled sociopolitical forces. But we choose to describe only a small subset of the antipatterns set that is particularly applicable in requirements engineering.

Environmental Antipatterns

Divergent Goals

Everyone must pull in the same direction. There is no room for individual or hidden agendas that don't align with those of the business. The divergent goals antipattern exists when there are those who pull in different directions.

There are several direct and indirect problems with divergent goals.

- Hidden and personal agendas divergent to the mission of an organization starve resources from strategically important tasks.
- Organizations become fractured as cliques form to promote their own self-interests.
- Decisions are second-guessed and subject to "review by the replay official" as staff try to decipher genuine motives for edicts and changes.
- Strategic goals are hard enough to attain when everyone is working toward them, without complete support they become impossible and introduce risk to the organization.

There is a strong correspondence between stakeholder dissonance and divergent goals, so be very aware of the existence of both.

Since divergent goals can arise accidentally and intentionally, there are two sets of solutions or refactorings.

Dealing with the first problem of comprehension and communication involves explaining the impact of day-to-day decisions on larger objectives. This was the case with the corporate executives of the box company described in Chapter 2. The executives forgot the bigger picture. There is more than providing a coffee mug with the mission statement, however. Remember that the misunderstanding is not because the staff are not aware of the mission or goals; organizations are generally very good at disseminating them. It is that they don't understand that their decisions have any impact on those goals. They have a very narrow perspective on the organization and that must be broadened.

The second problem of intentionally charting an opposing course is far more insidious, however, and requires considerable intervention and oversight. The starting point is to recognize the disconnect between their personal goals and those of the organization. Why do they feel that the stated goals are incorrect? If the motives really are personal, that they feel their personal success cannot come with success of the organization, radical changes are needed. Otherwise, the best recourse is to get them to buy into the organizational goals. This is most easily achieved if every stakeholder is represented in the definition and dissemination of the core mission and goals, and subsequently kept informed, updated, and represented.

Process Clash

A process clash is the friction that can arise when advocates of different processes must work together without a proven hybrid process being defined. The dysfunction appears when organizations have two or more well-intended but noncomplementary processes; a great deal of discomfort can be created for those involved. Process clash can arise when functional groups or companies (with different processes) merge, or when management decides to suddenly introduce a new process to replace an old one. Symptoms of this antipattern include poor communications—even hostility, high turnover, and low productivity.

The solution to a process clash involves developing a hybridized approach—one that resolves the differences at the processes' interfaces. Retraining and cross-training could also be used. For example, by training all engineering groups in requirements engineering principles and practices, mutual understanding can be achieved. Another solution is to change to a third process that resolves the conflict.

Management Antipatterns

Metric Abuse

The first management antipattern that might arise in requirements engineering is metric abuse, that is, the misuse of metrics either through incompetence or with deliberate malice (Dekkers and McQuaid 2002).

At the core of many process improvement efforts is the introduction of a measurement program. In fact, sometimes the measurement program is the process improvement. That is to say, some people misunderstand the role measurement plays in management and misconstrue its mere presence as an improvement. This is not a correct assumption. When the data used in the metric are incorrect or the metric is measuring the wrong thing, the decisions made based upon them are likely the wrong ones and will do more harm than good.

Of course, the significant problems that can arise from metric abuse depend on the root of the problem: incompetence or malice. Incompetent metrics abuse arises from failing to understand the difference between causality and correlation; misinterpreting indirect measures; underestimating the effect of a measurement program. Here's an example of the origin of such a problem. Suppose a fire control system for a factory is required to dispense fire retardant in the event of a fire. Fire can be detected in a number of ways—based on temperature, the presence of smoke, the absence of oxygen on the presence of gases from combustion, and so on. So, which of these should be measured to determine if there is a fire? Selecting the wrong metric can lead to a case of metrics abuse.

Malicious metrics abuse derives from selecting metrics that support or decry a particular position based upon a personal agenda. For example, suppose a manager institutes a policy that tracks the number of requirements written per engineer per day and builds a compensation algorithm around this metric. Such an approach is simplistic and does not take into account the varying difficulties in eliciting, analyzing, agreeing, and writing different kinds of requirements. In fact, the policy may have been created entirely to single out an individual who may be working meticulously, but too slowly for the manager's preference.

The solution or refactoring for metrics abuse is to stop the offending measurements. Measuring nothing is better than measuring the wrong thing. When data are available, people use them in decision-making, regardless of their accuracy.

Once the decks have been cleared Dekkers and McQuaid suggest a number of steps necessary for the introduction of a meaningful measurement program (Dekkers and McQuaid 2002):

1. **Define Measurement Objectives and Plans**: perhaps by applying the goal-question-metric (GQM) paradigm.
2. *Make Measurement Part of the Process*: don't treat it like another project that might get its budget cut or that one day you hope to complete.
3. **Gain a Thorough Understanding of Measurement**: be sure you understand direct and indirect metrics, causality vs. correlation, and, most importantly, that metrics must be interpreted and acted upon.
4. **Focus on Cultural Issues**: a measurement program will affect the organization's culture; expect it and plan for it.
5. **Create a Safe Environment to Collect and Report True Data**: remember that without a good rationale people will be suspicious of new metrics, fearful of a time-and-motion study in sheep's clothing.
6. **Cultivate a Predisposition to Change**: the metrics will reveal deficiencies and inefficiencies so be ready to make improvements.
7. **Develop a Complementary Suite of Measures**: responding to an individual metric in isolation can have negative side effects. A suite of metrics lowers this risk.

If you believe that you are being metric mismanaged, then you can try to instigate the above process by questioning management about why the metrics are being collected, how they are being used, and whether there is any justification for such use. You can also offer to provide a corrective understanding of the metrics with opportunities of alternate metrics and appropriate use or more appropriate uses of the existing metrics.

Mushroom Management

Mushroom management is a situation in which management fails to communicate effectively with staff. Essentially, information is deliberately withheld in order to keep everyone "fat, dumb, and happy." The name is derived from the fact that mushrooms thrive in darkness and dim light but will die in the sunshine. As the old saying goes "keep them in the dark, feed them dung, watch them grow … and then cut off their heads when you are done with them."

The dysfunction occurs when members of the team don't really understand the big picture; the effects can be significant, particularly with respect to requirements engineering when stakeholders get left out. It is somewhat insulting to assume that someone working on the front lines doesn't have a need to understand the bigger picture. Moreover, those who are working directly with customers, for example, might have excellent ideas that may have a sweeping impact on the company. So, mushroom management can lead to low employee morale, turnover, missed opportunities, and general failure.

Those eager to perpetuate mushroom management will find excuses for not revealing information, strategy, and data. To refactor this situation some simple strategies to employ include:

- Take ownership of problems that allow you to demand more transparency.
- Seek out information on your own. It's out there. You just have to work harder to find it and you may have to put together the pieces. Between you and the other mushrooms, you might be able to see most of the larger picture.
- Advocate for conversion to a culture of open-book management. With all refactoring, courage and patience are needed to affect change.

Other Paradigms for Requirements Management

Requirements Management and Improvisational Comedy

Improvisational comedy can provide some techniques for collaboration (and after all requirements engineering is the ultimate in collaboration) and for dealing with adversity in requirements management. Anyone who has ever enjoyed improvisational comedy (e.g., the television show *Whose Line Is It Anyway?*) has seen the exquisite interplay of persons with very different points of view and the resolution of those differences. The first author has studied improvisational comedy and has observed a number of lessons that can be taken away from that art.

- Listening skills are really important—both to hear what customers and other stakeholders are saying and to play off your partner(s) in the requirements engineering effort.
- When there is disagreement or partial agreement the best response is "yes, and ..." rather than "yes, but ..." That is, build on, rather than tear down ideas.
- Things will go wrong—both improvisational comedy and requirements engineering are about adapting.
- You should have fun in the face of adversity.
- Finally, you should learn to react by controlling only that which is controllable (usually, it is only your own reaction to certain events, not the events themselves).

You can actually practice some techniques from improvisational comedy to help you develop your listening skills, emotional intelligence, and ability to think on your feet, which in turn will improve your practice as a requirements engineer and as a team player.

For example, consider one improvisational skill-building game called "Zip, Zap, Zup (or Zot)."[2] Here is how it works. Organize four or more people (the more, the better) to stand in a circle facing inside. One person starts off by saying either zip, zap, or zup. If that person looks at you, you look at someone else in the circle and reply in turn—zip, zap, or zup. Participants are allowed no other verbal communication than the three words. The game continues until all participants begin to anticipate which of the three responses is going to be given. The game is a lot harder to play than it seems, and the ability to anticipate responses can take several minutes (if it is attained at all). The point of this game/exercise is that it forces participants to "hear" emotions and pick up on other nonverbal cues.

In another game, Dr. Know-it-all, a group of three or more participants answers questions together, with each participant providing just one word of the answer at a time. So, in a requirements engineering exercise, we would gather a collection of participants from stakeholder group A and ask them the following question. "Please complete the following sentence: the system should provide the following …" then participants provide the answer one word each, in turn. This is a very difficult experience, and it is not intended as a requirements elicitation technique—it is a thought building and team-building exercise, and it can help to inform the requirements engineer and the participants about the challenges ahead.

One final exercise involves the requirements engineer answering questions from two other people at the same time. This experience helps the requirements engineer to think on their feet and also simulates what they will often experience when interacting with customers where they may need to respond simultaneously to questions from two different stakeholders/customers.

While this author prefers comedy as the appropriate genre for requirements elicitation, others (e.g., Mahaux and Maiden 2008) have studied improvisational theater as a paradigm for requirements discovery. In any case, it seems clear that our brains tend to suppress our best ideas and improvisation helps you to think spontaneously and generously. So, try these exercises for fun and to develop these important skills.

Requirements Management as Scriptwriting

In many ways, the writing of screenplays for movies (scripts) has some similarities to requirements engineering. For example, Norden (2007) describes how requirements engineers can learn how to resolve different viewpoints by observing the screenwriting process. There are other similarities between the making of movies and software. These include:

■ Movies are announced well in advance while building certain expectations, which may not be delivered upon as the movie evolves (e.g., changes in actors, screenwriters, directors, plots, and release date). Software and systems are often announced in advance of their actual release and with subsequent changes in announced functionality and release date.
■ Egos are often a significant factor in movies, software, and systems.
■ There are often too many hands involved in making movies, software, and systems.
■ Sometimes the needs of movies exceed the technology (and new technology has to be developed as a result). The same is true for software and systems.
■ Movies often exceed budget and delivery expectations. Need we say more about software?
■ Movies are filmed out of order, requirements are given, and software and some systems may be built this way, too.
■ Movies are shot out of sequence and then assembled to make sense later.
    Software is almost always developed this way, too. Some systems are, too.
■ A great deal of work ends up getting thrown away—in movies, it ends up as film left on the cutting room floor; in software, as throwaway prototypes and one-time-use tools. While systems components are not usually built to be thrown away, many test fixtures are.

What can we learn from big picture production? In a short vignette within Norden's article, Jones provides the following tips for requirements engineers from screenwriting:

- Preparation is everything. Don't leap straight into the detail, but do your homework first.
- Expect to work through many drafts. Remember that different versions might each need to support planning and enable collaboration between stakeholders.
- Think in detail about the future system users—their backgrounds, attitudes, habits, likes, and dislikes.
- Don't just think about the users' physical actions. Remember what they might think (their cognitive actions) and feel.
- Remember your requirements story's arc. Requirements should tell the complete story of achieving user goals.
- Hold your audience—remember that someone must read your requirements document. Perhaps there is a place for tension and dramatic irony (Norden 2007).

VIGNETTE 10.1 Requirements Management: Frequently Asked Questions and Experience

We are often asked questions about requirements management by students and our real-world clients who hire us to evaluate their requirements engineering practices. Often, we found that there are some good requirement engineering practices in place, but the real problems are related to deeper organizational problems. Here are some observations from those experiences.

1. What are the clear boundaries between requirements management and organizational management?

   In theory, it's nice to think you can separate the requirements management process from overall organizational behavior and flaws, but you cannot. In practice, there is a blurry line between the two. The analogy is often used that putting clean water through dirty pipes is just as bad as dirty water. In the same way, good requirements engineering practices can be thwarted by bad management practices.

2. What is the extent of the authority requirements engineer usually have?

   This depends entirely on the company in question. But usually, the requirements engineer does not have a great deal of authority. After all, they are supposed to respond to the customer and guide them. They are not like a doctor who can prescribe (or refuse to prescribe) medication or a procedure. They are more like a decorator—they can make strong suggestions, but in the end the customer/sponsor makes the decisions.

3. Is the goal of requirements management not solving the management/environment problems but, instead, working around them to ensure the successful delivery of requirements?

The requirements engineer is not there to fix the organization—but they need to understand the organization and its problems. Several times we have been hired by companies to "fix" their requirements engineering. But often it turns out it is not a requirements engineering problem, but larger organizational problems. This is why we study certain antipatterns— to understand the organizational and managerial climate.

We offer the following general recommendations to help the requirements engineer deal with managerial and organizational/environmental dysfunction:

1. Develop the official requirements engineering process document collaboratively with internal stakeholders to increase buy-in.

2. Designate one or more individuals with requirements engineering responsibility for all projects (e.g., a "director of requirements engineering").

3. Adopt an appropriate requirements management tool consistently throughout the organization.

4. Consider conducting a dry run requirements engineering process for a fictitious customer in order to identify requirements engineering versus managerial and organizational problems.
5. Provide continuing education about requirements engineering throughout the organization.

Ultimately any organizational dysfunction must be corrected in order for the requirements engineering management processes and the overall requirements engineering processes to be effective.

Standards for Requirements Management[3]

There are many different international standards that provide reference process models, management philosophies, or quality standards that can be used to inform the requirements engineering effort. These standards are not mutually exclusive in that they may be used in a complementary manner. Figure 10.3 summarizes the relevant software engineering-related standards and how they are related.

In Figure 10.3, ISO standards are shaded and relevant standards are grouped into the following categories:

1. Improvement standards (metamodels) such as ISO 15504 (capability maturity determination and process assessment standard) and ISO 9001 (quality management).
2. Quality management and improvement frameworks, such as CMMI (capability maturity model integration) with SCAMPI (standard CMMI appraisal method for process improvement); SPICE (software process improvement and capability determination); ISO 15504; COBIT (control objectives for information and related technology) for IT organizations; Six Sigma; TL 9000, which details requirements and measurements to telecommunication quality management systems; ISO/TS (technical standard) 16949, which provides requirements to quality management in automotive systems; and AS (aerospace standard) 9100, which enhances ISO 9001 with specific requirements for the aerospace industry.

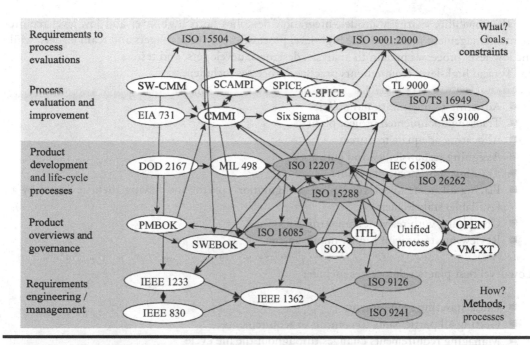

Figure 10.3   Standards applicable to requirements engineering. (From Ebert, C., Requirements engineering: Management. In P. Laplante (Ed.), *Encyclopedia of Software Engineering,* **pp. 932–948, Taylor & Francis, 2010. With permission.)**

3. Process and life cycle models include DoD (USA Department of Defense) 2167 MIL (military standard) 498, ISO 15288 (system life cycle), and ISO 12207 (software life cycle).
4. Process implementation and governance regulations and policies include the Sarbanes–Oxley Act (SOX), standards such as ISO 16085 (risk management), de facto standards such as SWEBOK and PMBOK (project management body of knowledge), and various frameworks (Ebert 2010).

We have already covered SWEBOK in Chapter 1, and IEEE Standard 830 was subsumed by IEEE Standard 29148, which was discussed in Chapters 4 and 6. We will explore some of the other standards in the following sections.

Capability Maturity Model Integration

The capability maturity model integration (CMMI) is a systems and software quality model consisting of five levels. The CMMI is not a life cycle model, but rather a system for describing the principles and practices underlying process maturity. CMMI is intended to help software organizations improve the maturity of their processes in terms of an evolutionary path from ad hoc, chaotic processes to mature, disciplined processes.

Developed by the Software Engineering Institute at Carnegie Mellon University, the CMMI is organized into five maturity levels. Predictability, effectiveness, and control of an organization's software processes are believed to improve as the organization moves up these five levels. While not truly rigorous, there is some empirical evidence that supports this position.

The capability maturity model integration describes both high-level and low-level requirements management processes. The high-level processes apply to managers and team leaders, while the low-level processes pertain to analysts, designers, developers, and testers.

Typical high-level requirements practices/processes include:

- Adhering to organizational policies
- Tracking documented project plans
- Allocating adequate resources
- Assigning responsibility and authority
- Training appropriate personnel
- Placing all items under version or configuration control and having them reviewed by all (available) stakeholders
- Complying with relevant standards
- Reviewing status with higher management

Low-level best practices/processes include:

- Understanding requirements
- Getting all participants to commit to requirements
- Managing requirements changes throughout the life cycle
- Managing requirements traceability (forward and backward)
- Identifying and correcting inconsistencies between project plans and requirements

These practices are consistent with those that have been discussed throughout the text.

Achieving level 3 and higher for the CMMI requires that these best practices be documented and followed within an organization.

The CMM family of quality models defines requirements management as:

> establishing and maintaining an agreement with the customer on the requirements for the software project. The agreement forms the basis for estimating, planning, performing, and tracking the software project's activities throughout the software life cycle. (Paulk et al. 1993)

For maximum flexibility, CMM does not prescribe which tools or techniques to use to achieve these goals.

ISO 9001

ISO Standard 9000, Quality Management, is a generic, worldwide standard for quality improvement. The standard, which collectively is described in four volumes, ISO 9000, ISO 9001 (Quality Management Systems—Requirements), ISO 9004 (Managing for the Sustained Success of an Organization), and ISO 9011 (Guidelines for Auditing Management Systems) is designed to be applied in a wide variety of environments. ISO 9000 applies to enterprises according to the scope of their activities. These ISO standards are process-oriented, common sense practices that help companies create a quality environment.

ISO 9001, Quality Management Systems—Requirements, provides some guidance on requirements management within the discussion for product realization. The standard calls for the following to be determined:

1. Quality objectives and requirements for the product
2. The need to establish processes, documents, and provide resources specific to the product
3. Required verification, validation, monitoring, inspection, and test activities specific to the product and the criteria for product acceptance
4. Records needed to provide evidence that the realization processes and resulting product meet requirements (ISO 9000 2015)

ISO 9001 does not provide specific process guidance for requirements management. The aforementioned recommendations, however, are helpful as a checklist, especially when used in conjunction with appropriate requirements metrics (some of these are discussed in Chapter 9). In order to achieve certification under the ISO standard, significant documentation is required.

ISO/IEEE 12207

ISO 12207: Standard for Information Technology—Software Life Cycle Processes, describes five primary processes: acquisition, supply, development, maintenance, and operation. ISO 12207 divides the five processes into activities, and the activities into tasks, while placing requirements upon their execution. It also specifies eight supporting processes—documentation, configuration management, quality assurance, verification, validation, joint review, audit, and problem resolution—as well as four organizational processes—management, infrastructure, improvement, and training.

The ISO standard intends for organizations to tailor these processes to fit the scope of their particular projects by deleting all inapplicable activities, and it defines ISO 12207 compliance as being in terms of tailored performance of those processes, activities, and tasks.

ISO 12207 provides a structure of processes using mutually accepted terminology, rather than dictating a particular life cycle model or software development method. Since it is a relatively high-level document, 12207 does not specify the details of how to perform the activities and tasks comprising the processes. Nor does it prescribe the name, format, or content of documentation. Therefore, organizations seeking to apply 12207 need to use additional standards or procedures that specify those details.

The IEEE recognizes this standard with the equivalent numbering: IEEE/ EIA 12207.0–1996, IEEE/EIA Standard Industry Implementation of International

Standard ISO/IEC12207:1995, and (ISO/IEC 12207) Standard for Information Technology—Software Life Cycle Processes. ISO/IEC 15504 Information Technology—Process Assessment, also termed SPICE, is a derivative of 12207.

Six Sigma

Developed by Motorola, Six Sigma is a management philosophy based on removing process variation. Six Sigma focuses on the control of a process to ensure that outputs are within six standard deviations (six sigma) from the mean of the specified goals. Six Sigma is implemented using define,

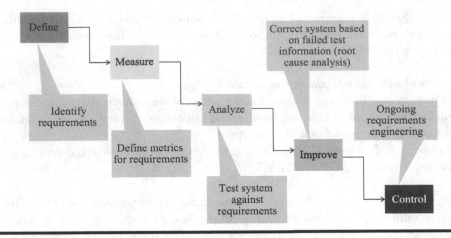

Figure 10.4   Six sigma process for requirements engineering.

measure, analyze, improve, and control DMAIC. An overview of a Six Sigma model for requirements management is depicted in Figure 10.4.

In Six Sigma, "define" means to describe the process to be improved, usually through using
some sort of business process model. In the case of requirements engineering, we can think of
this as identifying the required features that need improvement. "Measure" means to identify
and capture relevant metrics for each aspect of the process model, in our case, the identified
requirements. The goal-question-metric paradigm is helpful in this regard. "Analyze" means to
study the captured metrics for opportunities for improvement. "Improve" means to change some
aspect of the process so that beneficial changes are seen in the associated metrics, usually by
attacking the aspect that will have the highest payback. Finally, "control" means to use ongoing
monitoring of the metrics to continuously revisit the model, observe the metrics, and refine the
process as needed.

Six Sigma is more process yield-based than the CMMI so the CMMI process areas can be used
to support DMAIC (e.g., by encouraging measurement). And, while CMMI identifies activities,
Six Sigma helps optimize those activities.

Six Sigma can also provide specific tools for implementing CMMI practices (e.g., estimation
and risk management).

Some organizations use Six Sigma as part of their software quality practice. The issue here,
however, is in finding an appropriate business process model for the software production process
that does not devolve into a simple, and highly artificial, waterfall process. If appropriate metrics
for requirements can be determined (e.g., those discussed in Chapter 9), then Six Sigma can be
used to improve the RE process.

VIGNETTE 10.2   FBI Virtual Case File

The U.S. Federal Bureau of Investigation's (FBI) disastrous virtual case file
(VCF) system was a spectacular failure of requirements management, among
other things. This discussion is largely based on Goldstein's report (2005),
Alfonsi's analysis (2005), and Israel's project retrospective (2012).

Begun in 2000, VCF was supposed to automate the FBI's paper-based work environment and allow agents and intelligence analysts to share investigative information. The system would also replace the obsolete automated case support (ACS), which was based on 1970s software technology. After 12 years and $451 million, the project is not complete and may never be completed. The VCF may eventually become fully operational, but by all accounts, the project has been a financial and political failure. What went wrong with VCF? Quite a bit, some of it political, some of it due to the complexities of a mammoth bureaucracy, but some of the failures can be attributed to bad requirements engineering. Goldstein noted that the project was poorly defined from the beginning and that it suffered from slowly evolving requirements (an 800-page document to start). Even though JAD sessions were used for requirements discovery (several 2-week sessions over 6 months), these were used late in the project and only after a change of vendors. Apparently, there was no discipline to stop requirements from snowballing. An ever-changing set of sponsors (changes in leadership at the FBI as well as turnover in key congressional supporters) caused requirements drift. Moreover, apparently, there was too much design detail in the SRS, literally describing the color and position of buttons and text. And many requirements failed the IEEE 29148 rules—often they were not "clear, precise, and complete." Israel (2012) noted that the project should have incorporated extensive prototyping, but pressures to deliver working functionality made prototyping impossible.

Finally, there were also cultural issues at the FBI—sharing ideas and challenging decisions are not the norm. All of these factors, in any environment, would jeopardize the effectiveness of any requirements engineering effort.

A subsequent review of the project noted that there was no single person or group to blame—this was a failure on every level. But an independent fact-finding committee singled out FBI leadership and the prime contractor as being at fault for the reasons previously mentioned (Goldstein 2005).

The project was a complex one and according to former FBI CIO Israel was beyond the capabilities of the organization to manage (Israel 2012). The project went through multiple changes of CIOs and project leaders. Originally begun as an internal project, the FBI realized that it did not have the necessary expertise to complete it and, in 2007, handed the project to Lockheed Martin. But by 2010, after many setbacks, the FBI reclaimed the project and at this writing is expecting to complete the project using an agile approach (Israel 2012).

Aside from the political ramifications, and the usual problems when designing large complex systems, what can be learned from the failure of VCF? From a requirements engineering standpoint, there are three very specific lessons. First, don't rush to get a system out before elicitation is completed. Obviously, there are pressures to deliver, particularly for such an expensive and high-profile system, but these pressures need to be resisted by strong leadership. Second, the system needs to be defined completely and correctly from the beginning. While this was highly problematic for the VCF due to the various changes in FBI leadership during the course of its evolution, a well-disciplined process could have helped to provide pushback against late requirements changes. Finally, for very large systems, it is helpful to have an architecture in mind. We realize that it has been said that the architecture should not be explicitly incorporated in the requirements specification, but having an architectural target can help inform the elicitation process, even if the imagined architecture is not the final one.

Exercises

10.1 Should a request to add or change features be anonymous? Why or why not?

10.2 For the following specification, identify any requirements which are likely to become obsolete in a period of 2 years:

10.2.1 The smart home system in Appendix A

10.2.2 The wet well control system in Appendix B

10.3 How could metrics abuse begin to develop in an organization?

10.4 What are some sources of process clash in organizations?

10.5 Give an example of a process clash, from your own experience.

10.6 Give an example of metrics abuse, from your own experience.

10.7 What are some sources of divergent goals in organizations?

10.8 Give an example of divergent goals, from your own experience.

10.9 How can CMMI be used to identify and reconcile process clash?

10.10 With a group of friends, classmates, or teammates use Wideband Delphi to select a restaurant to share a meal from the following choices:

10.10.1 Olive Garden

10.10.2 Carrabas

10.10.3 Red Lobster

10.10.4 Golden Corral

10.10.5 PF Chang's

10.10.6 Ruth's Chris Steakhouse

10.11 Select a set of five requirements management tools. From information published on their websites, create a list of features provided that support the management aspects of requirements engineering.

10.12 Investigate and use the analytical hierarchy process (AHP) to resolve the decision-making problem in 7.

Notes

1 Some of this discussion is excerpted from Neill et al. (2012) with permission.

2 The authors admit that this is also a traditional drinking game.

3 This section is excerpted from Laplante (2006) and Ebert (2010) with permission.

References

Alfonsi, B. (2005). FBI's virtual case file living in limbo. *Security & Privacy*, 3(2): 26–31.

Andriole, S. (1998). The politics of requirements management. *IEEE Software*, 15: 82–84.

Berander, P., Damm, L. O., Eriksson, J., Gorschek, T., Henningsson, K., Jönsson, P., Kågström, S., Milicic, D., Mårtensson, F., Rönkkö, K., Tomaszewski, P. (2005). Software quality attributes and trade-offs. *Blekinge Institute of Technology*, 97(98): 19.

Bhat, J. M., Gupta, M., & Murthy, S. N. (2006). Overcoming requirements engineering challenges: Lessons from offshore outsourcing. *IEEE Software*, 23: 38–44.

Brown, W. J., Malveau, R. C., McCormick, H. W., and Mowbray, T. J. (1998). *AntiPatterns: Refactoring Software, Architectures, and Projects in Crisis*. Wiley, New York.

Dekkers, C. A., & McQuaid, P. A. (2002). The dangers of using software metrics to (mis) manage. *IT Professional*, 4(2): 24–30.

Ebert, C. (2010). Requirements engineering: Management. In P. Laplante (Ed.), *Encyclopedia of Software Engineering* (pp. 932–948). Taylor & Francis, Boca Raton, FL.

Goldstein, H. (2005). Who killed the virtual case file? *Spectrum*, 42(9): 24–35.

Hull E., Jackson K., Dick J. (2002). Management Aspects of Requirements Engineering. In: Requirements Engineering. Practitioner Series. Springer, London. https://doi.org/10.1007/978-1-4471-3730-6_8.

ISO 9000 Quality Management Systems. (2015). http://www.iso.org (accessed January 2017).

Israel, J. W. (2012). Why the FBI can't build a case management system. *Computer*, 45(6): 73–80.

Kassab, M. (2013, May). An integrated approach of AHP and NFRs framework. In *IEEE 7th International Conference on Research Challenges in Information Science (RCIS)* (pp. 1–8). IEEE, Paris, France.

Kassab, M. (2014, June). Early effort estimation for quality requirements by AHP. In *14th International Conference on Computational Science and Its Applications* (pp. 106–118). Springer International Publishing, Guimaraes, Portugal.

Kassab, M., & Kilicay-Ergin, N. (2015). Applying analytical hierarchy process to system quality requirements prioritization. *Innovations in Systems and Software Engineering*, 11(4): 303–312.

Laplante, P. A. (2006). *What Every Engineer Needs to Know about Software Engineering*. CRC/Taylor & Francis, Boca Raton, Florida, United States.

Lawrence, B. (1996). Unresolved ambiguity. *American Programmer*, 9(5): 17–22.

Mahaux, M., & Maiden, N. (2008). Theater improvisers know the requirements game. *Software*, 25: 68–69.

Neill, C. J., Laplante, P. A., & DeFranco, J. (2012). *Antipatterns: Identification, Refactoring, and Management*, 2nd edition. CRC Press, Boca Raton, FL.

Norden, B. (2007). Screenwriting for requirements engineers. *Software*, 24: 26–27.

Paulk, M., Weber, C. V., Garcia, S. M., Chrissis, M.-B. C., & Bush, M. (1993). *CMM Practices Manual*. CMU/SEI-93-TR-25. https://www.sei.cmu.edu/reports/93tr024.pdf (accessed February 2017).

Reinhart G., Meis J. (2012). Requirements Management as a Success Factor for Simultaneous Engineering. In: ElMaraghy H. (eds). Enabling Manufacturing Competitiveness and Economic Sustainability. Springer, Berlin, Heidelberg. https://doi.org/10.1007/978-3-642-23860-4_36

Saaty T.L. (1988). What is the Analytic Hierarchy Process?. In: Mitra G., Greenberg H.J., Lootsma F.A., Rijkaert M.J., Zimmermann H.J. (eds). Mathematical Models for Decision Support. NATO ASI Series (Series F: Computer and Systems Sciences), vol 48. Springer, Berlin, Heidelberg. https://doi.org/10.1007/978-3-642-83555-1_5

Schmid K. (2014). Challenges and Solutions in Global Requirements Engineering – A Literature Survey. In: Winkler D., Biffl S., Bergsmann J. (eds) Software Quality. Model-Based Approaches for Advanced Software and Systems Engineering. SWQD 2014. Lecture Notes in Business Information Processing, vol 166. Springer, Cham. https://doi.org/10.1007/978-3-319-03602-1_6

Sinha, V., & Sengupta, B. (2006). Enabling collaboration in distributed requirements management. *Software*, 23: 52–61.

Wiegers, Karl E. (2003). *Software Requirements*, 2nd edition. Microsoft Press, Redmond, Washington, United States.

Chapter 11

Value Engineering of Requirements

What, Why, When, and How of Value Engineering

Up until now, we have had a very little discussion of the costs of certain features that customers may wish to include in the system requirements. Part of the reason is that it made sense to separate that very complex issue from all of the other problems surrounding requirements elicitation, analysis, representation, validation, verification, and so on. Another reason for avoiding the issue is that it is a tricky one.

In the early 1980s, Barry Boehm proposed a model for software projects called the funnel curve that helped explain why software effort and cost prediction were so hard (Boehm 1981). The model had been around for some time, but Boehm first proposed its applicability to software (and later systems) projects. Steve McConnell refined the idea, calling the curve the "cone of uncertainty" (McConnell 1997). The model is very much relevant today, and it applies both to software-only systems and complex systems that may have little or no software in them. More significantly, the model helps explain why early requirements understanding is so important, which is why we recast the model as the requirements cone of uncertainty (Figure 11.1). It's easiest to understand the model by reading it from right to left.

Because requirements change, it is not until the project is delivered that the final list of requirements is known with certainty. Up until that time, the requirements can change. In fact, early in the life cycle of the project, at requirements definition time, in particular, the requirements are highly volatile, that is, uncertain. We show the estimation variability of the requirements at the project's initial conception on a scale of 4 to 1, that is, at this point in time, our estimates of time to complete, and hence, the cost of requirements could be as much as four times too conservative to four times too optimistic. Of course, this factor of four is entirely arbitrary.

The purpose of the cone model is to emphasize that it is only over time and experience that we gain precision in our ability to estimate the cost of the true requirements of the project, and so the variability of our estimates converge to their true values over time.

DOI: 10.1201/9781003129509-11

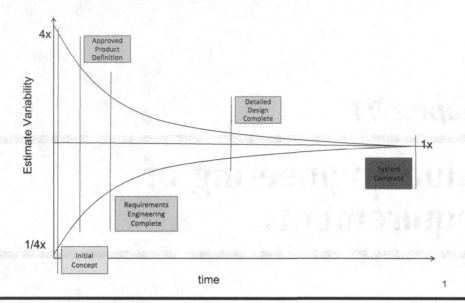

Figure 11.1   The requirements cone of uncertainty.

What Is Value Engineering?

There is a fundamental relationship between time, cost, and functionality. Project managers sometimes refer to this triad as the three legs of the project management stool. That is, you can't tip one without affecting the others. For example, you have already finished writing and agreeing to the specification and have provided an estimated cost to complete in response to some RFP (request for proposal). If the customer asks you to incorporate more features into the system, shouldn't the price and estimated time to complete increase? If not, then you were likely padding the original proposal and the customer will not like that. Conversely, if you propose to build an agreed-upon system for a price, say $1 million, and the customer balks, should you then respond by giving them a lower price, say $800,000? Of course not, because if you lower the price without reducing the feature set or increasing the time to complete, the customer will think that you gave them an inflated price originally. The correct response, in this case, would be to agree to build the system for $800,000, but with fewer features (or taking a much longer time).

When dealing with affordability during requirements elicitation, analysis, agreement, and negotiation, both the vendor and customer have to balance certain factors. The requirements of the system determine the functional and nonfunctional features or performance of the system. A schedule has to be determined and agreed upon. The vendor and customer have to identify various risk factors in both the actual system features and the processes to develop the system. The requirements, schedule, and risk factors determine the cost to produce, and thus, the price to the customer. The vendor is entitled to a fair profit, but at the same time, the customers' expectations need to be managed.

To properly manage customer expectations, deal with trade-offs between functionality, time, and cost, it is, therefore, necessary to have some way of estimating costs for system features. Making such estimations is not easy to do accurately at the early stages of a project, such as during the requirements elicitation activities. However, it is necessary to make such cost and effort estimations. The activities related to managing expectations and estimating and managing costs are called value or affordability engineering (Reeves et al. 2010).

When Does Value Engineering Occur?

Value engineering occurs throughout the system life cycle and is typically considered a project management activity. But for the requirements engineer, value engineering has to take place to help manage customer expectations concerning the final costs of the delivered system and the feasibility or infeasibility of delivering certain features.

During the early stages of elicitation, it is probably worthwhile to entertain some value engineering activities—but not too much. For example, you may wish to temper a customer's expectations about the possibility of delivering an elaborate feature if it will break the budget. On the other hand, if you are too harsh in providing cost assessments early, you can curtail the customer's creativity and frustrate them. Therefore, be cautious when conducting value engineering when working with user-level requirements.

The best time to conduct the cost analysis is at the time when the systems-level requirements are being put together. It is at this time that better cost estimates are available, and this is a time when trade-off decisions can be discussed more successfully with the customer, using some of the negotiating principles previously mentioned.

A simple model, then a reasonable place to start with an affordability engineering process for requirements, is shown in Figure 11.2.

First, working with the customer and other stakeholders, we elicit, analyze, and create requirements using standard approaches. Next, we use some estimating techniques, such as some constructive cost modeling tools to calculate the effort needed to produce those requirements. The constructive cost modeling systems tool, COSYSMO, can be used to estimate the effort.

Expert opinion, analogy, Wideband Delphi, or a combination of these can be used. Next, we generate a cost profile for these requirements based on the effort estimates just calculated. If the cost of these requirements is too high, then we need to revise the requirements set based on some trade-off analysis, for example, by using a risk matrix.

We recalculate the effort and cost for the revised requirements set until the vendor and customer are satisfied.

Finally, once the development effort commences, we can use standard project monitoring techniques, such as earned value analysis (EVA), to ensure that as requirements are built out, they meet their target cost structure. If we perceive problems with schedule or cost, the vendor and customer can work together to modify the requirements to maintain the schedule and budget.

This ideal model is attractive, but there are challenges in realizing this model.

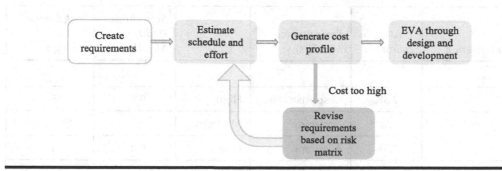

Figure 11.2　Simple affordability engineering for requirements.

Challenges to Simple Cost vs. Risk Analysis

A typical approach to balance requirements cost, risk, schedule, and so forth is to put these factors in a matrix-like one shown in Table 11.1 and employ some kind of optimization technique, even a heuristic one, to minimize cost, risk, or schedule.

But there are several problems with this somewhat simplistic representation. For example, computing true costs of requirements can be quite difficult early on, as the cone of uncertainty suggests. Further, the matrix does not take into account any dependencies between requirements. These could be accounted for by adding an additional column in the matrices to list dependent requirements, but then the formulation of the problem is different. Another problem arises in the quantification of risk and identification. Different customers and stakeholders will rate risk differently. The risk matrix approach also does not account for the consistency of rating risk factors between projects. And while you can use consensus-building techniques to deal with all of these problems, this is a time-consuming and, hence, expensive approach.

There are additional problems when trying to estimate the cost of requirements early in the software or system's life cycle.

It has already been noted that different customers and stakeholders will rate requirements risk differently. But they will also decompose functionality differently. For example, one customer might view a requirement for a certain style user screen to be a minor task, while another might view the implementation of this feature as a major undertaking.

Customers and other stakeholders have different levels of commitment to a project and hence will invest more or less thought into their inputs. An employee who must live with any requirements decisions made will certainly behave differently than an outside consultant who does not need to cope with the decisions made during requirements analysis.

Those participating in the requirements elicitation process will have different personal involvement or "skin in the game" with respect to the eventual project outcome, and this fact will influence the quality of their participation in requirements elicitation and agreement.

Be aware, too, of the effects of differing personal agendas of the participants. For example, a customer representative who is leaving the company soon will care less about the correctness of a feature description than the prospective manager of the project.

Finally, cost-based requirements engineering can encourage individuals to set up some "plausible deniability," that is, the ability to give an excuse if the project does not go as planned.

Table 11.1 Risk Matrix Approach to Affordability Optimization

Requirement #	Cost Estimate	Performance Metric(s)	Rank	Schedule (Staff Months)	Risk Factor 1=low, 5=high
...					
3.1.2	2.5K	Response time	High	0.1	5
3.1.3	5K	Accuracy	Medium	0.3	3
3.1.4	7.5K	Accuracy	Low	0.4	1
...					

You can identify these individuals, usually, because they are unwilling to attach their name to certain documentation, or they use hedging words during requirements negotiation.

In the remainder of the chapter, we'll look at some simple approaches to assist in the value engineering activity for requirements engineers. These sections provide only an introduction to these very complex activities that many project management, cost accounting, and system engineering. Really, a set of experts with these skills is needed to provide the most accurate information for decision-making.

Estimating Software Effort

Software size is the base measure that is used to calculate project effort, duration, and cost. One way to respond to the need to deal comprehensively and objectively with the software project effort is in terms of their corresponding size.

Software size estimation is the process of predicting the size of a software product. Accurate size estimation is critical to effectively managing the software development process. The project planner must understand the scope of the software to be built and generate an estimate of its size before a project estimate can be made (Pressman 2015).

Software size can be described in terms of length, complexity, and functionality. These three aspects of size are discussed next.

The Aspect of Software Size

Internal product attributes describe a software product in a way that is dependent only on the product itself. One of the most useful attributes is the size of a software product, which can be measured statically without executing the system. In the context of project planning, size refers to the quantifiable outcome of the software project.

Since other physical objects are easily measurable, it might be assumed that measuring the size of software products should be straightforward. In practice, however, size measurement can be difficult. Simple measures of size are often rejected because they do not provide adequate information. Those who reject a measure because it does not provide enough information may be expecting too much of a simple measure.

For example, if a human size is measured as a single attribute such as weight, then we can determine the number of people who can safely ride in an elevator at one time. However, we cannot determine whether passengers will bump their heads on the elevator door. If a human size is measured in terms of two attributes such as weight and height, then we can determine both the number of people who can safely ride in an elevator at one time and whether passengers will bump their heads on the elevator door.

Similarly, if software size is measured in terms of the number of LOC, the fact that it is not useful in measuring quality does not negate its value. Rather this might indicate a requirement for more information.

It is therefore often useful to define an external attribute such as size in terms of more than one internal attribute. Applying measures to different goals does not invalidate them for their original purpose. Ideally, we want to define a set of views for software size. Each view should capture a key aspect of software size. Pressman (2015) suggests that software size can be described with three views: length, complexity, and functionality.

Software Length

Length is the physical size of the product. There are three major development products whose size would be useful to know: the specification, the design, and the code. The length of the specification can indicate how long the design is likely to be, which in turn is a predictor of code length.

The most commonly used measure of source code program length is the number of LOC.[1] Many different approaches to counting LOC have been proposed. The Software Engineering Institute, for example, has developed a set of guidelines to help in deciding how to measure LOC (Park 1992). This recommendation is flexible in that it allows you to tailor the definition of LOC for your needs.

Another aspect of software length is the length of specifications and design. Specification and design documents may use text, graphs, or mathematical diagrams and symbols to express information. In measuring code length, an atomic object must be identified to count (LOC, executable statements, source instructions, operators, and operands). Similarly, for specification and design documents, one or more objects are identified and counted.

In the case of dataflow diagrams, objects such as processes (bubble nodes), external entities (box nodes), data stores (line nodes), and data flow (arcs) are counted. In the case of a class diagram, objects such as classes are counted. It is common in the industry to use the number of pages to measure length for documents containing text and graphs.

Software Complexity

Complexity can be interpreted in different ways. In the context of software size, complexity refers to algorithmic complexity and problem complexity. Problem complexity (also called computational complexity) is a branch of the theory of computation in computer science that focuses on classifying problems according to their inherent difficulty. Here, a problem is understood in the narrow sense of a task that is in principle amenable to be solved by a computer. Informally, a problem is regarded as inherently difficult if solving the problem requires a large number of resources, independent of the algorithm used for solving it. The theory formalizes this intuition, by introducing mathematical models of computation and casting computational tasks mathematically as decision problems. The degree of difficulty can be quantified in the number of resources needed to solve these problems, such as time and storage. In particular, the theory seizes the practical limits on what computers can and cannot do.

On the other hand, algorithmic complexity reflects the complexity of the algorithm used to solve the problem. A key distinction between computational complexity theory and analysis of the algorithm is that the latter is devoted to analyzing the number of resources needed by a particular algorithm to solve a concrete problem, whereas the former asks a more general question. Namely, it targets classifying problems that can, or cannot, be solved with appropriately restricted resources. A mathematical notation called big-O notation is used to define an order relation on functions. The big-O form of a function is derived by finding the dominating term $f(n)$. Big-O notation captures the asymptotic behavior of the function. Using this notation, the efficiency of algorithm A is $O(f(n))$, where, for input size n, algorithm A required at most $O(f(n))$ operations in the worst case.

For example, the function $f(n) = 3n^2 + 2n + 26$ is big-O n^2 written as $O(n^2)$. The algorithm will therefore require at most $O(n^2)$ operations.

According to the ISO/IEC 14143-6:2012 (Functional Size Measurement), the methods to measure the length and complexity aspects of the size have the following limitations:

■ These methods cannot always be applied in the early phases of software development life cycles.
■ These methods cannot always be understood by the user of the software.

To overcome the above-mentioned limitations, methods that are not based on length or complexity have been proposed. Most of the methods that are used today to measure the size of the software are based upon the "Functionality" of the software. These methods measure the size of the software by measuring the functionality that it provides to the customer.

Functional Size

Functional size methods (FSMs) have shifted the focus from measuring the technical characteristics of the software toward measuring the functionality of the software that is required by the intended users of the software. It is important to note that functional size is the only standardized way to measure the software size. This method is independent of the development tools and the programming languages. It is also independent of the technical requirements of the software.

The first method, named function points (FPs), which calculates the functionality of the software is designed in 1979 by Albrecht. The Function Point Analysis (FPA) method served as the basis for the first FSM industrial method. Over the years, different variations and varieties of FSM methods have emerged. A preview evolution of FSM methods is presented in Table 11.2. We next provide a brief overview of two of these methods: FPs and the COSMIC.

Estimating Using Function Points

Function points were introduced in the late 1970s as an alternative to product metrics based on source line count. This aspect makes FPs especially useful to the requirements engineer. The basis of function points is that as more powerful programming languages were developed; the number of source lines necessary to perform a given function decreased. Paradoxically, however, the cost/LOC measure indicated a reduction in productivity, as the fixed costs of software products were largely unchanged (Albrecht 1979).

The solution to this effort estimation paradox is to measure the functionality of software via the projected number of interfaces between modules and subsystems in programs or systems. A big advantage of the function point metric is that it can be calculated during the requirements engineering activities.

Function Point Cost Drivers

The following five software characteristics for each module, subsystem, or system represent the function points or cost drivers:

■ Number of inputs to the application (I)
■ Number of outputs (O)
■ Number of user inquiries (Q)

Table 11.2 Concepts, FSM Methods, and Description

Year	Method Name
1979	Albrecht FPA/IFPUG FPA
1982	DeMarco's Bang Metrics
1986	Feature Points
1988	MK II FPA
1990	NESMA FPA
1990	Asset-R
1992	3-D FP
1994	Object Points
1994	FP by Matson, Barret, and Mellichamp
1997	Full Function Points (FFP)
1997	Early FPA (EFPA)
1998	Object-Oriented FP
1999	Predictive OP
1999	COSMIC FFP
2000	Early and Quick COSMIC FFP
2000	Kammelar's Component OP
2001	Object-Oriented Method FP
2004	FiSMA FSM

- Number of files used (F)
- Number of external interfaces (X)

In addition, the FP calculation takes into account weighting factors for each aspect that reflect their relative difficulty in implementation, and the function point metric consists of a linear combination of these factors, as shown in Equation 11.1.

$$FP = w_1 I + w_2 O + w_3 Q + w_4 F + w_5 X \qquad (11.1)$$

where the w_i coefficients vary depending on the type of application system. Then, complexity factor adjustments are applied for different types of application domains. The full set of coefficients and corresponding questions can be found by consulting an appropriate text on software metrics. The International Function Point Users Group maintains a web database of weighting factors and function point values for a variety of application domains.

For the purposes of cost and schedule estimation, and project management, FPs can be mapped to the relative lines of source code—in particular, programming languages. A few examples are given later in Table 11.3.

Estimating Using COSMIC

The FSM method developed by the Common Software Measurement International Consortium (COSMIC) has been adopted as an international standard (ISO 19761) and is referred to as the COSMIC method. This approach has been designed to measure the functional size of management information systems, real-time software, and multilayer systems. Its design conforms to all ISO requirements for FSM methods and was developed to address some of the major weaknesses of earlier methods. For example, FPA dates back almost 30 years to a time when the software was much smaller and much less varied.

COSMIC focuses on the "user view" of functional requirements and is applicable throughout the development life cycle, right from the requirements phase to the implementation and maintenance phases. Before starting to measure using the COSMIC method, it is imperative to carefully define the purpose, the scope, and the measurement viewpoint. This may be considered as the first step of the measurement process. The measurer defines why the measurement is being undertaken, and/or what the result will be, as well as the set of functionalities to be included in a specific FSM exercise. Measurements taken using the COSMIC method with a different purpose and scope and a different measurement viewpoint may therefore give quite a different size.

In the measurement of software functional size using the COSMIC method, the software functional processes and their triggering events must be identified. In COSMIC, the unit of measurement is a data movement, which is a base functional component that moves one or more data attributes belonging to a single data group. Data movements can be of four types: Entry, Exit, Read, or Write. The functional process is an elementary component of a set of user requirements triggered by one or more triggering events either directly or indirectly via an actor. It comprises at least two data movement types: an Entry plus at least either an Exit or a Write. The triggering event is an event occurring outside the boundary of the measured software and initiates one or more functional processes. The subprocesses of each functional process are sequences of events. An Entry moves a data group, which is a set of data attributes, from a user across the boundary into the functional process, while an Exit moves a data group from a functional process across the boundary to the user requiring it. A Write moves a data group lying inside the functional process to persistent storage, and a Read moves a data group from persistent storage to the functional process.

The Relationship between Functional Size and Effort

Software cost and effort estimation play a significant role in the successful completion of any software. Resources are assigned according to the effort required to complete the software. Accurate effort estimation leads to the completion of the software project on the scheduled time. Many models and approaches have been developed in the past 40 years to estimate the effort. Most of the models take software size as a basic input to estimate the effort. We have already discussed that it is better to use functional size instead of the length of code to estimate effort. The effort is usually calculated by using the functional size of the software. There is a strong relationship between functional size and effort. Valid measured functional size has the potential to improve effort estimation and reduce the "cone of uncertainty" effect on the project planning. It is critical to correctly establish a relationship between functional size and effort so that we could be able to estimate effort accurately. There are many project and product factors that affect positively or negatively this relationship. Environmental factors, technical factors, and operating constraints are some of them.

Many significant attempts have been taken to explore the relationship between the size and effort and also to identify the subset of those factors which may affect this relation.

For example, A study carried out by Maxwell and Forselius (2000) in Finnish companies to explore the factors that affect productivity and effort estimation reveal the following main factors:

- Application Programming Language
- Application Type (e.g., a management information system)
- Hardware Platform
- User Interface
- Development Model
- DBMS Architecture
- DB Centralization
- Software Centralization
- DBMS Tools
- Case Cools
- Operating System
- Company where project was developed
- Business Sector (banking, insurance, etc.)
- Customer Participation
- Staff Availability
- Standard Use, Method Use, Tool Use
- Software Logical Complexity
- Requirement Volatility
- Quality Requirement
- Efficiency Requirement, Installation Requirement
- Staff's Analysis Skills
- Staff's Tools Skills, Staff's Team Skills
- Staff's Application Knowledge

Angelis et al. (2001) have also made an important contribution toward finding the different factors that affect size and effort relationships. These authors study the projects in the International Software Benchmarking Standards Group (ISBSG). The ISBSG database contains data about recently developed projects characterized mostly by attributes of categorical nature such as the project business area, organization type, application domain, and usage of certain tools or methods. The authors found seven important factors that affect the relationship between size and effort:

- Development Type
- Development Platform
- Language Type

- Used Methodology
- Organization Type
- Business Area Type
- Application Type

A third study by Liebchen and Shepperd (2005) that aims at reporting on an ongoing investigation into software productivity and its influencing factors confirmed the intuitive notion that different industry sectors exhibit differences in productivity. It is due to the fact that industry sectors also affect productivity. The following factors were reported:

- The Degree of Technical Innovation, Business Innovation, Application Innovation
- Team Complexity
- Client Complexity
- Degree of Concurrency
- Development Team Degree of Experience with Tools, Information Technology, Hardware, or with Adopted Methodology,
- The Project Management Experience

A fourth study in different Swedish companies (Magazinovic and Pernstå120 2008) shows that the following factors affect the effort estimation:

- Requirement Volatility (Unclear and Changing Requirement).
- Un//availability of Templates.
- Lack of coordination between product development and other parts of the project.

Estimating Using COCOMO and Its Derivatives

One of the most widely used software cost and effort estimation tools is Boehm's COCOMO model, first introduced in 1981. COCOMO is an acronym for the constructive cost model, which means that estimates are determined from a set of parameters that characterize the project. There are several versions of COCOMO including the original (basic), intermediate, and advanced models, each with increasing numbers of variables for tuning the estimates. The latest COCOMO models better accommodate more expressive modern languages as well as software generation tools that tend to produce more code with essentially the same effort. There are also COCOMO derivatives that are applicable for web-based applications and software-intensive (but not pure software) systems.

COCOMO

COCOMO models are based on an estimate of lines of code, modified by a number of factors. The equations for project effort and duration are

$$Effort = A \prod_{i=1}^{n} cd_i \left(size\right)^{p_1} \tag{11.2}$$

$$Duration = B \left(Effort\right)^{p_2} \tag{11.3}$$

where A and B are a function of the type of software system to be constructed, *size* represents the number of lines of code, cd_i represents different cost drivers (this number could vary), and P_1 and P_2 are dependent on certain characteristics of the application domain.

If the system is organic, that is, one that is not heavily embedded in the hardware (e.g., the pet store POS system), then we would set $A=3.2$ and $B=1.05$. If the system is embedded, that is, closely tied to the underlying hardware, for example, as in the visual inspection subsystem of the baggage handling system, then the following parameters are used: $A=2.8$ and $B=1.20$. Finally, if the system is semidetached (e.g., the smart home system), that is, partially embedded, then we would set $A=3.0$ and $B=1.12$. Note that the exponent for the embedded system is the highest, leading to the longest time to complete for an equivalent number of delivered source instructions.

As mentioned, cd_i are cost drivers based on a number of factors, including:

- Product reliability and complexity
- Platform difficulty
- Personnel capabilities
- Personnel experience
- Facilities
- Schedule constraints
- Degree of planned reuse
- Process efficiency and maturity
- Precedentedness (i.e., novelty of the project)
- Development flexibility
- Architectural cohesion
- Team cohesion

These have qualitative ratings on a Likert scale ranging from very low to very high, that is, numerical values are assigned to each of the responses.

Finally, P_1 and P_2 represent certain attributes of the application domain, for example, characterization of economies of scale, ability to avoid rework, novelty of the application, and team cohesion. Incidentally, *Effort* represents the total project effort in person-months, and *Duration* represents calendar months. These figures are necessary to convert the COCOMO estimate to an actual cost for the project.

In the advanced COCOMO models, a further adaptation adjustment factor is made for the proportion of code that is to be used in the system, namely design modified, code modified, and integration modified. The adaptation factor, A, is given by Equation 11.4.

$$A = 0.4 \ (\%\text{design modified}) + 0.3 \ (\%\text{code modified}) + 0.3(\%\text{integration modified}) \quad (11.4)$$

COCOMO is widely recognized and respected as a software project management tool. It is useful even if the underlying model is not really understood. COCOMO software is commercially available and can even be found on the web for free use.

An important consideration for the requirements engineer, however, is that COCOMO bases its estimation on lines of code, which are not easily estimated at the time of requirements engineering. Other techniques, such as function points, are needed to provide line of code estimates based on feature sets that are available when developing the requirements.

Here is an example of how this might work. A customer asks for a cost estimate for the desired feature. Based on the details of that feature, and the various weighting factors needed to calculate

Table 11.3 Programing Language and Lines of Code per Function Point

Language	Lines of Code per Function Point
C	128
C++	64
Java	64
SQL	12
Visual Basic	32

Source: Adapted from Jones, C., *IEEE Comput.*, 29: 103–104, 1996.

FP, the FP metric is computed. That number is converted to lines of code count using the conversion shown in Table 11.3 (or another appropriate conversion). These lines of code count numbers, along with the various other aspects of the project, are plugged into a COCOMO estimator, which yields an estimate of time and effort (meaning person-hours) to complete the project.

This estimate of the number of person-months that is generated from COCOMO can then be converted into an appropriate cost estimate for the customer, a task that would probably be done through the sales department.

Quick Access to COCOMO II - Constructive Cost Model calculator
https://phil.laplante.io/requirements/COCOMO.php

WEBMO

WEBMO is a derivative of COCOMO that is intended specifically for project estimation of web-based projects, where COCOMO is not always as good. WEBMO uses the same effort and duration equations as COCOMO but is based on a different set of predictors, namely:

- Number of function points
- Number of XML, HTML, and query language links
- Number of multimedia files
- Number of scripts
- Number of Web building blocks

These have qualitative ratings on a Likert scale ranging from very low to very high, and numerical equivalents are shown in Table 11.4 (Reifer 2002).

Similar tables are available for the cost drivers in the COCOMO model but are embedded in the simulation tools, so the requirements engineer only has to select the ratings from the Likert scale.

In any case, the net result of a WEBMO calculation is a statement of effort and duration to complete the project in person-months and calendar months, respectively.

Table 11.4 WEBMO Cost Drivers and Their Values (Reifer 2002)

	Ratings				
	Very Low	Low	Nominal	High	Very High
Cost Driver					
Product reliability	0.63	0.85	1.00	1.30	1.67
Platform difficulty	0.75	0.87	1.00	1.21	1.41
Personnel capabilities	1.55	1.35	1.00	0.75	0.58
Personnel experience	1.35	1.19	1.00	0.87	0.71
Facilities	1.35	1.13	1.00	0.85	0.68
Schedule constraints	1.35	1.15	1.00	1.05	1.10
Degree of Planned					
Reuse	—	—	1.00	1.25	1.48
Teamwork	1.45	1.31	1.00	0.75	0.62
Process efficiency	1.35	1.20	1.00	0.85	0.65

COSYSMO

COSYSMO (COnstructive SYstem engineering MOdel) is a COCOMO enhancement for systems engineering project cost and schedule estimation (Valerdi 2008). COSYSMO is intended to be used for cost and effort estimation of mixed hardware/software systems based on a set of size drivers, cost drivers, and team characteristics. The formulation of the COSYSMO metrics is similar to that for COCOMO, using Equations 11.2 and 11.3.

Like COCOMO, COSYSMO computes effort (and cost) as a function of system functional size and adjusts it based on a number of environmental factors related to systems engineering (Figure 11.3).

The main steps in using COSYSMO are:

1. Determine the system of interest.
2. Decompose system objectives, capabilities, or measures of effectiveness into requirements that can be tested, verified, or designed.
3. Provide a graphical or narrative representation of the system of interest and how it relates to the rest of the system.
4. Count the number of requirements in the system/marketing specification or the verification test matrix for the level of design in which systems engineering is taking place in the desired system of interest.
5. Determine the volatility, complexity, and reuse of requirements (Valerdi 2008).

In COSYSMO, the size drivers include the counts of the following items as taken right from the SRS document:

- Total system requirements
- Interfaces

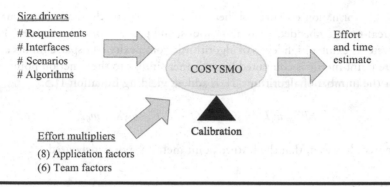

Figure 11.3 COSYSMO operational concept.

- Operational scenarios
- The unique algorithms that are defined

Other drivers include:

- Requirements understanding
- Architecture complexity
- Level of service requirements
- Migration complexity
- Technology maturity

which are ranked using a Likert scale in a similar manner as the COCOMO size and cost drivers. Finally, cost drivers based on team characteristics include:

- Stakeholder team cohesion
- Personnel capability
- Personnel experience/continuity
- Process maturity
- Multisite coordination
- Formality of deliverables
- Tool support

These are also rated on a Likert scale. The decomposition of requirements into an appropriate level of detail or granularity is particularly important, but not very easy to do. An effective approach to requirements decomposition can be found in Liu et al. (2010).

The COSYSMO model contains a calibration data set of more than 50 projects provided by major aerospace and defense companies such as Raytheon, Northrop Grumman, Lockheed Martin, SAIC, General Dynamics, and BAE Systems (Valerdi 2008).

Feature Points

Feature points are an extension of function points developed by Software Productivity Research, Inc. in 1986. Feature points address the fact that the classical function point metric was developed for management information systems and therefore is not particularly applicable to many other systems, such as real-time, embedded, communications, and process control software. The motivation is that these systems exhibit high levels of algorithmic complexity, but sparse inputs and outputs.

The feature point metric is computed in a manner similar to the function point, except that a new factor for the number of algorithms, A, is added, yielding Equation 11.5.

$$FP' = w_1I + w_2O + w_3Q + w_4F + w_5X + w_6A \tag{11.5}$$

It should be noted, however, that the feature point metric is not widely used.

Use Case Points

Use case points (UCP) allow the estimation of an application's size and effort from its use cases. This is a particularly useful estimation technique during the requirements engineering activity when use cases are the basis for requirements elicitation.

The use case point equation is based on the product of four variables that are derived from the number of actors and scenarios in the use cases, and various technical and environmental factors. The four variables are:

1. Technical complexity factor (TCF)
2. Environment complexity factor (ECF)
3. Unadjusted use case points (UUCP)
4. Productivity factor (PF)

These lead to the basic use case point equation:

$$UCP = TCF \times ECF \times UUCP \times PF \tag{11.6}$$

This metric is then used to provide estimates of project duration and staffing from data collected from other projects. As with function points, additional technical adjustment factors can be applied based on project, environment, and team characteristics (Clemmons 2006).

Artificial neural networks (Nassif et al. 2012a) and stochastic gradient boosting techniques (Nassif et al. 2012b) have been used to enhance the predictive accuracy of use case points. Use case points are clearly beneficial for effort estimation in those projects where the requirements are created predominantly via use cases.

Considerations for Nonfunctional Requirements in Size Measurements

The majority of the existing FSMs have been primarily focused on sizing the functionality of a software system. Size measures are expressed as single numbers (FP), while nonfunctional requirements (NFRs) are only be considered with respect to the task of adjusting the (unadjusted) FP counts to the project context or the environment in which the system is supposed to work.

Despite the above challenges that arise when dealing with NFRs, it is important to be able to make decisions about the scope of software by given resources and budget based on a proper estimation of building both FRs and NFRs. As the effort is a function of size, one way to respond to the need to deal comprehensively with the effect of NFRs on the effort of completing the software project is to measure their corresponding functional size.

Kassab et al. (2007) provided the nonfunctional requirements size measurement (NFSM) method as a solution to deal with the problem of quantitatively assessing the NFR modeling process early in the project by extending the COSMIC functional size measurement method. The method relies on classifying the NFRs into two broad categories: those that can be satisfied through functional operationalizations, and those that are satisfied through other architectural decisions. The total size of the NFR that can be satisfied into functionality within the system is then calculated by measuring the total changes in the functional size of functionalities triggered by introducing these NFRs.

More recently, in October 2019, IEEE published IEEE 2430, "Trial-Use for Software Non-Functional Requirements Sizing Measurements," based on software nonfunctional assessment process (SNAP) points developed by the International Function Point Users Group (IFPUG). IEEE 2430 presents the design of a measurement method for the sizing of nonfunctional software requirements using four nonfunctional categories (as in data operations, interface design, technical environment, and architecture software), which are further subdivided into 16 subcategories. The IEEE 2430 measurement drew some criticism though (e.g., Abran 2020) as its design fails primary school mathematics and produces numerical noise rather than a number with metrological properties required in engineering.

Software Effort Estimation in Practice

Stay Updated: For the up-to-date data from the RE state of practice survey.
https://phil.laplante.io/requirements/updates/survey.php

In the 2020 survey on RE state of practices (Kassab and Laplante 2022), the majority of the surveyed (65%) reported performing some requirements estimation activity. This number has slightly increased from the previous edition of the survey conducted in 2013 (Kassab et al. 2014). In the 2020 Survey, "Story Points" was the most popular technique reported (Figure 11.4). This is a shift from 2013 when "Expert Judgements" was the most popular back then. Interestingly, Story Points was also present in 17 percent of the projects that selected only Waterfall as a development approach. Function points and COCOMO remained unpopular techniques in both Agile and Waterfall in 2020 as in the past survey. The majority of projects (87%) also reported consideration for Non-Functional Requirements (NFRs) during the effort estimation. This percentage was consistent among the Agile and Waterfall samples too. This was a particularly interesting finding considering the common belief that Agile approaches tend to deal less with NFRs in comparison to the Waterfall. One reason for this belief is that NFRs are not always apparent when the requirements are presented through user stories which is a common representation technique in Agile. We further linked the question of estimation techniques to whether the duration of the project was within schedule and whether the cost was within budget. Overall, we observed that when a project relied more on "Story Points" and "Group estimation" techniques, there was a higher level of agreement that it was maintained within schedule and budget. Since Waterfall projects reported

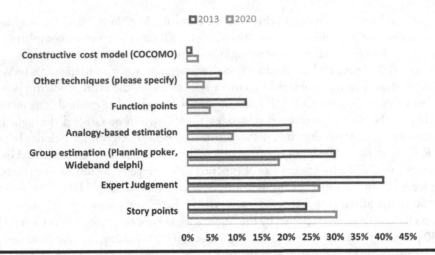

Figure 11.4 Effort Estimation techniques employed—2020 vs. 2013

more reliance on experts' judgments in comparison to Agile projects that reported more reliance on Group estimation and Story Points, it was interesting to find that 50% of the Agile projects also reported a level of agreement that the project duration was within schedule (compared to 43% for Waterfall). In addition, 43% of Agile projects reported an agreement that the project cost was within budget (compared to 37% for Waterfall).

VIGNETTE 11.1   Boston Big Dig Project

The Boston Big Dig project was the most complex highway project ever undertaken in the United States. The project consisted of several bridges, tunnels, and extensions that traversed the city and connected major highways, the airport, and downtown. Initially budgeted for 2.8 billion dollars and completion in 1998, it was finished in 2007 at a final cost of $14.6–$22 billion dollars—no one knows the true cost. The project overran the schedule by 9 years. Not long after completion problems such as metal corrosion, concrete failure, and tunnel collapse began to appear.

While instances of poor workmanship and inferior materials were identified, the root causes for the failure of this project were lack of accountability, risk planning and oversight, and the use of too many contractors (Sigmund 2007). But it is also clear that a comprehensive and correct requirements affordability program was not undertaken.

Shane et al. (2009) implied that "during the planning and design phase of project development external factors such as local government concerns and requirements, fluctuations in the rate of inflation, scope change, scope creep, and market conditions [led to] underestimation of project costs." They also suggested that a design/build situation, where the engineers both design and build the structure, would have increased accountability. While design/build is common for complex hardware and software-intensive systems, it is not usually the case for a highway construction project—one firm or the government does the design work, and the project is put out to bid for construction firms to respond.

The Big Dig project was a failure that spanned more than requirements engineering—there were project management failures, political failures, and construction errors—but the disaster started with the poor requirements engineering process.

Requirements Feature Cost Justification

Consider the following situation. A customer has the option of incorporating feature A into the system for $250,000 or forgoing the feature altogether. Currently, $1,000,000 is budgeted for activity associated with feature A. It has been projected that incorporating the feature into the software would provide $500,000 in immediate cost savings by automating several aspects of the activity.

Should the manager decide to forgo including feature A into the system, new workers would need to be hired and trained before they can take up the activities associated with feature A. At the end of 2 years, it is expected that the new workers will be responsible for $750,000 cost savings. The value justification question is "should the new feature be included in the system or not?" We can answer this question after first discussing several mechanisms for calculating the value of certain types of activities, including the realization of desired features in the system.

Return on Investment

Return on investment (ROI) is a rather overloaded term that means different things to different people. In some cases, it means the value of the activity at the time it is undertaken. To others, it is the value of the activity at a later date. In other cases, it is just a catchword for the difference between the cost of the system and the savings anticipated from the utility of that system. To still others, there is a more complex meaning.

One traditional measure of ROI for any activity or system is given as

$$ROI = \text{average net benefits/initial costs} \qquad (11.7)$$

The challenge with this model for ROI is the accurate representation of average net benefit and initial costs. But this is an issue of cost accounting that presents itself in all calculations of costs vs. benefits.

Other commonly used models for the valuation of some activity or investment include net present value (NPV), internal rate of return (IRR), profitability index (PI), and payback. We'll look at each of these shortly.

Other methods include Six Sigma and proprietary Balanced Scorecard models. These kinds of approaches seek to recognize that financial measures are not necessarily the most important component of performance. Further considerations for valuing software solutions might include customer satisfaction, employee satisfaction, and so on, which are not usually modeled with traditional financial valuation instruments.

There are other more complex accounting-oriented methods for valuing software. Discussion of these techniques is beyond the scope of this text. The references at the end of the chapter can be consulted for additional information (e.g., Raffo et al. 1999; Morgan 2005).

Net Present Value

NPV is a commonly used approach to determine the cost of software projects or activities. The NPV of any activity or thing is based on the notion that a dollar today is worth more than that same dollar tomorrow. You can see the effect as you look farther into the future, or back to the past (some of you may remember when gasoline was less than $1 per gallon). The effect is due to the fact that, except in extraordinary circumstances, the cost of things always escalates over time.

Here is how to compute NPV. Suppose that *FV* is some future anticipated payoff either in cash or in anticipated savings. Suppose *r* is the discount rate and *Y* is the number of years that the cash or savings is expected to be realized. Then, the net present value of that payoff is

$$NPV = FV/(1+r)^Y \qquad\qquad (11.8)$$

NPV is an indirect measure because you are required to specify the market opportunity cost (discount rate) of the capital involved.

To see how you can use this notion in value engineering of requirements, suppose that you expect a certain feature, *B*, to be included in the system at a cost of $60,000. You believe that the benefits of this feature are expected to yield $100,000 2 years in the future. If the discount rate is 3%, should the feature be included in the new system?

To answer this question, we calculate the NPV of the feature using Equation 11.7, taking into account the cost of the feature

$$NPV - cost = 100,000/1.03^2 - 60,000 = 34,259$$

Since NPV less its cost is positive, yes, the feature should be included in the requirements.

Equation 11.7 is useful when the benefit of the feature is realizable after some discrete period of time. But what happens if there is some incremental benefit of adding a feature; for example, after the first year, the benefit of the feature doubles (in future dollars). In this case, we need to use a variation of Equation 11.7 that incorporates the notion of continuing cash flows.

For a sequence of cash flows, CF_n, where $n=0, \ldots k$ represents the number of years from initial investment, the net present value of that sequence is

$$NPV = \sum_{n=0}^{k} \frac{CF_n}{(1+r)^n} \qquad\qquad (11.9)$$

CF_n could represent, for example, a sequence of related expenditures over a period of time, such as features that are added incrementally to the system or evolutionary versions of a system.

Internal Rate of Return

The internal rate of return (IRR) is defined as the discount rate in the NPV equation that causes the calculated NPV, minus its cost, to be zero. This can be done by taking Equation 11.7 and rearranging to obtain

$$0 = FV/(1+r)^Y - c$$

where *c* is the cost of the feature and FV its future value. Solving for *r* yields

$$r = \left[FV/c \right]^{1/Y} - 1 \tag{11.10}$$

For example, to decide if we should incorporate a feature or not, we compare the computed IRR of including the feature with the financial return of some alternative, for example, to undertake a different corporate initiative or add a different feature. If the IRR is very low for adding the new feature, then we might simply want to take this money and find an equivalent investment with lower risk. But if the IRR is sufficiently high for the new feature, then the decision might be worth whatever risk is involved in its implementation.

To illustrate, suppose that a certain feature is expected to cost $50,000. The returns of this feature are expected to be $100,000 of increased output 2 years in the future. We would like to know the IRR on adding this feature.

Here, NPV–cost = $100,000/(1 + r)^2$–50,000. We now wish to find the r that makes the NPV = 0, that is, the break even value. Using Equation 11.9 yields

$$r = \left[100,000 / 50,000 \right]^{1/2} - 1$$

This means $r = 0.414 = 41.4\%$. This rate of return is very high for this feature, and we would likely choose to incorporate it into the system.

Profitability Index

The profitability index (PI) is the NPV divided by the cost of the investment, I:

$$PI = NPV / I \tag{11.11}$$

PI is a "bang-for-the-buck" measure, and it is appealing to managers who must decide between many competing investments with positive NPV when there are financial constraints. The idea is to take the investment options with the highest PI first until the investment budget runs out. This approach is not bad but can suboptimize the investment portfolio.

One of the drawbacks of the profitability index as a metric for making resource allocation decisions is that it can lead to a suboptimization of the result. To see this effect, consider the following situation. A customer is faced with some hard budget-cutting decisions and needs to remove some features from the proposed system. To reach this decision, an analysis is prepared based on the profitability index shown in Table 11.5.

Suppose the capital budget is $500K. The PI ranking technique would cause the customer to pick A and B first, leaving inadequate resources for C. Therefore, D will be chosen, leaving the overall NPV at $610K. However, using an integer programming approach will lead to a better decision (based on maximum NPV) in that features A and C would be selected, yielding an expected NPV of $620K.

The profitability index method is helpful in conjunction with NPV to help optimize the allocation of investment dollars across a series of feature options.

Payback Period

To the project manager, the payback period is the time it takes to get the initial investment back out of the project. Projects with short paybacks are preferred, although the term "short" is

Table 11.5 A Collection of Prospective Software Features, Their Project Cost, NPV, and Profitability Index

Feature	Projected Cost (in $10s of thousands)	NPV (in $10s of thousands)	PI
A	200	260	1.3
B	100	130	1.3
C	300	360	1.20
D	200	220	1.1

completely arbitrary. The intuitive appeal is reasonably clear: the payback period is easy to calculate, communicate, and understand.

Payback can be used as a metric to decide whether to incorporate a requirement in the system or not. For example, suppose implementing feature D is expected to cost $100,000 and result in a cost savings of $50,000 per year. Then, the payback period for the feature would be 2 years.

Because of its simplicity, payback is the ROI calculation least likely to confuse managers. However, if the payback period is the only criterion used, then there is no recognition of any cash flows, small or large, to arrive after the cutoff period. Furthermore, there is no recognition of the opportunity cost of tying up funds. Since discussions of payback tend to coincide with discussions of risk, a short payback period usually means a lower risk. However, all criteria used in the determination of payback are arbitrary. And from an accounting and practical standpoint, the discounted payback is the metric that is preferred.

Discounted Payback

The discounted payback is the payback period determined on discounted cash flows rather than undiscounted cash flows. This method takes into account the time (and risk) value of the sunken cost. Effectively, it answers the questions, "how long does it take to recover the investment?" and "what is the minimum required return?"

If the discounted payback period is finite in length, it means that the investment plus its capital costs are recovered eventually, which means that the NPV is at least as great as zero. Consequently, a criterion that says to go ahead with the project if it has *any* finite discounted payback period is consistent with the NPV rule.

To illustrate, in the payback example just given, there is a cost of $100,000 and annual maintenance savings of $50,000. Assuming a discount rate of 3%, the discounted payback period would be longer than 2 years because the savings in year 2 would have an NPV of less than $50,000 (figure out the exact payback period for fun). But because we know that there is a finite discounted payback period, we know that we should go ahead and include feature D.

Putting It All Together

Suppose the smart home was being built for some special customer. This customer presumably conducted their own ROI analysis for this system using one of the methods previously introduced and now understands how much they would like to pay for the system (or even parts of the system). For example, if this was a one-of-a-kind home (think of Bill Gates' smart home), then the investment in this functionality would need to add at least as much as the cost of the features to the resale price.

Let's say that the homeowner felt that this feature would add $250K to the asking price of the resold home and has decided that an ROI of 1.25 is acceptable. Therefore, they would be willing to pay $200K for these features.

Suppose we analyzed these requirements and used COSYSMO to derive a cost estimate of $210K. The customer may indicate that they are only willing to pay $200K. In order to lower the price, we can suggest eliminating certain requirements in order to meet the budgetary constraints.

For example, suppose we omit requirements 8.1.10 and 8.1.13, we then get a new estimate of $200K, and COSYSMO predicts that it will take 10 calendar months to build and that the customer can resell the system immediately upon completion. Now we can compute a profitability index as well as an internal rate of return.

Let's assume a discount rate of 3%. The NPV of this activity, using $250K over 10 months at 3% (using simple interest), is

$$NPV = \$250{,}000 \times (1.03)^{(10/12)} - \$200{,}000 = \$43{,}917.14$$

Continuing, the profitability index on this set of requirements is

$$PI = \$243{,}917 \, / \, \$200{,}000 = 1.22$$

The internal rate of return on this investment is

$$IRR = \left[\$250{,}000 \, / \, \$200{,}000\right]^{(12/10)} - 1 = 30.7\%$$

So we resolved a cost issue for our customer by revising the effort and generated a good investment.

Share your Opinion: What kinds of cost/benefit approaches (if any) does your organization use during requirements engineering?
https://phil.laplante.io/requirements/cost.benefit.php

VIGNETTE 11.2   Buying versus Leasing Specialty Equipment

Sometimes satisfying system requirements may necessitate the acquisition of some sort of expensive, specialized equipment. It could be a testing framework for an unusual application domain, ultra-precision instruments, or some kind of build equipment (such as a large crane for heavy construction work). For example, supercomputers are quite expensive and few companies own them, choosing to lease them (or the computational time) as project needs demand. But other companies make the strategic decision to avoid leasing altogether and may own one or more supercomputers (even if they sit idle some of the time). Purchasing a multimillion-dollar piece of equipment for a single (even a high-value project) might not make sense but leasing it can carry significant costs too. Therefore, a cost-benefit (or value) analysis has to be made to determine whether to buy or lease any major piece of equipment if it is not already owned (or owned, but unavailable because it is assigned to another project). The questions that must be considered are project, opportunity, and financially based. Some of the questions to be considered include:

- If purchasing, how will it be financed?
  - What is the lead time on delivering the equipment? Will it be a cash purchase or paid via an equipment loan?
  - What are the terms for the loan (are they favorable, given your company's credit situation)?
  - Could the cash for purchase be used for other, more important needs?
  - Can the company lease the equipment to others even competitors when not needed?
  - What is the total cost of ownership (energy, maintenance, repair, insurance, costs)?
  - Is there a favorable resale value for the used equipment?
- If leasing,
  - What will be the availability of the equipment (certain leased equipment can be scarce at key times)?
  - Are the leasing terms favorable given your company's credit profile?
  - What kind of support does the leasing firm provide (e.g., training, maintenance)?
- General questions:
  - What are the required skill set to use the equipment?
  - Are these skill sets already available in-house or can someone be hired or trained or will temporary employees be brought in for the duration of the project.
- Will this piece of equipment be for one-time use versus ongoing use going forward?
- What kind of downtime will this equipment experience?
- What are the tax advantages of depreciating the equipment, versus recording the leasing costs as an expense?

The buy versus lease decision is a very complicated one that is situation dependent. Be very thorough in making such a decision because omitting an important cost consideration can be disastrous. Also, in the absence of all the facts, don't be quick to judge those who choose to buy versus lease (or vice versa).

Exercises

11.1 Why is it so important to determine the cost of features early, but not too early in the requirements engineering process?

11.2 What factors determine which metric or metrics a customer can use to help make meaningful cost-benefit decisions of proposed features for a system to be built?

11.3 How does the role of ranking requirements help in feature selection cost-benefit decision-making?

11.4 What changes (if any) would you need to make to the COCOMO, COSYSMO, or feature/function point equation calculations to incorporate ranking of requirements?

11.5 Complete the derivation of Equation 11.8 from Equation 11.6 by setting the NPV equation (less the cost of the investment) to zero and solving for r.

11.6 Use COSYSMO (find a Web implementation) to conduct an effort and schedule analysis for the smart home system in Appendix A.

11.7 Use COSYSMO to conduct an effort and schedule analysis for the wet-well control system in Appendix B.

*11.8 Investigate the use of other decision-making techniques, such as integer programming, in helping to decide on the appropriate feature set for a proposed system.

Note

1 Or KLOC (1000 LOC). Another related metric is source lines of code (SLOC). The major difference being that a single source line of code may span several lines. For example, a long if-then-else statement could be a single SLOC but many delivered source instructions.

References

Abran, A. (2020). IEEE 2430 non-functional sizing measurements: A numerical placebo. *IEEE Software*, 38(3): 113–120.

Albrecht, J. (1979, October). Measuring application development productivity. In *Proceedings of the IBM Applications Development Symposium* (pp. 14–17). Monterey, CA.

Angelis, L., Stamelos, I., & Morisio, M. (2001). Building a software cost estimation model based on categorical data. In *Proceedings of the 7th International Symposium on Software Metrics* (pp. 4–15) METRICS. IEEE Computer Society, Washington, DC.

Boehm, B. W. (1981). *Software Engineering Economics*. Prentice Hall, Englewood Cliffs, NJ.

Clemmons, R. K. (2006). Project estimation with use case points. *The Journal of Defense Software Engineering*, 2: 18–22.

ISO/IEC 14143-6:2012-Functional Size Measurement. (2012). https://www.iso.org/standard/60176.html

ISO/IEC 19761:2011 Software Engineering — COSMIC: A Functional Size Measurement Method. (2011). https://www.iso.org/standard/54849.html

Jones, C. (1996). Activity-based software costing. *IEEE Computer*, 29: 103–104.

Kassab, M., & Laplante, P. (2022). The current and evolving landscape of requirements engineering state of practice. *IEEE Software*. Accepted.

Kassab, M., Neill, C., & Laplante, P. (2014). State of practice in requirements engineering: Contemporary data. *Innovations in Systems and Software Engineering*, 10(4): 235–241.

Kassab, M., Ormandjieva, O., Daneva, M., & Abran, A. (2007). Non-functional requirements size measurement method (NFSM) with COSMIC-FFP. In *Software Process and Product Measurement* (pp. 168–182). Springer, Berlin, Heidelberg.

Liebchen, G. A., & Shepperd, M. (2005). Software productivity analysis of a large data set and issues of confidentiality and data quality. In *Proceedings of the 11th IEEE International Software Metrics Symposium* (September 19–22, 2005, p. 46). METRICS, IEEE Computer Society, Washington, DC.

Liu, K., Valerdi, R., & Laplante, P. A. (2010). Better requirements decomposition guidelines can improve cost estimation of systems engineering and human systems integration. In *8th Annual Conference on Systems Engineering Research*. Hoboken, NJ.

Magazinovic, A., & Pernstål, J. (2008). Any other cost estimation inhibitors? In *Proceedings of the Second ACM-IEEE International Symposium on Empirical Software Engineering and Measurement* (pp. 233–242), Kaiserslautern, Germany, ESEM '08. ACM, New York.

Maxwell, K. D., & Forselius, P. (2000). Benchmarking software development productivity. *IEEE Software*, 17(1): 80–88.

McConnell, S. (1997). *Software Project Survival Guide.* Microsoft Press, Redmond, WA.

Morgan, J. N. (2005). A roadmap of financial measures for IT project ROI. *IT Professional*, 7: 52–57.

Nassif, A. B., Capretz, L. F., & Ho, D. (2012a, December). Estimating software effort using an ANN model based on use case points. In *2012 11th International Conference on Machine Learning and Applications (ICMLA)* (Vol. 2, pp. 42–47). IEEE, Boca Raton, Florida, USA.

Nassif, A. B., Capretz, L. F., Ho, D., & Azzeh, M. (2012b, December). A Treeboost model for software effort estimation based on use case points. In *2012 11th International Conference on Machine Learning and Applications (ICMLA)* (Vol. 2, pp. 314–319). IEEE, Boca Raton, Florida, USA.

Park, R. E. (1992). *Software Size Measurement: A Framework for Counting Source Statements.* Carnegie-Mellon University, Software Engineering Institute, Pittsburgh, PA.

Pressman, R. S. (2015). *Software Engineering: A Practitioner's Approach.* McGraw-Hill Higher Education. New York, USA.

Raffo, D., Settle, J., & Harrison, W. (1999). *Investigating Financial Measures for Planning of Software IV&V.* Research Report #TR-99-05. Portland State University, Portland, OR.

Reeves, J., DePasquale, D., & Lim, E. (2010, August). Affordability engineering: Bridging the gap between design and cost. In *AIAA SPACE 2010 Conference & Exposition* (p. 8904), Anaheim, California.

Reifer, D. (2002). Estimating web development costs: There are differences. *Crosstalk*, 15(6): 13–17.

Shane, J. S., Molenaar, K. R., Anderson, S., & Schexnayder, C. (2009). Construction project cost escalation factors. *Journal of Management in Engineering*, 25(4): 221–229.

Sigmund, P. (2007). Triumph, tragedy mark Boston's big dig project. In *Construction Equipment Guide.* http://www.constructionequipmentguide.com/triumph-tragedy-mark-bostons-big-dig-project/8751 (accessed February 2017).

Software Non-Functional Assessment Process (SNAP), Assessment Practices Manual, Release 2.4. (2017). International Function Points Users Group, Princeton Junction, NJ.

Trial-Use for Software Non-Functional Requirements Sizing Measurement. (2019). IEEE Standard 2430, IEEE Standards Association, Piscataway, NJ.

Valerdi, R. (2008). *The Constructive Systems Engineering Cost Model (COSYSMO): Quantifying the Costs of Systems Engineering Effort in Complex Systems.* VDM Verlag, Berlin.

Chapter 12

Requirements Engineering: A Road Map to the Future

Shaping Factors of the Future Landscape of Requirements Engineering

The Greek philosopher Heraclitus is famous for his conviction that constant change is a fundamental truth of the universe. Since requirements engineering is integral to software/system development, and with fingers firmly on the pulse of changes to disruptive technologies and development approaches, one can envision the factors that will influence the landscape of requirements engineering in near future, say, 10 years from now.

There is no question that technological disruption (e.g., AI, IoT, Blockchain) will affect the way that businesses, consumers, or industries function. These same technologies will also influence the requirements engineering practices. We will discuss the relationship between these technologies and the requirements of engineering shortly.

Moreover, as systems become more and more distributed and divided into a set of components, so do the requirements as well. Globalization, as we discussed in Chapter 10, and outsourcing drive the implementation of market-specific features. Stakeholders can be geographically located in multiple certain countries and components sourced from even other countries.

Next, consider that systems are no longer always built from scratch. Companies are transitioning to offering products as services instead of stand-alone offerings. The use of commercial off-the-shelf (COTS) software and the move to microservices architectures and cross-organizational systems will also influence the traditional requirements engineering practices. For example, requirements will keep on being layered on top of one another due to the separation of program features into separate services, and these will need to have their own set of requirements as well as proper maintenance. Besides, requirements management will need to include an ongoing negotiation process with a continuously expanding set of stakeholders and dynamic architectures that needs to be adaptable and expandable.

Finally, the ongoing open-source movement will keep on influencing requirements engineering as well. As engineers keep on using modules from a myriad of other developers, more focus must be paid to ensuring licenses compliance and the potential of constant updates breaking the system and the packages' average lifetime.

DOI: 10.1201/9781003129509-12

The Changing Landscape of Requirements Engineering

In Kassab (2015) and Kassab and Laplante (2022), we reported on the results from four conducted surveys (2003, 2008, 2013, and 2020) of software development and requirements engineering practices by working professionals in the field. We presented different results views from these surveys throughout the book. The results reflected the changing landscape of software development as one would expect; the emergence of agile as the dominant development paradigm being the most obvious.

Some other salient observations that emerge from comparing the results from the four surveys are as follows:

- Overall the trend shows that the RE practices in general within a project had slightly lost some satisfaction level in the past few years. In the "2013 Survey", 82% reported satisfaction (agreement and neutral) with regards to RE practices in general applied within a project; compared to 70% in 2020. Despite the decline, the satisfaction level is still higher than what had been reported in the "2008 Survey" (67%).
- Bose et al. (2008) recommended that the projects using Agile "should learn from various elicitation techniques that have been in use in the RE of the Waterfall." We indeed observe that for Agile projects in 2020, "use cases" and "interviews," originally being popular techniques for the Waterfall, are becoming common techniques in Agile projects as well.
- It is quite interesting to see the opposite also holds true. Even though that "user stories" were developed specifically for Agile, it is finding its way to the Waterfall as well.
- A number of paradigms and techniques show no significant changes from the "2013 Survey," particularly with respect to requirements inspection, prototyping, and utilizing semiformal notations.
- While the "2013 Survey" indicated still significant budget and schedule performance issues, the data from the "2020 Survey" indicated an improvement in overall performance in maintaining the duration of the project within schedule and costs within budget estimates. There are obviously many factors at work here, including a) the changing in the landscape of the employed estimation techniques from "expert judgments" to "Story Points" and "Group Estimations " which are analytically proven to be more reliable; and b) the increasing rise of agile approaches that come with more pressures placed on development project timelines and budgets in comparison of the conventional Waterfall.
- Along the same vein, we notice that Agile projects outperformed the Waterfall projects in the "2020 Survey" in maintaining the projects within the budget and schedule. On the other hand, respondents of the "2020 Survey" expressed a very comparable level of satisfaction regarding the applied RE practices in their Agile projects compared to the Waterfall projects.
- Formal methods (e.g., formal languages, formal inspections, formal estimation techniques) remain largely not utilized.

Requirements Engineering for Small Businesses

Small businesses are defined as enterprises that employ fewer than 50 persons and whose annual turnover or annual balance sheet total does not exceed EUR 10 million, according to the European

Commission (2005). Hence, most startups may also be considered as small businesses. In the United States, 88% of employer firms are considered to be small businesses (JPMorgan, 2018). These businesses accounted for over half of net job creation in the U.S. in 2014. According to the U.S. Small Business Administration, the overwhelming majority of software development firms are also considered to be small businesses, and this will most likely continue to be the case for the next decade.

Even though there is ample information available on solid requirements engineering practices for software systems, the requirements engineering community has mostly overlooked the small business characteristics and their requirements engineering practices (e.g., requirements elicitation, analysis, validation, estimation, tools support). This situation exists is even though using correct practices can mean the difference between success and failure of the small business itself. Azar et al. (2007) identified three reasons for the lack of requirements engineering focus in small businesses:

- Due to the small staff sizes, often the requirements engineer is also the developer, architect, and/or the QA.
- The learning curve in implementing the industry's best practices for RE is steep and requires hiring talents with a solid educational and practical background which can be a financial burden on a small business.
- Financial constraints can also hamper the use of RE processes and tools. Typically, a small organization's attention to RE may come only after early successes that haven't relied heavily on it.

In Kassab (2021), observations from the four RE surveys are reported on how requirements engineering are conducted in small businesses:

- A wide diversity in the employed requirements engineering practices for small businesses was reported in the 2020 survey. We can rationalize that this diversity is the result of evolutionary adaptation, as these businesses have to adapt to their particular ecological niche. The context of these businesses is the essence of their adopted practices.
- Despite the diversity, we note that when requirements engineering practices are employed, there is a tendency to adopt easy, flexible, and inexpensive techniques (e.g., the majority opted for open-source RE management tools).
- The adoption of inexpensive requirements engineering techniques doesn't necessarily hinder the level of satisfaction of employing them in small businesses. This is evidenced in a third observation when we compared the level of satisfaction in regard to the employed RE activities between the small businesses sample and the overall sample. Even more satisfaction level was reported in small business sample in regards to the followed SDLC, requirements elicitation and analysis, and requirements traceability.
- A high degree of cultural cohesion that the small businesses exhibit with the majority of responses reporting an agreement that the team size was adequate for the RE challenges that face the project, and an agreement that the ability and previous software development experience of the software development team was adequate. The high level of agreement may imply a homogeneity in the professional environment of the analyzed sample, which contributes to efficiency in communication and sharing an understanding of the requirements within a small team.

Requirements Engineering and Disruptive Technologies

It is impossible to address all the diverse facets of requirements engineering potential research areas or to comprehensively explore the disruptive impact of change in those areas, or even to hint at the potential for innovative creativity brought about by those same disruptions. Instead, this section will highlight a selection of disruptive technologies and their commensurate opportunities and challenges in requirements engineering.

Requirements Engineering and Blockchain

Blockchain is a type of distributed ledger technology that is characterized by the CoDIFy-Pro characteristics: Consensus, Decentralization, Immutability, Finality, and Provenance, explained as follows.

A blockchain consists of a set of nodes connected through a peer-to-peer network. Each node in the network maintains an exact copy of the blockchain creating a decentralized structure. When a new block is added to the chain, all the nodes of the network need to reach a consensus on the validity of that block. The consensus mechanisms are protocols that ensure all nodes on the network are synchronized with each other and agree on which transactions are valid (and only those are added to the blockchain). New entries to the blockchain are added by appending them to the end of the ledger. Once recorded, data can't be altered, and the transaction history is combined into a chain structure without the possibility of additional branches of alternative transactions emerging or wedging into the middle of a chain. Each block contains a cryptographic hash and a

---

VIGNETTE 12.1   3-D Printing

Architects and engineers have used paper, cardboard, and wood models of buildings, bridges, and other structures for centuries to help in requirements visualization. Automobile and airplane engineers have used clay, carved wood, and, later, plastic models to visualize new products from almost the beginning of their disciplines.

We've already discussed the value of using mockups and prototypes for requirements engineering, but in the past, skilled model makers were needed to build these. Today, relatively inexpensive 3-D printers can be used to build some or all of the structural elements of a prototype system for the purposes of requirements discovery, validation, and reconciliation. Large size systems can be built to scale, though 3-D printing of, for example, a prototype automobile body is completely feasible at full size. Even small companies can use advanced 3-D printing for prototyping and requirements discovery using public makerspaces (Mertz).

So, expect to see more prototyping and modelmaking using advanced 3-D printing as a regular practice in many application domains.

REFERENCE

L. Mertz, "Dream It, Design It, Print It in 3-D: What Can 3-D Printing Do for You?," in *IEEE Pulse*, vol. 4, no. 6, pp. 15–21, Nov. 2013, doi: 10.1109/MPUL.2013.2279616.

---

timestamp creating an immutable record of all the transactions in the network. The finality that characterizes the blockchain means that a single and shared ledger provides one unique place to determine ownership of an asset or completion of a transaction, while the provenance implies that network participants have access to the knowledge of where an "asset" came from and how its ownership changed over time.

Blockchain technology's transformative capabilities have been rapidly recognized as a turning point in many use case scenarios beyond the financial sector. A study by Chakraborty et al. (2018) reported that 3,000 blockchain software projects were hosted on Github in March 2018. A similar search we conducted on Github in June 2021 yielded more than 90,000 projects. This emerging technology's impetus is now being utilized in multiple ways, from global payments to managing the COVID-19 vaccinations supply chain to even tracking diamond sales.

The World Economic Forum estimates 10% of the global GDP to be stored using blockchain by 2027. A renowned scientific study and market consulting firm, Gartner, estimated investment decisions worth $3.1 trillion in blockchain technology projected by 2030 (Kandaswamy and Furlonger 2018). In synchronous with the growing sphere of blockchain, numerous data points testify to the increasing demand for blockchain developers and engineers. As of June 2021, a search that we conducted for the blockchain-related jobs in the U.S. retrieved 5,603 results on "Indeed.com." The skyrocketing demand for blockchain-related jobs has also translated into a significant salary bump. According to global statistics provided by Hired.com, the median salary for a blockchain-related job opening is approximately 1.3 times higher than the standard salary for software engineers. Unlike traditional software development, blockchain engineers need to secure an immutable and decentralized database hosted on distributed nodes connected through a peer-to-peer network without a pre-existing trust relationship (Kassab et al. 2021).

Although blockchain technology has the potential to disrupt traditional business models in many business domains, surprisingly the interrelation between blockchain and software engineering has received little attention and only recently (Demi 2020). The significant differences between blockchain-oriented development and traditional software development motivated the blockchain community to propose a new development paradigm named Blockchain-Oriented Software Engineering (BOSE) (Destefanis et al. 2018).

Requirements Engineering must be included in any blockchain-based project. While there is still a significant incongruity regarding how the blockchain's unique five characteristics can be mapped to the elicited requirements (this area is open to research), Kassab (2021) presented a first attempt toward exploring NFRs for blockchain-oriented systems in particular in respect to (i) privacy, (ii) performance and scalability, (iii) interoperability, (iv) usability, (v) compliance with regulations, and (vi) operating and financial constraints. Future research agenda on the link between blockchain and requirements engineering include:

■ Identification of the design and architectural requirements that emerge due to the five blockchain qualities. Specifically, how these qualities can be satisfied during the high-level and low-level design phases, and what are the new requirements that could be introduced due to this?

■ Exploring the new emerging NFRs (not listed in the traditional qualities taxonomies) for blockchain-based systems when used as a platform for more focused domain-related applications (e.g., healthcare, education, retail). What are these, and how they will be satisfied?

Requirements Engineering and Internet of Things

The term "Internet of Things" (IoT) has recently become popular to emphasize the vision of a global infrastructure that connects physical objects/things, using the same Internet Protocol, allowing them to communicate and share information (Sula et al. 2013). The term "IoT" was coined by Kevin Ashton in 1999 to refer to "uniquely identifiable objects/things and their virtual representations in an internet-like structure" (Han 2011; Uzelac et al. 2015).

According to the industry analysis firm Gartner, 8.4 billion "things" were connected to the Internet in 2017, excluding laptops, computers, tablets, and mobile phones. According to the GSM Association, an industry organization that represents the interests of mobile network operators worldwide, the number of IoT devices is expected to grow to 25.1 billion by 2025.

While there is still no universally accepted and actionable definition exists, Jeff Voas, a computer scientist at the US National Institute of Standards and Technology (NIST), recommends using the acronym NoT (Network of Things) interchangeably to refer to IoT systems: "IoT is an instantiation of a NoT, more specifically, IoT has its 'things' tethered to the Internet. A different type of NoT could be a Local Area Network (LAN), with none of its 'things' connected to the Internet. Social media networks, sensor networks, and the Industrial Internet are all variants of NoTs. This differentiation in terminology provides ease in separating out use cases from varying vertical and quality domains (e.g., transportation, medical, financial, agricultural, safety-critical, security-critical, performance-critical, high assurance, to name a few). That is useful since there is no singular IoT, and it is meaningless to speak of comparing one IoT to another."

According to Voas, a NoT can be described by five primitives proposed in (Voas 2016):

1. **Sensor** is an electronic utility (e.g., proximity sensors, pressure sensors, temperature sensors, gas sensors, image sensors, acoustic sensors) that measure physical properties such as sound, weight, humidity, temperature, and acceleration. Properties of a sensor could be the transmission of data (e.g., Radio-Frequency IDentification [RFID]), Internet access, and/or be able to output data based on specific events.

2. **A Communication Channel** is "a medium by which data are transmitted (e.g., physical via Universal Serial Bus, wireless, wired, and verbal)." Since data is the "blood" of a NoT, communication channels are the "veins" and "arteries," as data moves to and from intermediate events at different snapshots in time.

3. **Aggregator** is "a software implementation based on mathematical function(s) that transforms groups of raw data into intermediate, aggregated data. Raw data can come from any source." Aggregators have two actors for consolidating large volumes of data into lesser amounts:
   a. *Cluster* is "an abstract grouping of sensors (along with the data they output) that can appear and disappear instantaneously."

4. b. *Weight* is "the degree to which a particular sensor's data will impact an aggregator's computation."

5. **Decision Trigger** "creates the final result(s) needed to satisfy the purpose, specification, and requirements of a specific NoT." Decision trigger is a conditional expression that triggers an action and abstractly defines the end purpose of a NoT. A decision trigger's outputs can control actuators and transactions.

6. **External Utility** (eUtility) is "a hardware product, software, or service, which executes processes or feeds data into the overall dataflow of the NoT."

There are various challenges associated with building requirements for IoT systems. First, it is a relatively new domain, and capturing the requirements based on the proper domain knowledge is necessary before designing and developing IoT-based systems. Secondly, when specifying the functionality for IoT applications, attention is naturally focused on concerns such as fitness of purpose, wireless interoperability, energy efficiency, and so on. Conventional requirements elicitations techniques such as domain analysis, joint application development (JAD), and quality function deployment (QFD) among others are usually adequate for these kinds of requirements. But in some domains, such as healthcare or education, where IoT applications can be deployed, some quality requirements are probably of greater concern.

For example, given the increased communication and complexity of IoT technology, there is an increase in security-related concerns (Georgescu and Popescu 2015). Many of the devices used in a provisioned, specialized IoT will collect various data whether that surveillance is known or not" (Laplante et al. 2015). But why are these data being collected? Who owns the data? And where does the data go? These are questions that need to be answered by the legal profession and government entities that will oversee the deployment of IoT systems in various domains.

On the other hand, by embedding sensors into front field environments as well as terminal devices, an IoT network can collect rich sensor data that reflect the real-time environment conditions of the front field and the events/activities that are going on. Since the data is collected in the granularity of elementary event level in a 7X24 mode, the data volume is very high and the data access pattern also differs considerably from traditional business data. The related "scalability" requirements will need to be addressed.

Finally, deploying IoT systems opens the doors for new quality attributes to emerge. For example, there are questions on the moral role that IoT may play in human lives, particularly concerning personal control. Applications in the IoT involve more than computers interacting with other computers. Fundamentally, the success of the IoT will depend less on how far the technologies are connected and more on the humanization of the technologies that are connected. IoT technology may reduce people's autonomy, shift them toward particular habits, and then shift power to corporations focused on financial gain. For deploying IoT technologies in classrooms, for instance, this effectively means that the controlling agents are the organizations that control the tools used by the academic professionals but not the academic professionals themselves (Gubbi et al. 2013). The dehumanization of humans in interacting with machines is a valid concern. Many studies indicate that face-to-face interaction between students will not only benefit a child's social skills but also positively contributes toward character building. The issue that may arise from increased IoT technologies in education is the partial loss of the social aspect of going to school. Conversely, using IoT in virtual learning environments can be of special support to students of special needs (e.g., dyslexic and dyscalculic needs (Lenz et al. 2016)).

Similarly, when constructing IoT systems for the healthcare domain, it is important to engage all stakeholders when trying to define a notion of "caring" for a new healthcare system, and it is critically important to engage systems engineers, computer scientists, doctors, nurses, and most important patients during requirements discovery. Laplante and Kassab (2017) presented a structured approach based on the above discussed NoT primitives for describing IoT for healthcare systems while illustrating their approach for three use cases and discussing relevant quality issues that arise, in particular, the need to consider "caring" as an emerging quality requirement.

VIGNETTE 12.2   Virtual Reality

Virtual reality (VR) is a computer-generated simulation where the user can interact with a virtual environment. In fully immersive systems, the viewer sees the virtual world through a virtual reality headset or a projection-based device and interacts with the 3D objects with controllers. Augmented reality is a VR simulation where real-world images are augmented by computer-generated ones; for example, in the case of inspecting a bridge, schematics of the under-lying materials can be viewed as if using X-rays. Augmented reality is widely used in surgical situations where images from ultrasound, CAT scans, MRI scans, etc. can be virtually superimposed on the image of the patient's body as it is viewed through special optical equipment.

You are likely familiar with VR from the many gaming systems available today, and for entertainment and training (e.g., flight simulation, surgery), VR is widely used. But VR can also be used for requirements elicitation, valida-tion, and reconciliation. For example, architects and homebuilders can use virtual walkthroughs through simulated buildings to help customers experi-ence the building before it is built.

These kinds of walkthroughs help customers experience size, scale, and position and to remove the "aha, I didn't want that" moments that come after the building or system has been constructed. VR is especially useful for RE in novel applications where usability information is hard to get—by putting users in VR worlds to simulate the use of a new product the require-ments engineer can see how they use the product to determine needed or unneeded features, identify hazards, and improve the user experience before the product is built.

REFERENCES

C. C. d. Santos, M. E. Delamaro and F. L. S. Nunes, "The Relationship between Requirements Engineering and Virtual Reality Systems: A Systematic Literature Review," *2013 XV Symposium on Virtual and Augmented Reality*, 2013, pp. 53–62, doi: 10.1109/SVR.2013.52.

Aman Bhimani, "Feasibility of Using Virtual Reality in Requirements Elicitation Process," Master's Thesis, Kennesaw State University, pp. 11–28, 2017, https://digitalcommons.kennesaw.edu/cgi/viewcontent.cgi?article=1001& context=msse_etd

---

Requirements Engineering and Artificial Intelligence (AI)

McCarthy (2007) defines AI as the science and engineering of developing computer programs which when fed into a machine make the machine exhibit the intelligence of humans. AI has been well researched for use in requirements engineering. Over the years, many AI techniques have been employed to capture, represent, and analyze requirements.

Requirements elicitation is probably the most investigated requirements engineering task. One prominently explored task within elicitation is ambiguity detection, which is typically sup-ported via a combination of natural language processing and machine learning (e.g., Chechik 2019 and Ferrari and Esuli 2019). Another sample AI-elicitation approach is the work from Peclat and Ramos (2018) who examined using semantic analysis for identifying security concerns from collections of unstructured textual documents. The automated analysis of user feedback and emotions (part of affective computing) to capture requirements and feedback directly and automatically from the end is also gaining traction (Groen et al. 2017).

Another topic that received great attention is using AI for requirements traceability. For example, Hayes et al. (2019) presented the use of metadata, such as readability indexes, as a resource for requirements traceability.

Automated classification remains also one of the most mature AI techniques applied to tackle requirements engineering problems. Stanik et al. (2019) presented an application for classifying reviews to better make sense and use of the user opinions shared in social media. They compared traditional machine learning with deep learning when classifying app reviews and users' tweets into problem reports (potential bugs), inquiries (feature requests), and being irrelevant (noisy feedback). Rashwan (2012) also presented a semantic analysis approach to classifying requirements into functional and nonfunctional requirements.

Requirements analysis practices have also been approached with AI. For example, del Sagrado and del Aguila (2018) presented an approach using Bayesian network requisites to predict whether the requirements specification documents have to be revised. Requisites provide an estimation of the degree of revision for a given requirements specification (i.e., SRS). Thus, it helps when identifying whether a requirements specification is sufficiently stable and needs no further revision.

While AI had the potential to revolute requirements engineering practices, many challenges still exist. Deep learning is data-hungry, which calls for a community effort in data sharing, curation, and provenance. More work on understanding the data is necessary. Also, as with other disruptive technologies, emerging quality requirements need to be fully understood. In the context of AI, there is a need to be aware of new quality requirements, such as explainability, freedom from discrimination, and specific legal requirements (particularly those regulating the deployment of AI in safety-critical contexts).

It is also worth mentioning that there is a distinction between "AI in RE" and "RE in AI." While the first focuses on how to develop machine learning and deep learning to improve requirements engineering tasks, the second focuses on how requirements engineering will contribute to AI technology. Dalpiaz and Niu (2020) provided a summary of the current research trends in each of these two directions.

Requirements Engineering and Cloud Computing

In industrial practice, cloud computing is becoming increasingly established as an option for increasing productivity, reducing labor and infrastructure costs, and improving agility. According to the definition from NIST, cloud computing is defined as "a model for enabling ubiquitous, convenient, on-demand network access to a shared pool of configurable computing resources (e.g., networks, servers, storage, applications, and services) that can be rapidly provisioned and released with minimal management effort or service provider interaction" (Mell and Grance (2011).

Despite the obvious benefits, cloud-based solutions face questions relating to the best architectures, privacy concerns, difficulty complying with regulations, performance problems, and availability issues. Wind and Schrödl (2010) examined selected requirements engineering approaches to study their extent to accommodate specific requirements of cloud-based solutions. They recommend that the following characteristics need to be examined when selecting a requirements engineering approach for cloud computing systems:

■ The architectural capacity of a requirements engineering approach to describe the connecting element between the individual application components.

■ The agility of the requirements engineering approach in relation to the description of architecture such as the integration of multidiscipline components from different domains, or different requirements sources, affects the requirements engineering process.

■ The structured elicitation of infrastructure requirements. These infrastructure requirements must be allocated into areas of service quality, security, and economic dimension.

■ Comprehensive inclusion of the customer into the entire development process during every phase. Even where this can be difficult, due to different language biases and differing levels of understanding by developers, this must not be abandoned.

■ The preparation for an optimally functioning change management system for the phase following delivery, in order to be able to implement any modifications in the service area.

Requirements Engineering and Affective Computing

Affective computing is an interdisciplinary field spanning computer science, psychology, and cognitive science, and focuses on the study and development of systems that can recognize, interpret, process, and simulate human affects. Awareness of emotions to be experienced by end-users is of utmost importance in requirements engineering and therefore emotional requirements should be always considered. These emotions and stakeholders' cognitive states arise and evolve differently during the elicitations, prioritization, and negotiation processes. Hence, it is becoming increasingly important to understand these emotions and affects to be experienced by stakeholders within sociotechnical systems. The sentiment expressed together with opinions helps to assess important topics and creates actionable insights.

Trending research topics to further explore the relation between affective computing in requirements include:

■ Methods and artifacts for elicitation and modeling of emotional requirements, including the relevant approaches of participatory requirements engineering.

■ The potential and challenges of different types of personality characteristics of software engineers when conducting requirements engineers along with the limitation regarding the current state of the art.

■ Defining or adapting psychological models of affect to RE (e.g., understanding what may trigger positive or negative emotions during the requirement engineering process).

■ Exploration of biometric sensors emerging from hardware (e.g., smartwatch), which enable new measurement techniques to support the validation and verification of requirements.

References

Azar, J., Smith, R. K., & Cordes, D. (2007). Value-oriented requirements prioritization in a small development organization. *IEEE Software*, 24(1): 32–37.

Bose, S., Kurhekar, M., & Ghoshal, J. (2008). Agile methodology in requirements engineering. *SETLabs Briefings Online*, 13–21.

Chakraborty, P., Shahriyar, R., Iqbal, A., & Bosu, A. (2018, October). Understanding the software development practices of blockchain projects: A survey. In *Proceedings of the 12th ACM/IEEE International Symposium on Empirical Software Engineering and Measurement* (pp. 1–10), Oulu, Finland.

Chechik, M. (2019, September). Uncertain requirements, assurance and machine learning. In *2019 IEEE 27th International Requirements Engineering Conference (RE)* (pp. 2–3). IEEE, Jeju Island, South Korea.

Dalpiaz, F., & Niu, N. (2020). Requirements engineering in the days of artificial intelligence. *IEEE Software*, 37(4): 7–10.

del Sagrado, J., & del Aguila, I. M. (2018). Stability prediction of the software requirements specification. *Software Quality Journal*, 26(2): 585–605.

Demi, S. (2020, August). Blockchain-oriented requirements engineering: A framework. In *2020 IEEE 28th International Requirements Engineering Conference (RE)* (pp. 428–433). IEEE, Zurich, Swizerland.

Destefanis, G., Marchesi, M., Ortu, M., Tonelli, R., Bracciali, A., & Hierons, R. (2018, March). Smart contracts vulnerabilities: A call for blockchain software engineering? In *2018 International Workshop on Blockchain Oriented Software Engineering (IWBOSE)* (pp. 19–25). IEEE, Gothenburg, Sweden.

European Commission. (2005). *The New SME Definition: User Guide and Model Declaration Section*. Office for Official Publications of the European Communities, Brussels.

Ferrari, A., & Esuli, A. (2019). An NLP approach for cross-domain ambiguity detection in requirements engineering. *Automated Software Engineering*, 26(3): 559–598.

Georgescu, M., & Popescu, D. (2015). How could Internet of Things change the E-learning environment. *The 11th International Scientific Conference eLearning and Software for Education*, Bucharest, Romania.

Groen, E. C., Seyff, N., Ali, R., Dalpiaz, F., Doerr, J., Guzman, E., ... Stade, M. (2017). The crowd in requirements engineering: The landscape and challenges. *IEEE Software*, 34(2): 44–52.

Gubbi, J., Buyya, R., Marusic, S., & Palaniswami, M. (2013). Internet of Things (IoT): A vision, architectural elements, and future directions. *Future Generation Computer Systems*, 29(7): 1645–1660.

Han, W. (2011). Research of intelligent campus system based on IOT. In *Advances in Multimedia, Software Engineering and Computing* (Vol. 1, pp. 165–169). Springer.

Hayes, J. H., Payne, J., & Leppelmeier, M. (2019, September). Toward improved artificial intelligence in requirements engineering: Metadata for tracing datasets. In *2019 IEEE 27th International Requirements Engineering Conference Workshops (REW)* (pp. 256–262). IEEE, Jeju Island, South Korea.

JPMorgan Chase Co. (2018). Small Businesses Are an Anchor of the US Economy. https://www.jpmorgan-chase.com/institute/research/smallbusiness/small-business-dashboard/economic-activity.

Kandaswamy, R., & Furlonger, D. (2018). Blockchain-based transformation: A Gartner trend insight report. https://www.gartner.com/en/doc/3869696-blockchain-based-transformation-a-gartner-trend-insight-report

Kassab, M. (2015, August). The changing landscape of requirements engineering practices over the past decade. In *2015 IEEE Fifth International Workshop on Empirical Requirements Engineering (EmpiRE)* (pp. 1–8). IEEE, Ottawa, Canada.

Kassab, M. (2021). How requirements engineering is performed in small businesses? *Proceeding of the Workshop on Requirement Engineering for Software Startups and Emerging Technologies in Conjunction with 29th IEEE International Requirements Engineering Conference 2021*. Vitual.

Kassab, M., & Laplante, P. (2022). The current and evolving landscape of requirements engineering state of practice. *IEEE Software*. DOI: 10.1109/MS.2022.3147692.

Kassab, M., Destefanis, G., DeFranco, J., & Pranav, P. (2021, May). Blockchain-Engineers wanted: An empirical analysis on required skills, education and experience. In *2021 IEEE/ACM 4th International Workshop on Emerging Trends in Software Engineering for Blockchain (WETSEB)* (pp. 49–55). IEEE.

Laplante, P. A., Kassab, M., Laplante, N. L., & Voas, J. M. (2017). Building caring healthcare systems in the Internet of Things. *IEEE Systems Journal*, 12(3): 3030–3037.

Laplante, P. A., Laplante, N., & Voas, J. (2015). Considerations for healthcare applications in the Internet of Things. *Rel. Dig.*, 61(4): 8–9.

Lenz, L., Pomp, A., Meisen, T., & Jeschke, S. (2016). How will the Internet of Things and Big Data analytics impact the education of learning-disabled students? A concept paper. In *3rd MEC International Conference on Big Data and Smart City (ICBDSC)* (pp. 1–7). IEEE, Muscat, Oman.

McCarthy, J. (2007). What Is Artificial Intelligence? Computer Science Department, Stanford University, Stanford, CA. http://www-formal.stanford.edu.ezaccess.libraries.psu.edu/jmc/whatisai.pdf

Mell, P., & Grance, T. (2011). The NIST definition of cloud computing. https://csrc.nist.gov/publications/detail/sp/800-145/final

Peclat, R. N., & Ramos, G. N. (2018). Semantic analysis for identifying security concerns in software procurement edicts. *New Generation Computing*, 36(1): 21–40.

Rashwan, A. (2012, May). Semantic analysis of functional and non-functional requirements in software requirements specifications. In *Canadian Conference on Artificial Intelligence* (pp. 388–391). Springer, Berlin, Heidelberg.

Stanik, C., Haering, M., & Maalej, W. (2019, September). Classifying multilingual user feedback using traditional machine learning and deep learning. In *2019 IEEE 27th International Requirements Engineering Conference Workshops (REW)* (pp. 220–226). IEEE, Jeju Island, South Korea.

Sula, A., Spaho, E., Matsuo, K., Barolli, L., Miho, R., & Xhafa, F. (2013). An IoT-based system for supporting children with autism spectrum disorder. In *2013 Eighth International Conference on Broadband and Wireless Computing, Communication and Applications* (pp. 282–289), IEEE, Compiegne, France.

U.S. Small Business Administration. https://www.sba.gov

Uzelac, A., Gligoric, N., & Krco, S. (2015). A comprehensive study of parameters in physical environment that impact students' focus during lecture using Internet of Things. *Computers in Human Behavior*, 53: 427–434.

Voas, J. M. NIST SP 800-183 Networks of 'Things'. http://dx.doi.org/10.6028/NIST.SP.800-183, http://nvlpubs.nist.gov/nistpubs/SpecialPublications/NIST.SP.800-183.pdf, 2016

Wind, S., & Schrödl, H. (2010, December). Requirements engineering for cloud computing: A comparison framework. In *International Conference on Web Information Systems Engineering* (pp. 404–415). Springer, Berlin, Heidelberg.

Appendix A: Software Requirements Specification for a Smart Home

1 Introduction

1.1 Purpose: Mission Statement

Making residential enhancements that will pave the way for an easy and relaxed transition into retired life.

Document prepared for the Smith family home, a pre-existing building.

1.2 Scope

The "Smart Home" system, herein referred to as "The System," will be a combination of hardware and software which will provide an escape from daily routines and mundane tasks. This product seeks out the items that consume the most time but do not need to. The goal is to automate that which does not really need human interaction, to free the occupants to enjoy themselves in their retirement. The system will not free the mundane household chores from any human interaction, but it will require only as little as needed.

1.3 Definitions, Acronyms, and Abbreviations

pH—see http://en.wikipedia.org/wiki/PH RFID—Radio Frequency Identification SH—Smart Home

SAN—Storage Area Network SRS—System Requirements Specification WPA—Wi-Fi Protected Access WEP—Wired Equivalent Privacy USB—Universal Serial Bus

1.4 References

802.11 IEEE Specification. https://standards.ieee.org/findstds/standard/802.11-2016.html (accessed June 2017).

1.5 Overview

Requirements have been divided into key functional areas, which are decomposed into features within those functional areas. Functional and nonfunctional requirements exist within the laid-out document format. The order of the leading sections and the corresponding requirements should be interpreted as priority.

2 Overall Description

2.1 Product Perspective

This system consists of many standalone devices. At this time, it is not known if these devices exist in the commercial market or not. The document has a holistic approach, intermingling the demand for devices with the functions of the system. The first step in continuing with this project would be to decide on feasibility and do some cost analysis for some of the requirements contained herein.

This document seeks to lay out areas where all interfaces are abstracted. There are areas where there would clearly need to be communication between the various interfaces, but since there is no targeted device, there is no known protocol for speaking with that device.

2.2 Product Functions

The functions of this product will be divided into six categories. Accessibility is the first and most highly desired by the customer. This functional category seeks to improve the user experience of the system by providing various benchmarks for deciding on usability. The second major functional area is environmental considerations. The aim of this area is to ensure that the inhabitants have a safe environment to live and that the SH system enhances this environment instead of adding risks or hazards to the environment. The most important aspects for this category within this document will be monitoring and helping the quality of air and water. The third category is energy efficiency. It is desired by the system and the customers to not only enhance their lives while living in this SH but also live in an efficient way. This system will not only monitor the occupants' energy usage but also seek to improve the occupants' abilities to save on energy costs. Fourth we have security. Security is important to the customers as they want their home to be safe. The security system in the SH will provide added layers of protection from various crimes as they are happening, but also add layers to help prevent crimes from happening in the first place. The security will also give the occupants more peace of mind as they will have far greater control and oversight should they ever need to go for some extended trip away from their SH. The fifth section deals with media and entertainment. The goal of this section of the system is to decentralize home entertainment and make it available to the occupants wherever they desire. Finally, there will be automation. This is the section that will get into the guts of what is meant by taking as much of the human element out of routine tasks as possible. The summation and harmonization of all the six categories of the SH will provide for a truly rewarding living experience for the occupants of the SH.

2.3 User Characteristics

The primary users of this system will be two older adults entering retirement. One of the adults spent his work life doing IT support and has a mild degree of electronic and computer expertise.

The other was a school teacher and is not very familiar or comfortable with electronic and computing devices. These individuals are both of sound physical abilities, although one is a little shorter and suffers from sporadic hip pains.

2.3.1 User/Stakeholder Profiles

Stakeholder	Interests	Constraints
Local building codes	Ensuring the safety of the building for the inhabitants	Multiple business codes, especially around electrical interfaces
Robert and Elizabeth Smith	Inhabitants interested in easing their lives	None
Interior designer	Ensuring the functionality of the system does not deter from the esthetic	None
Building architect	Ensuring the existing structure can support the improvements	None
Construction workers	Making sure the construction details are clearly identified	None
Developers	Making sure interfaces are defined	None
Tim Smith [son]	Ease of use for occasional user	None
Cats	Safety	None
Relatives	Ease of use and comfort	One relative in wheelchair
House sitters	Easily understanding limited sets of functionality	None
Guests	Ease of use and comfort	None
Maid service	Minimal functional understanding	None
Utility company	Negative for alternative energy, sharing alternate use policies	None
Internet services provider	Making sure bandwidth and services are available	None
Tivo	Negative, will lose business	None

2.4 Constraints

IEC 61508—Providing for functional safety.
None others at this time, as feasibility and cost estimation activities are out of scope.

2.5 Assumptions and Dependencies

All hardware is available.
All devices will present the data listed below.
Occupants will provide feed elements for the devices that require them.

3 Core System Requirements

This section will list all of the core functional requirements for the SH system.

3.1 Central Processing Requirements

3.1.1 System shall operate on a system capable of multiprocessing.
3.1.2 System shall operate on a system capable of near-real-time execution of instructions.
    3.1.2.1 System shall service triggers or stimuli in no more than 500 ms.
3.1.3 System shall operate in a highly available and fault-tolerant manner.
    3.1.3.1 System shall have a reported uptime of 99.99% (4 NINES).
    3.1.3.2 System shall recover from locked state in less than 1 second.
3.1.4 System shall have a database associated with it that can handle transaction processing at a rate of 1,000 transactions per minute.
3.1.5 System shall have redundant databases for fail over purposes.
3.1.6 System shall perform periodical offsite and onsite backups of all configuration and reporting data.
3.1.7 System shall support wireless encryption protocols WPA [1–2] and WEP.
3.1.8 System shall support wired Ethernet for 1 gigabit/second and contain cat 6e cabling.
3.1.9 System may contain a separate SAN device for storage flexibility.
3.1.10 System may contain a separate video recorder/processor for process distribution.
3.1.11 If a system supports recording more than three television shows simultaneously, then the system shall have a separate video recorder.

4 Accessibility

Accessibility is defined as the need for the SH system to be usable by all persons, including those with any physical impairments or those with difficulty operating and/or understanding complex electronic systems. Priority = High.

4.1 Use of SH Features

4.1.1 SH system shall be usable by those with slight eye loss.
    4.1.1.1 System shall not have any buttons smaller than 1 inch square.

4.1.1.2 System shall have all consoles and controlling devices between 4 and 5 feet from ground level.

4.1.1.3 System shall have backlighting on all buttons for nighttime ease of use.

4.1.1.4 System shall have options to increase and decrease font sizes on web interfaces and all console and controlling devices.

4.1.1.5 System shall have liquid layouts for all graphical interfaces for display on many different types of display devices.

4.1.2 System shall be easy to use.

4.1.2.1 System shall be understood by users of all levels of understanding with no more than 4 hours of training.

4.1.2.2 System shall have a help function associated with all user entry possibilities.

4.1.2.3 System shall have text-to-speech capabilities to allow the user to receive vocal instructions for help and how to menu items.

5 Environment

Environment encompasses air quality controls but also includes other environmental elements such as lighting levels and water purification. Priority = High.

5.1 Water and Air Purification

Water purification and air quality are key factors to a good environment within the SH. This system seeks to improve by monitoring both air quality and water purification. Priority = High.

5.1.1 SH shall have a reverse osmosis water purification system.

5.1.2 SH shall have a nonfiltered water system.

5.1.3 System shall store how much water passes through the filtration unit each day.

5.1.4 System may send notifications to users about how much water passes through the filtration system.

5.1.5 System shall have the option to send notifications to users when water filtration unit needs cleaning.

5.1.6 System shall incorporate a water softener system into the water system.

5.1.7 System shall monitor the salt in the water softener.

5.1.8 System shall accept user input for desirable levels of salt in the water softener device.

5.1.9 System shall send notifications to users when salt in softener gets below user-defined levels.

5.1.10 System shall monitor the air filter.

5.1.11 System shall send a notification to users when the air filter needs to be cleaned and/or changed.

5.1.12 System shall provide monitors for measuring air quality.

5.1.13 System shall accept user input for air quality thresholds.

5.1.14 System shall notify users when air quality reaches levels outside the user-defined thresholds.

5.2 Health and Safety Detectors

This section describes the role and interfaces for various common household detectors. The detectors carry out their common functionality, but reactions are automated, and historical data are logged. Priority = High.

5.2.1 SH shall have at least one (1) multipurpose detector for detecting smoke and carbon monoxide on each floor.

5.2.2 System shall not interfere in any way with detector's manufacturer's operating procedures.

5.2.3 System shall accept user input for dangerous levels of smoke and carbon dioxide.

5.2.4 System shall trigger warning and require additional confirmation when users select levels outside of the manufacturer's settings for dangerous levels of smoke and carbon dioxide.

5.2.5 System shall notify proper authorities when levels above user-defined thresholds of smoke or carbon monoxide are detected.

5.2.6 System shall tie into detectors to send remote alert messages to users when elevated levels of smoke or carbon monoxide are detected.

5.2.7 System shall utilize radon detector in the basement.

5.2.8 System shall allow users to set defined ceiling for radon levels.

5.2.9 System shall accept input from users for notification events for radon level detections.

5.2.10 System shall send notifications based on user-defined notification events to interested parties when radon levels are more than the user-defined ceiling.

5.2.11 System shall activate basement fan system when radon levels report above-defined ceiling.

5.2.12 System shall record radon levels routinely.

5.2.13 System shall allow users to view reports on radon levels.

5.2.14 System shall persist radon level data for no less than 90 days.

6 Energy Efficiency

Energy efficiency covers the extent to which the SH system enables the users to monitor and enhance energy efficiency of the house through "smart" and adaptive controls and interfaces. Priority = High.

6.1 Air Conditioner/Heating

Controlling and adapting the air conditioning and heating are important aspects of using energy efficiently. Not only does the SH seek to improve the ease of use of traditional thermostats, but also provides intelligence in order to optimize the use of the system. Priority = High.

6.1.1 SH shall be divided up into zones for heating and cooling.

6.1.2 System shall accept desired temperature settings for each zone, for no less than four periods in the day.

6.1.3 System shall accept input for desired room temperature when room is unoccupied.

6.1.4 System shall detect motion to determine if a room is occupied and make proper adjustments to the temperature.

6.1.5 System shall differentiate between pets and occupants for motion detection and temperature adjustment.

6.1.6 System shall monitor outdoor temperature and humidity.

6.1.7 System shall shut down air conditioning and open windows if temperature outside is cooler than the inside temperature.

6.1.8 System shall not open or close any windows if there is something in the desired path of the window (see Figure A.1).

6.1.9 System shall reverse directions of windows if they encounter any resistance.

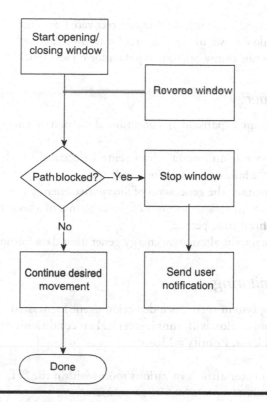

Figure A.1 Window movement flow chart.

6.1.10 System shall send notifications to users if windows need to reverse path or windows cannot complete a desired action (open or close).

6.2 Time-of-Day Usage

Time-of-day usage refers to common utility programs that provide reduced rates for utilities used during off-peak time periods.

6.2.1 Appliances shall be configured if they need to be used as time-of-day devices.

6.2.2 System shall accept definitions for the range for time-of-day savings.

6.2.3 System shall queue up appliance(s) to run when time-of-day period starts if it is a time-of-day device.

6.2.4 System shall allow a user to override the time-of-day setting to run the device/appliance immediately.

6.2.5 System may send a notification when the device has completed its work.

6.3 Water Recovery

Promote reusable resources by capturing rainwater to use for irrigation. Priority = Medium.

6.3.1 System shall have a water recovery system for rainwater.

6.3.2 System shall use water recovered from rain for lawn irrigation.

6.3.3 System shall record amounts of rainwater recovered on a monthly basis.

6.3.4 System shall allow a user to view reports for rainwater recovery.

6.3.5 System shall persist rainwater recovery data for no less than 24 months.

6.4 Alternative Energy

Interface to allow for future expansion and addition of alternative energy sources. Priority = Low.

6.4.1 System shall provide an interface into central electrical supply for alternative energy to supply power to the house (i.e., solar or wind).

6.4.2 System shall monitor the generation of alternative energy.

6.4.3 System shall present reports to users of the amounts of alternative energy generated during some user-defined time period.

6.4.4 System shall maintain alternative energy generation data for no less than 2 years.

6.5 Air Flow Monitoring

The airflow monitoring system serves as a detection agent for wasted energy. The monitoring of airflow, especially in conjunction with running central air conditioning or heating, will lead to the discovery of drafts and leaks. Priority = Medium.

6.5.1 System shall monitor airflow in various rooms within the SH.

6.5.2 System shall present reports for airflow to users.

6.5.3 System shall persist air flow data for no less than 3 years.

6.5.4 System shall accept input for thresholds for detecting drafts or leaks of air within the house.

6.5.5 Assuming the house is running centralized air or heat, system shall send notifications if drafts are detected that exceed the user-defined threshold.

7 Security

Security includes aspects of home security like alerts in the event of a break-in, video monitoring of premises or areas of interest, as well as unattended monitoring while the occupant(s) is away. Priority = High.

7.1 Home Security

Home security centers around controlling the access points to the SH as well as providing many cameras to provide views into areas of the SH. The SH will provide automated and human commanded responses to various security situations. Priority = High.

7.1.1 System shall have biometric and keypad door locks for all points of entry into the house.

7.1.2 System shall allow users to encode a one-time use door code for expected visitors.

7.1.2.1 System shall allow users to remotely code a one-time use code for visitors (i.e., over the phone, Internet, or some other mobile device).

7.1.3 System shall record all entries based on code entered or biometrics presented.

7.1.4 System shall present a report to users for all entries.

7.1.5 System shall persist home entry data for no less than 10 years.

7.1.6 System shall allow for RFID tags to open garage doors.

7.1.7 System shall allow for biometric and keypad entry to open garage doors.

7.1.8 System shall allow user to configure maximum duration for garage door to remain open.

7.1.9 System shall shut open the garage door if it is open past the user-defined maximum.

7.1.10 System shall allow a user to override automatically shutting the garage door, that is, "Hold Open option."

7.1.11 Garage door shall reverse course if something is blocking its path.

7.1.12 System shall notify user if the garage door is unable to safely close.

7.1.13 System shall allow users to configure entry routine for all RFID, biometric, and key codes, that is, upon entry for user X through the garage door, turn on garage light, hall light, and kitchen light.

7.2 Unattended Home

Unattended home is a set of responses to various triggers throughout the home and immediate responses to those triggers. This will aid in security of the home. Priority = High.

7.2.1 System shall allow users to set an away mode.

7.2.2 System shall define away mode as time and date range when users will be away from their house.

7.2.3 System shall allow users to configure lights in any room to go on for some defined duration while they are away.

7.2.4 System shall deploy motion detectors to always be on for the duration while the user is away.

7.2.5 System shall differentiate between motion detected for pets and humans.

7.2.6 System shall send a notification to users if any motion detectors are triggered while the user is away.

7.2.7 User shall be presented options to view various cameras via the web or some other mobile device when motion detectors are triggered.

7.2.8 User shall be given the option to alert the authorities when motion detectors are triggered.

7.2.9 System shall turn on user-defined lights in and outside the house when motion detectors are triggered.

7.3 Monitoring Anywhere

Occupants and users of the SH's system should be able to monitor the home from anywhere they wish. This includes many different cameras within the SH as well as various points of entry and other triggers placed throughout the home. This will give the occupants more freedom to travel while feeling their home is secured and well cared for. Priority = Medium.

7.3.1 System shall show camera data streams to any television in the house.

7.3.2 System shall incorporate cameras at points of entry with doorbells to allow users to view visitor.

7.3.3 System shall allow user to remotely unlock the door to permit entry to the visitors.

7.3.4 System shall allow users to notify emergency personnel of possible intruder.

7.3.5 System shall permit users to view security cameras from a secure website or mobile device for remote viewing of property.

8 Media/Entertainment

Media and entertainment include the ability to create, store, and access multiple forms of media and entertainment such as audio and video anywhere in the house. Priority = Medium.

8.1 Recording Television Shows

Recording television shows allows the users to throw out the VCR and gives them a more automated and intelligent solution for recording all their favorite shows or movies that play through the television. Priority = Medium.

8.1.1 System shall allow user to record any show on television.

8.1.2 System shall present a web interface with a grid listing similar to the TV guidebook for users to select shows to record.

8.1.3 System shall allow user to record a minimum of two television shows simultaneously.

8.1.4 System shall make storage for recorded shows expandable.

8.1.5 System shall free storage space as needed by first-in first-out (FIFO) or some other defined priority schedule.

8.1.6 System shall provide a search feature to search through television shows to select which one to record.

8.1.7 System shall provide user the ability to record all occurrences of a specified show.

8.1.8 System shall provide user the ability to record only new instances of a specified show.

8.1.9 System shall provide telephone menu options for customer to dial in and select channel, time, and duration to record.

8.1.10 System shall present users option to select quality for recording.

8.1.11 System shall present user option to not automatically overwrite the television recording.

8.1.12 System shall give user the option to only store X number of episodes from a certain series at a time.

8.1.13 System may skip commercials when system is able to detect the commercial.

8.1.14 System shall monitor storage space for future recordings.

8.1.15 System shall send a notification when resources get low enough where recordings will be overwritten.

8.1.16 System shall permit users to not automatically delete a show or a series.

8.1.17 System shall not record any new shows if there is space available for recovery.

8.1.18 System shall send notifications to users if there is no longer space available to record new shows.

8.2 Video Entry

Video entry is the mechanism by which various formats of video data are able to be loaded into the repository for video playback. Priority = Medium.

8.2.1 System shall allow for video input into digital library.

8.2.2 System shall allow for storage of video metadata such as category, genre, title, and rating.

8.2.3 System shall provide an interface to users to edit and update video metadata.

8.2.4 System shall accept one-button touch support for incorporating VHS tape into digital library.

8.2.5 System shall accept one-button touch support for incorporating DVD videos into digital library, where law and technology provide.

8.3 Video Playback

Video playback allows the users of the SH to be able to enjoy video, both recorded and preloaded content from anywhere within the house. Priority = Medium.

8.3.1 System shall allow for recorded video playback at any television in the house.

8.3.2 System shall allow for other video media to be available for playback in any room with a television.

8.3.3 System shall allow all common features of a VCR player or DVD player, such as fast forward, rewind, and chapter skip.

8.3.4 System shall allow user to skip commercials where commercials are detected.

8.3.5 System shall prevent user from playing back identical media on multiple televisions at the same time.

8.3.6 System shall allow user to remove the recording from storage when they are done watching.

8.3.7 System shall allow user to remove other video media from the storage.

8.4 Audio Storage and Playback

The audio storage and playback are important features of the smart home. This section will describe how the audio is imported into the digital library as well as what capabilities there are for distributing or sharing the audio either to various rooms in the SH or to external media. Priority = Medium.

8.4.1 System shall accept input into digital audio library from CD.

    8.4.1.1 System shall allow user to enter a CD into a tray and immediately rip the CD.

    8.4.1.2 System shall collect all available metadata from the CD from the Internet for categorization.

    8.4.1.3 System shall store audio binary in a lossless format.

8.4.2 System shall accept input into the digital audio library from a USB device.

8.4.3 System shall provide interface for users to manually place an audio file into the digital audio library.

8.4.4 System shall automatically normalize volume of all audio files loaded into the digital audio library.

8.4.5 System shall store information about audio files in some searchable entity.

8.4.6 System shall provide users the ability to alter metadata for any file in the collection.

8.4.7 System shall allow users to remove audio files from the digital audio library.

8.4.8 System shall allow for categorization of audio files by important fields such as genre, artist, and album.

8.4.9 System shall allow audio playback.

8.4.9.1 System shall allow for wired or wireless connection to any device capable of audio playback.

8.4.9.2 System shall allow centralized panel to play back various audio files in different rooms of the house.

8.4.9.3 System shall provide an access point in garage so digital audio can be downloaded to an automobile audio system.

8.4.10 System shall allow users to author new CDs.

8.4.10.1 System shall allow users to select tracks for newly authored CD from a playlist or from the complete library.

8.4.10.2 System shall allow user to select which format to use.

8.4.10.3 System shall allow user to select a CD burning drive.

8.4.10.4 System shall provide guidance to users for available space depending on the drive and media selected as well as the format chosen.

8.4.10.5 System shall verify proper media is in the selected drive.

8.4.10.6 System shall allow user to select order for the tracks.

8.4.10.7 System shall allow user to confirm track information to start authoring the CD.

8.4.10.8 System shall perform necessary audio conversion and burn the CD based on the user's authoring details.

8.4.10.9 System may notify user when the authoring process is complete.

8.4.11 System shall allow users to create, edit, and delete playlists.

8.4.11.1 Playlist shall consist of one to "N" number of tracks selected from the digital library.

8.4.11.2 A single track may reside in any number of playlists.

8.4.11.3 A single track may not reside in an individual playlist more than once.

8.4.11.4 System shall allow users to set a name and description for all playlists created.

8.4.12 System shall allow users to transfer music from the digital library to portable music players.

8.4.12.1 System shall allow track transfer according to both complete playlists as well as individual tracks.

8.4.12.2 System shall allow users to format or delete the current selection of tracks on the portable device before transferring.

8.4.12.3 System shall allow users to append additional tracks or playlists to the portable device as space permits.

8.4.12.4 System shall add tracks to the device in the order they were selected by the user until all tracks are transferred or the device is full.

9 Automation

Automation is the process of making something routine happen automatically, on some set schedule or by some defined trigger such that little or no human intervention is needed. Priority = Medium.

9.1 Pet Care

The pet care system will be added to ensure that proper unattended care of the pets in the household takes place. In this instance, cats are the pets within the household, but the system should extend to other future pets as well. The goal is to have a mostly automated system to take care of the pets' primary needs such as food, water, and waste disposal, but the system can also track health care needs such as appointments and vaccinations. The logic flow for the feeding functionality of the pet care function is shown in Figure A.2. Priority = High.

9.1.1 System shall handle providing water for the pets.

9.1.1.1 Pet watering bowls shall be tied into the water filtration system [ref. requirement 5.1.1].

9.1.1.2 System shall monitor the consumption of water on a daily basis for pet bowls.

9.1.1.3 System may send time-defined notifications to users detailing the water consumption by pets.

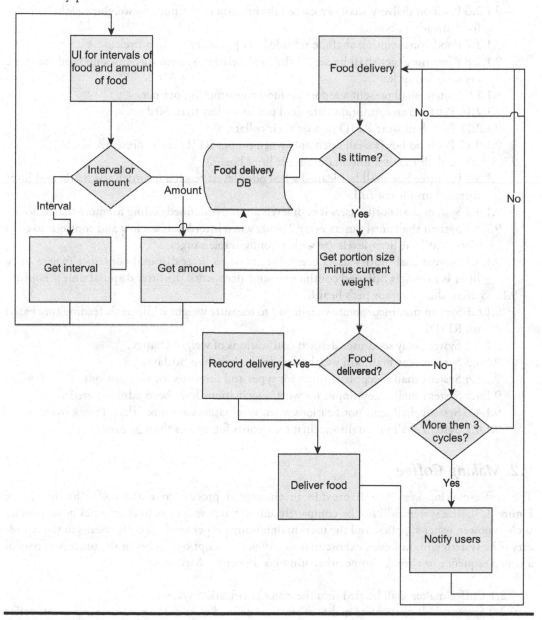

Figure A.2 Pet feeding flow diagram.

9.1.1.4 System shall present a report for water consumption by pets.

9.1.1.5 Pet water consumption data shall persist for no less than 30 days.

9.1.2 System shall provide food for pets.

9.1.2.1 System shall accept user input for intervals to deliver food.

9.1.2.2 System shall notify users when food in storage reaches low levels, as users will be required to fill the storage depot.

9.1.2.3 Pet's food shall be delivered to their bowls at every user-defined interval.

9.1.2.4 System shall allow user to set portion weight for every bowl for the system.

9.1.2.5 Pet food delivery shall be portioned to user-defined weight.

9.1.2.6 Pet food delivery shall not exceed the amount of the portion weight; weight includes food already in bowl.

9.1.2.7 Food consumption shall be recorded per pet every feeding cycle.

9.1.2.8 Alert messages shall be sent if the food delivery system dispenses no food for three consecutive cycles.

9.1.2.9 System shall present a report for food consumption per pet.

9.1.2.10 Pet food consumption data shall persist no less than 30 days.

9.1.2.11 Pets shall wear RFID tags on their collars.

9.1.2.12 Pet food bowls shall open only when proper RFID tag is present.

9.1.3 System shall monitor and maintain pet litter box(es).

9.1.3.1 Pet litter box shall be cleaned when odor levels reach a user-defined mark, and litter disposal unit is not full.

9.1.3.2 System shall notify users if odor levels are above defined ceiling for more than 8 hours.

9.1.3.3 System shall notify users every 2 hours when litter levels are low and continue to send alerts until the litter levels are within configurable ranges.

9.1.3.4 System shall notify users every 4 hours when litter disposal container (where dirty litter is stored) is full, and continue to send alerts until the litter disposal unit is not full.

9.1.4 System shall monitor pet's health.

9.1.4.1 System may incorporate weight pad to measure weight while pet is feeding (pet based on RFID).

9.1.4.2 System may send user-defined notifications of weight change.

9.1.4.3 System may maintain weight data for no less than 30 days.

9.1.4.4 System shall accept user input for types and intervals for vaccinations.

9.1.4.5 System shall accept input for when vaccinations have been administered.

9.1.4.6 System shall send notifications when vaccinations are more than 1 week overdue.

9.1.4.7 System shall maintain vaccination records for no less than 5 years.

9.2 Making Coffee

The coffee-making system will provide an automated process to make coffee for users (see Figure A.3). The system will not be completely autonomous as it does have external dependencies such as power, water supplies, and the users maintaining proper levels of coffee beans in the repository. The system will, however, expose many configuration options to begin the process as part of a timed sequence or through some other stimulus. Priority = Medium.

9.2.1 Coffee maker shall be tied into the water purification system.

9.2.2 System shall start coffee maker at any user-defined time as long as water is present, coffee bean levels are sufficient, and unit is powered.

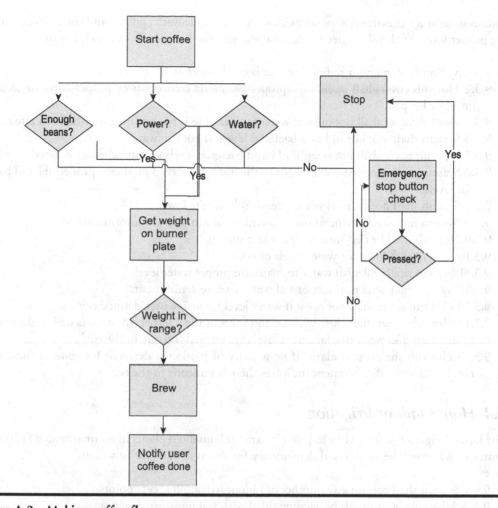

Figure A.3 Making coffee flow.

9.2.3 System shall send a notification when bean levels are low.

9.2.4 When bean levels are too low to make coffee, system shall send an alert and coffee maker shall blink a warning indicator.

9.2.5 Coffee maker shall use a reusable filter.

9.2.6 System shall send a notification when filter should be cleaned or changed.

9.2.7 Coffee maker shall shut off if weight measured by burner plate is less than the weight of the carafe or more than the weight of a full carafe.

9.2.8 Coffee maker shall have an emergency stop button clearly visible and accessible.

9.2.9 Coffee maker shall stop within 1 microsecond when button is pressed.

9.2.10 Coffee maker shall turn off if carafe is not removed from burner plate for 30 consecutive minutes.

9.3 Hot Tub Maintenance

Hot tub maintenance is important to the users as this process will enable them to have little interaction in the daily needs of maintaining the hot tub. The process once again is not completely autonomous as it has external dependencies such as available water, power, and the users maintaining proper levels of chemicals needed for maintaining the hot tub. Priority = Medium.

9.3.1 System shall monitor temperature of hot tub water at all times.

9.3.2 Hot tub cover shall open with proper biometric credentials or proper entry of code on numeric keypad.

9.3.3 Hot tub cover shall not open if water temperature is not within a user-defined safe range.

9.3.4 System shall monitor pH and bacterial levels of hot tub water.

9.3.5 Hot tub cover shall not open if pH and bacterial levels are outside user-defined norms.

9.3.6 System shall administer chemicals to the hot tub water to maintain proper pH and bacterial levels.

9.3.7 System shall notify users when chemical levels are low.

9.3.8 System may send notifications of chemical administration information.

9.3.9 Hot tub shall be tied into central water system.

9.3.10 System shall monitor water levels of hot tub.

9.3.11 System shall replenish water to maintain proper water level.

9.3.12 System may send notifications of water level replenishment.

9.3.13 Hot tub cover shall not open if water level is outside user-defined norms.

9.3.14 Hot tub cover shall close with button press or if no activity/motion is detected for some time range, and water displacement levels are normal (no one in the tub).

9.3.15 Hot tub shall sound alarm if no activity or motion is detected for some defined time range and water displacement indicates there is someone in the tub.

9.4 Home Indoor Irrigation

The home irrigation system aims to provide care for household plants in an unattended fashion. A source of water will be an external dependency for this system. Priority = Low.

9.4.1 System shall control any number of indoor irrigation access points.

9.4.2 All access points shall be accompanied with soil moisture detector.

9.4.3 System shall allow user to define desired moisture level and watering intervals for each watering access point in the house.

9.4.4 System shall control watering units at each access point to keep soil at a steady moisture level.

9.4.5 If plants are on an interval, system shall bring the moisture level up to defined range during each watering interval.

9.4.6 System shall record water used and average soil moisture levels per access point.

9.4.7 System shall maintain indoor watering information for no less than 45 days.

9.4.8 System may accept input for soil moisture warning levels.

9.4.9 System shall send notifications if moisture levels drop below user-defined floors for more than 4 hours.

9.5 Outdoor Irrigation

The outdoor irrigation system will promote a healthy lawn as well as landscaping. The system will depend on water supplies but will be able to automate the lawn watering process. Priority = Low.

9.5.1 Irrigation system shall be plugged into the water system.

9.5.2 System shall control any number of yard irrigation devices.

9.5.3 System shall allow user to configure any irrigation device.

9.5.4 Irrigation devices shall be configurable for type of stream, amounts of water, and turn rotation during cycle.

9.5.5 System shall test irrigation devices independently for setting configurations.

9.5.6 System shall run a test cycle of all irrigation devices simultaneously to test configuration and coverage.

9.5.7 System shall have access to any number of devices reporting soil moisture.

9.5.8 System may base water cycles on soil moisture or by a set schedule.

9.5.9 System shall allow user to set groups of irrigation devices or individual irrigation devices.

9.5.10 System shall not run irrigation devices if rain is detected.

9.5.11 System shall be able to retrieve weather forecast from the Internet.

9.5.12 System may be configured to skip a user-defined number of watering cycles if rain is in the immediate forecast (i.e., rain is 60% likely over the next 2 days).

9.5.13 System shall record the amount of water deployed through each individual irrigation device.

9.5.14 System shall present users with reports for water deployment through the lawn irrigation system.

9.5.15 System shall maintain data for lawn irrigation for no less than 30 days.

9.5.16 System shall allow for motion detectors to be present in specified areas in lawn (i.e., flowerbeds or flower pots).

9.5.17 System shall allow for users to configure settings for when to activate lawn motion detectors.

9.5.18 System shall deploy countermeasures (i.e., loud sound, scent repellent) when motion detectors are tripped during user-defined time periods (for scaring off animals trying to eat plants).

9.6 Outside Building Cleaning

The outdoor building cleaning system will allow automatic, periodic cleanings of all exterior surfaces of the building to promote better curb appeal. This system will depend on ready water supplies and users maintaining proper levels of chemical or cleaning agents used by the system. The system is made to be abstract enough to enable users to clean virtually any exterior surface. Priority = Medium.

9.6.1 SH shall have reservoirs for cleaning different surfaces outside the home (i.e., windows and siding).

9.6.2 System shall monitor levels of all materials needed to clean exterior surfaces.

9.6.3 System shall send notifications when materials are low.

9.6.4 System shall accept any number of cleaning devices to control.

9.6.5 System shall allow users to assign category to the type of device under the system's control.

9.6.6 System shall accept input on what type of schedule should be used to deploy devices for cleaning various exterior surfaces.

9.6.7 System shall deploy cleaning devices according to the user inputted schedule.

9.6.8 System shall store and report information about cleaning material usages on a daily basis.

9.6.9 System shall maintain cleaning material usage data for no less than 30 days.

9.7 Ability to Configure Routines

The ability to configure routines will enhance the lives of the occupants, especially those portions of their lives which are routine. While life is mostly variable, there are some situations where routines are the mode of operation. All systems able to be controlled by the SH shall be presented as options to configure and set new routines within this system. It is adaptable and changeable as the lives and routines of the occupants change over time. Priority = Medium.

9.7.1 System shall allow users to configure routines.

9.7.2 System shall allow users to set alarm or wake up calls for various occupants within the house, including visitors.

9.7.3 System shall allow users to control certain activities as a result of a trigger. Example trigger-based routines would be alarm at some time, 5 minutes afterward turn on bedroom TV, 10 minutes after the alarm turn on the shower, 15 minutes after the alarm ensure coffee maker is operational or coffee is warm.

9.8 Voice Activation

The voice activation system currently will consist of a finite set of commands to which the SH will programmatically respond. In the future, this should be extended such that any commands can be programmed to control any device or system interfaced by the SH. Priority = High.

9.8.1 System shall support voice activation in major living areas (i.e., living room, kitchen).

9.8.2 System shall support commands to raise the current target temperature of the thermostat.

9.8.3 System shall support commands to lower the current target temperature of the thermostat.

9.8.4 System shall support command to draw a bath in the master bathroom

9.8.5 Master bed shall have heating element capable of warming the bed.

9.8.6 System shall support command to begin prewarming the bed in the master bedroom.

9.8.7 System shall support command to prepare the hot tub for use.

9.8.8 System shall support commands to dim or switch off lights in any room in the house.

9.8.9 System shall support commands to turn air conditioning or heating on and off.

9.8.10 System shall support commands to open windows and/or blinds on various levels of the house.

9.8.11 System shall support command to lock all points of entry.

9.8.12 System shall support command to secure the house, which would lock all points of entry and close all windows and blinds.

9.9 Driveway

The system is geared to provide ease and safety in the winter months to attempt to prevent snow accumulation on the driveway surface, and more importantly the formation of ice. Priority = Low.

9.9.1 Driveway shall have heating element installed underneath it.

9.9.2 System shall constantly monitor driveway surface temperature.

9.9.3 System will turn on driveway heating if the surface temperature of the driveway is conducive to freezing water.

9.9.4 Driveway heating element will shut off or not run if the driveway is above 40° F.

9.9.5 System will monitor and record when driveway heating element is in use.

9.9.6 System may be set to only run heating surface at night or based on time-of-day settings.

9.10 Kitchen Food Stocking

The kitchen food stocking program will provide a way for the occupants to control and view inventory from anywhere in the world. This will be helpful when shopping for groceries as well as when deciding what options may be available for dinner. Priority = Low.

9.10.1 System shall allow users to enter food associated with RFID tag into the kitchen inventory system.

9.10.2 System shall present reports to users of food inventory.

9.10.3 System shall allow users to call in (i.e., from grocery store) to check on stock of certain items in the kitchen's inventory.

9.10.4 System shall monitor and track the usage of certain items.

9.10.5 System shall present users with reports on item usage (i.e., for diets, and food spending forecasting).

9.10.6 System shall maintain item inventory and usage for no less than 18 months.

9.10.7 System shall provide an interface for recipe center [ref. 9.11] to provide feedback on the stock of items needed for recipe.

9.10.8 System may provide an intelligent interface to create shopping list templates based on average food usage.

9.11 Kitchen Recipe Center

The kitchen recipe center will provide users with an easy way to recall and cook recipes while operating in the kitchen. The system will provide easy access to recipes and provide voice-automated help and limited automation for backing functions. The recipe center also provides some publishing mechanisms to share recipes with family and friends. Priority = Medium.

9.11.1 System shall allow users to enter recipes.

9.11.2 System shall allow users to define categories for recipes stored within the recipe center (i.e., appetizer, beef main course, dessert).

9.11.3 System shall provide a touchpad interface in the kitchen for users to search, recall, and view recipes.

9.11.4 System shall provide an interface for users to add, modify, and delete recipes from the repository.

9.11.5 System shall provide an interface to the food stock to create grocery lists of what items may be needed for some arbitrary number of recipes.

9.11.6 System shall provide users with recipes in a specified category where all items are currently in stock. (i.e., "What can I make tonight?").

9.11.7 System shall provide users the ability to send recipes to friends electronically (i.e., email, micro webpages).

9.11.8 System shall provide users the ability to create/print a categorized cookbook of all recipes currently within the system.

9.11.9 System shall allow users to store image file linked to any recipe within the system.

9.11.10 System shall allow user to enter assisted baking mode.
9.11.10.1 System shall automatically preheat the oven.
9.11.10.2 System shall verbalize order of ingredients to add.
9.11.10.3 System shall accept verbal confirmation once item is added before instructing to add the next item.

9.12 Phone System

The phone system will be a unified approach to handling voice mail for the occupants of the house. The key functions are allowing for easier retrieval from anywhere as well as extending the system via multiple virtual inboxes. Priority = Medium.

9.12.1 System shall serve as an answering machine for household.
9.12.2 System shall allow users to configure number of rings before answering.
9.12.3 System shall allow users to configure any number of phone mailboxes for recipients.
9.12.3 System shall allow users to record greeting message that will be played after user-defined number of rings.
9.12.4 System shall allow users to configure greeting message to be played for individual mailboxes.
9.12.5 System shall record messages for recipients along with date, time stamp, and incoming phone number to nonvolatile memory.
9.12.6 System shall send out a notification to users when they have a new message in their mailbox (i.e., email, text message, pages).
9.12.7 System shall make messages available via authenticated web interface for user retrieval.
9.12.8 System may use voice-to-text engine to send the text representation of the message to user's email account.

9.13 Wall Pictures

The wall pictures allow occupants of the home to allow friends and family members to share pictures with them and have those pictures displayed on select wall monitors throughout the house. Priority = Low.

9.13.1 System shall provide wireless support for driving any number of wall-mounted monitors for picture display.
9.13.2 System shall provide web-based interface for authenticated users to publish new photos for display on wall monitors.
9.13.3 System shall allow users to configure which pictures get displayed.
9.13.4 System shall allow users to configure which remote users can submit pictures to which wall monitor.
9.13.5 System shall support the following playback modes: Random—display random photos. Slideshow—display photos in order for some user-defined time. Single—display only selected or most recently submitted photo.
9.13.6 System shall provide remote users with storage for up to 20 pictures in their repository or 100MB, whichever is greater.

9.14 Mail and Paper Notification

System to notify occupants of status and delivery events for both mail and newspaper boxes. Priority = Low.

9.14.1 System shall monitor any number of mail and newspaper boxes for motion and weight.

9.14.2 System shall allow users to set notification events for those boxes.

9.14.3 System shall send notifications when motion is detected coupled with a change in static weight of the box.

9.14.4 System shall allow user to turn off any notification events for a set period (i.e., when snow or something else may trigger the motion and weight sensors).

9.14.5 System shall permit user to query the status of any of the boxes. The status would be empty or occupied.

Appendix B: Software Requirements for a Wastewater Pumping Station Wet-Well Control System[1]

1 Concept of Operations

A wastewater pumping station is a component of the sanitary sewage collection system that transfers domestic sewage to a wastewater treatment facility for processing. A typical pumping station includes three components: a sewage grinder, a wet well, and a valve vault (Figure B.1). Unprocessed sewage enters the sewage grinder unit so that solids suspended in the liquid can be reduced in size by a central cutting stack. The processed liquid then proceeds to the wet well which serves as a reservoir for submersible pumps. These pumps then add the required energy/head to the liquid so it can be conveyed to a wastewater treatment facility for primary and secondary treatment. The control system specification that follows describes the operation of the wet well.

1.1 Purpose

This specification describes the software design requirements for the wet well control system of a wastewater pumping station. It is intended that this specification provide the basis of the software development process and as preliminary documentation for end users.

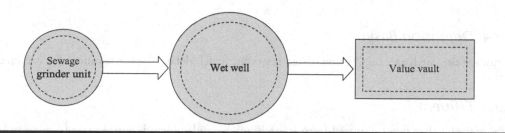

Figure B.1   Typical wastewater pumping station process.

1.2 Scope

The software system described in this specification is part of a control system for the wet well of a wastewater pumping station. The control system supports an array of sensors and switches that monitor and control the operation of the wet well. The design of the wet well control system shall provide for the safety and protection of pumping station operators, maintenance personnel, and the public from hazards that may result from its operation. The control system shall be responsible for the following operations:

1. Monitoring and reporting the level of liquid in the wet well
2. Monitoring and reporting the level of hazardous methane gas
3. Monitoring and reporting the state of each pump and whether it is currently running or not
4. Activating a visual and audible alarm when a hazardous condition exists
5. Switching each submersible pump on or off in a timely fashion depending on the level of liquid within the wet well
6. Switching ventilation fans on or off in a timely fashion depending on the concentration of hazardous gas within the wet well

Any requirements that are incomplete are annotated with TBD and will be completed in a later revision of this specification.

1.3 Definitions, Acronyms, and Abbreviations

The following is a list of definitions for terms used in this document.

1.3.1 Audible Alarm

The horn that sounds when an alarm condition occurs

1.3.2 Controller

Equipment or a program within a control system that responds to changes in a measured value by initiating an action to affect that value

1.3.3 DEP

Department of Environmental Protection

1.3.4 Detention Basin

A storage site, such as a small unregulated reservoir, which delays the conveyance of wastewater

1.3.5 Effluent

Any material that flows outward from something; examples include wastewater from treatment plants

1.3.6 EPA

Environmental Protection Agency

1.3.7 Influent

Any material that flows inward from something; examples include wastewater into treatment plants

1.3.8 Imminent Threat

A situation with the potential to immediately and adversely impact or threaten public health or safety

1.3.9 Manhole

Hole, with removable cover, through which a person can enter into a sewer, conduit, tunnel, etc., to repair or inspect

1.3.10 Methane

A gas formed naturally by the decomposition of organic matter

1.3.11 Overflow

An occurrence by which a surplus of liquid exceeds the limit or capacity

1.3.12 Precast

A concrete unit which is cast and cured in an area other than its final position or place

1.3.13 Pump

A mechanical device that transports fluid by pressure or suction

1.3.14 Remote Override

A software interface that allows remote administrative control of the pumping control system

1.3.15 Seal

A device mounted in the pump housing and/or on the pump shaft to prevent leakage of liquid from the pump

1.3.16 Security

Means used to protect against the unauthorized access or dangerous conditions. A resultant visual and/or audible alarm is then triggered.

1.3.17 Sensor

The part of a measuring instrument which responds directly to changes in the environment

1.3.18 Sewage Grinder

A mechanism that captures, grinds, and removes solids ensuring a uniform particle size to protect pumps from clogging

1.3.19 Submersible Pump

A pump having a sealed motor that is submerged in the fluid to be pumped

1.3.20 Thermal Overload

A state in which measured temperatures have exceeded a maximum allowable design value

1.3.21 Valve

A control consisting of a mechanical device for controlling the flow of a fluid

1.3.22 Ventilation

The process of supplying or removing air by natural or mechanical means to or from a space

1.3.23 Voltage

Electrical potential or electromotive force expressed in volts

1.3.24 Visible Alarm

The strobe light that is enabled when an alarm condition occurs

1.3.25 Wet Well

A tank or separate compartment following the sewage grinder which serves as a reservoir for the submersible pump

2 Overall Description

2.1 Wet-Well Overview

The wet well for which this specification is intended is shown in Figure B.2. The characteristics of the wet well described in this specification are as follows:

1. The wet-well reservoir contains two submersible pumps sized to provide a fixed capacity.
2. Hazardous concentrations of flammable gases and vapors can exist in the wet well.

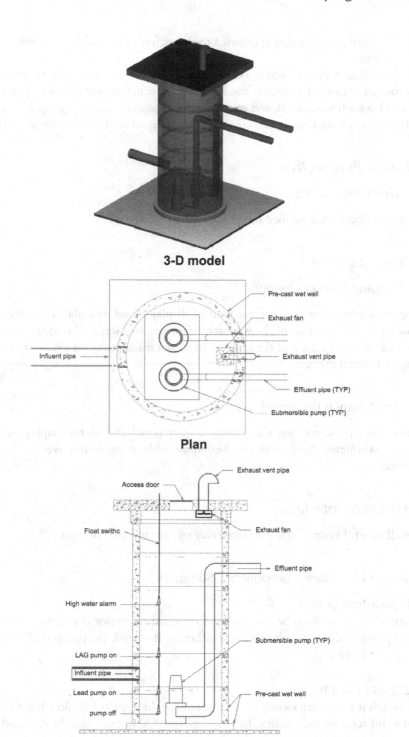

Figure B.2   Typical wet well (model, plan, and sectional diagram).

3. It has a ventilation fan that is oriented to direct fresh air into the wet well rather than just exhaust from the well.
4. An alarm and indicator light is located outside so that operators can determine if a hazardous condition exists. Hazardous conditions include, but are not necessarily limited to, a high gas level, a high water level, and pump malfunction.
5. A float switch is used to determine the depth of liquid currently in the wet well.

2.2 Product Perspective

2.2.1 System Interfaces

The system interfaces are described in the subsequent subsection.

2.2.2 User Interfaces

2.2.2.1 Pumping Station Operator

The pumping station operators use the control display panel and alarm display panel to control and observe the operation of the submersible pumps and wet well environmental conditions. Manipulation of parameters and the state of the submersible pumps is available when the system is running in manual mode.

2.2.2.2 Maintenance Personnel

The maintenance personnel use the control display panel and alarm display panel to observe the current parameters and state of the submersible pumps and wet well, and perform maintenance.

2.2.3 Hardware Interfaces

The wet-well control system hardware interfaces are summarized in Figure B.3.

2.2.3.1 Major Hardware Components: Summary

2.2.3.1.1 Moisture Sensor
Each submersible pump shall be equipped with a moisture sensor that detects the occurrence of an external pump seal failure. Should a seal failure be detected, the pump shall be turned off and alarm state set (Table B.1).

2.2.3.1.2 Float Switch
The float switch is a mercury switch used to determine the depth of liquid within the wet well and set the on or off state for each pump. Three switch states have been identified as lead pump on/off, lag pump on/off, and high water alarm.

2.2.3.1.3 Access Door Sensor
The access door sensor is used to determine the state, either opened or closed, of the wet well access door.

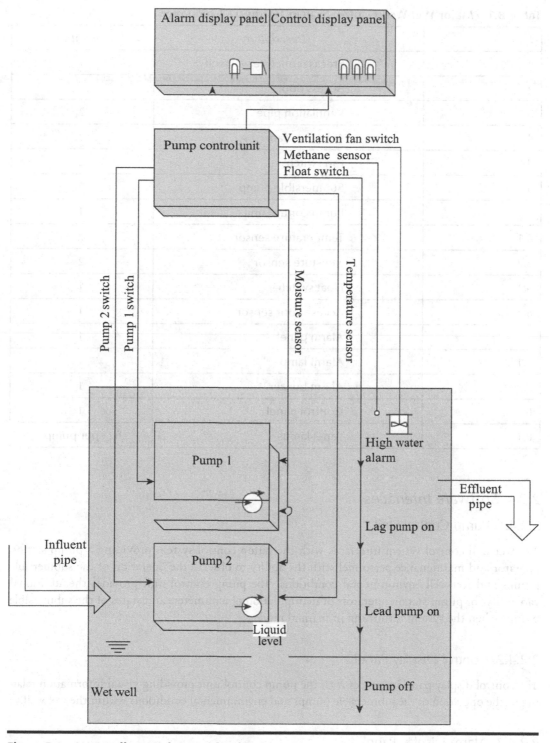

Figure B.3   Wet-well control system hardware.

Table B.1   Major Wet-Well Control System Hardware Components

Item	Description	Quantity
1	Precast concrete wet well	1
2	Access door	1
3	Ventilation pipe	2
4	Axial flow fan	2
4.1	Fan switch	2
5	Submersible pump	2
6	Pump control unit	1
6.1	Temperature sensor	2
6.2	Moisture sensor	2
6.3	Float switch	1
6.4	Access door sensor	1
7	Alarm panel	1
7.1	Alarm lamp	1
7.2	Alarm buzzer	1
8	Control panel	1
8.1	Panel lamps	6 (3 per pump)

2.2.4 Software Interfaces

2.2.4.1 Pump Control Unit

The wet-well control system interfaces with the pump control system providing a pump station operator and maintenance personnel with the ability to observe the operation of the submersible pumps and wet-well environmental conditions. The pump control unit provides the additional capability for pump station operators of manipulation of parameters and states of the submersible pumps when the system is running in manual mode.

2.2.4.2 Control Display Panel

The control display panel interfaces with the pump control unit providing visual information relating to the operation of the submersible pumps and environmental conditions within the wet well.

2.2.4.3 Alarm Display Panel

The alarm display panel interfaces with the pump control unit providing visual and audible information relating to the operation of the submersible pumps and the environmental conditions within the wet well.

2.2.5 Operations

The wet-well control system shall provide the following operations:

1. Automated operation
2. Local manual override operation
3. Local observational operation

2.3 Product Functions

The wet-well control system shall provide the following functionality:

1. Start the pump motors to prevent the wet well from running over and stop the pump motors before the wet well runs dry.
2. Keep track of whether or not each motor is running.
3. Monitor the pumping site for unauthorized entry or trespass.
4. Monitor the environmental conditions within the wet well.
5. Monitor the physical condition of each pump for the existence of moisture and excessive temperatures.
6. Display real-time and historical operational parameters.
7. Provide an alarm feature.
8. Provide a manual override of the site.
9. Provide automated operation of the site.
10. Equalize the run time between the pumps.

2.4 User Characteristics

2.4.1 Pumping Station Operator

Authorized personnel trained with the usage of the wet-well control system when it is in manual mode

2.4.2 Maintenance Personnel

Authorized personnel trained with the usage of the wet-well control system

2.5 Constraints

System constraints include the following items:

1. Regulatory agencies including but not limited to the EPA and DEP
2. Hardware limitations
3. Interfaces to other applications
4. Security considerations
5. Safety considerations

2.6 Assumptions and Dependencies

Assumptions and dependencies for the wet-well control system include the following items:

1. The operation of the sewage grinder unit is within expected tolerances and constraints at all times.

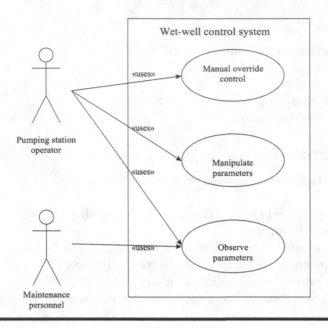

Figure B.4 Wet-well use case diagram.

2. A power backup system has been proved as a separate system external to the wet-well control system.
3. The operation of the controls within the valve vault is within expected tolerances at all times.

3 Specific Requirements

The following section defines the basic functionality of the wet-well control system (Figure B.4).

3.1 External Interface Requirements

3.1.1 User Interfaces

3.2 Classes/Objects

3.2.1 Pump Control Unit

1. The pump control unit shall start the submersible pump motors to prevent the wet well from running over and stop the pump motors before the wet well runs dry.

    LeadDepth represents the depth of liquid when the first pump should be turned on.

    LagDepth represents the depth of liquid when the second pump should be turned on.

    HighDepth represents the depth of liquid that the wet well should be kept below. Should the depth of liquid be equal to or exceed HighDepth, the alarm state is set.

AlarmState represents a Boolean quantity such that at any time *t*, the audible and visual alarms are enabled.

Depth represents the amount of liquid in the wet well at any time *t* in units of length.

Pumping represents a Boolean quantity such that at any time *t*, the pumps are either running or not.

> *Depth: Time → Length Pumping : Time → Bool*
> *High Depth > Lag Depth > Lead Depth Depth ≥ Lag Depth ⇒ Pumping*
> *Depth ≥ High Depth ⇒ Alarm State*

2. The pump control unit shall start the ventilation fans in the wet well to prevent hazardous levels of methane. The introduction of methane into the wet well shall cause the ventilation fan to turn on.
3. The pump control unit shall keep track of whether or not each motor is running.
4. The pump control unit shall keep track of whether or not each motor is available to run.
5. If a pump motor is not available to run and a request has been made for the pump motor to start, an alternative motor shall be started in its place.
6. An alarm state shall be set when the high water level is reached.
7. An alarm state shall be set when the high methane level is reached.
8. The starting and stopping of the pump motors shall be done in a manner that equalizes the run times on the motors.
9. Level switches shall be used to indicate when pump motors should be started.
10. The pump control unit shall be notified if excess moisture is detected in a pump motor.
11. The pump control unit shall be notified if a pump motor overheats and shall shut down the overheated motor.
12. The pump control unit shall be responsible for monitoring the pumping site.
13. The pump control unit shall be responsible for recording real time and historical operational parameters.
14. The pump control unit shall be responsible for providing the alarm feature.
15. There shall be an automatic and manual mode for the pump control unit. Each pumping station shall either be in automatic mode or manual mode.
16. Monitor and detect prohibited entry to the wet well through the access door by way of broken electrical circuit. Both audible and visible alarms shall be activated.
17. Monitor and detect the occurrence of a pump motor seal leak. If a leak has been detected, both an audible and visible alarm should be activated.

3.2.2 Control Display Panel

1. The control display panel shall have a digital depth of influent measured in feet.
2. Monitor and detect prohibited entry by way of a broken electrical circuit. Both audible and visible alarms shall be activated.
3. The pump control unit shall be responsible for displaying real-time and historical operational parameters.
4. Indicator lights shall be provided for pump running state.
5. Indicator lights shall be provided for pump seal failure state.
6. Indicator lights shall be provided for pump high-temperature failure state.
7. Indicator lights shall be provided for high wet-well level alarm state.

3.2.3 Alarm Display Panel

1. Indicator lights shall be enabled when an alarm state is activated.
2. A buzzer shall sound when an alarm state is activated.

3.2.4 Float Switch

1. When the depth of liquid is equal to or greater than the lead pump depth, the float switch shall set a state which causes the first pump to turn on.
2. When the depth of liquid is equal to or greater than the lag pump depth, the float switch shall set a state which causes the second pump to turn on.
3. When the depth of liquid is equal to or greater than the allowable high liquid depth, the float switch shall set an alarm state.

3.2.5 Methane Sensor

1. When the volume of methane is equal to or greater than the high methane volume, the methane sensor shall set a state which causes the ventilation fans to turn on.
2. When the volume of methane is equal to or greater than the allowable maximum methane volume, the methane sensor shall set an alarm state.

HighMethane represents the volume of methane which should cause the exhaust fans to turn on.

MaxMethane represents the volume of methane that the wet well should be kept below. Should the volume of methane be equal to or exceed MaxMethane the alarm state is set.

ExhaustFan represents a Boolean quantity such that at any time t, the exhaust fan is either running or not running.

AlarmState represents a Boolean quantity such that at any time t, the audible and visual alarms are enabled.

A partial formalization of the methane sensor operation is given as follows:

$$Max\ Methane > High\ Methane\ Exhaust\ Fan:\ Time \Rightarrow Bool\ Alarm\ State:\ Time \Rightarrow Bool$$
$$Methane \geq Max\ Methane \Rightarrow Exhaust\ Fan\ Methane < Max\ Methane \Rightarrow Exhaust\ Fan$$
$$Methane \geq Max\ Methane \Rightarrow Alarm\ State$$

Note

1 Reproduced from Laplante (2006), used with permission.

References

Laplante, P. A. (2006). *What Every Engineer Needs to Know about Software Engineering*. CRC Press, Boca Raton, FL.

Appendix C: Unified Modeling Language (UML)

Introduction

Unified modeling language (UML) is a system of concepts and notation for abstracting and representing discrete systems, particularly but not exclusively object-oriented software systems. It is widely used as a common notation for describing software systems in publications and design models. Although widely misunderstood as merely a pictorial notation, UML is more importantly a set of modeling concepts that are more general than those contained in most programming languages and widely applicable to real-world discrete systems as well as software systems. UML is "unified" because it combines a number of competing but similar modeling languages that were previously fighting for domination, eventually superseding all or most of them in public usage; "modeling" because its primary intended use is to construct abstract models of discrete systems that can subsequently be converted into a concrete implementation, such as a programming language, database schema, or real-world organization; and a "language" because it comprises both internal structural concepts (metamodel) and an external representation (visual notation). UML supports an engineering-based design approach in which models of software are constructed to understand and organize a system and fix problems before software is written. Although modeling is universally used by engineers, architects, and economists, software personnel have often been reluctant to use modeling despite the large number of errors often found in completed software.

UML comprises a number of loosely connected facets, often called submodel types or (in reference to the visual notation) diagram types. UML is intended to be a general-purpose modeling language, that is, it is applicable to all or at least most kinds of modeling problems. Unlike many academic modeling languages that emphasize elegance, a minimal set of basic concepts, and efficacy in theorem proving, UML is a pragmatic, "messy" modeling language with some redundant concepts and multiple ways of expressing them, similar to most mainline programming languages or natural languages. The various submodel types vary in elegance, level of abstraction, and usefulness. Because different submodel types were proposed and defined by different people as part of a consensus-based development process, they sometimes fit together a little awkwardly, like many major software applications.

UML is informed by object-oriented principles, although it can be used to model nonobject-oriented systems as well. It can be used at various levels of abstraction from very high-level system models to detailed models of programs.

Diagram Types

In this description, the term "diagram type" is used as a shorthand to represent a set of related concepts within a submodel type of UML and their visual notation. The most commonly used UML diagram types include class diagrams, use case diagrams, sequence diagrams, state machine diagrams, and activity diagrams.

Class Diagram

The class diagram is by far the most widely used kind of UML diagram. It describes the concepts within a system and their relationships. It is used to organize the information within a system into classes, understand the behavior of the classes, and generate their data structure. Figure C.1 shows a (much simplified) example of a class diagram showing the kinds of recording requests within a digital video recorder (DVR) system attached to a satellite television receiver.

A class is the description of a system concept that can be represented by data structure, behavior, and relationships. For example, "RecordRequest" is a class representing an entry in the DVR list of television shows to be recorded. An individual instance of a class is an object. A class has attributes, each of which describes a data value held by each object that is an instance of the class. For example, a RecordRequest has a numerical priority indicating its urgency if the DVR disk becomes full. Each object has its own particular data values for the attributes of its class, and the values of each object may differ from other objects of the same class. The data values of a single object can vary over time. An attribute includes a name, a data type, and optional constraints on the values that objects of the class may hold. A class can also have operations, each of which describes a behavior that objects of the class can perform. Individual objects do not have distinct operations; all the objects of the same class have the same behavior, which can be implemented by shared code, similar machinery, printed documents, or other ways appropriate to the kind of system.

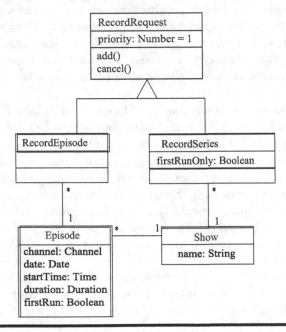

Figure C.1   Class diagram.

Classes are shown as rectangles with three sections, some of which may be optionally suppressed in a diagram. The top section shows the name of the class, and the middle section shows a list of attributes in the form "name: type = initial value."

In the example, the default priority of a record request is 1. The bottom section shows a list of operations in the form "name (parameter-list): returntype." For example, a new record request can be added to the recording queue, and an existing request can be canceled. Various other indicators may be attached to attributes and operations.

There are two major kinds of relationships among classes: associations and generalizations. Associations are semantic relationships between pairs (or, in general, tuples) of objects. The value of an association is a set of object tuples. Associations are often implemented as pointers within objects (in C, C++, and Java) or as sets or tables of tuples of object identifiers (in SQL and other databases). Modeling associations is of great importance because they are the glue that holds a system together. A binary association, by far the most common kind, is shown as a line between two classes. For example, a show contains episodes. A number near an end of an association line indicates the multiplicity of the association in that direction, that is, how many instances of the class at that end may be associated with any one instance of the class at the other end. The value may be a single number or an integer range. An asterisk represents many instances, that is, zero or more without limit. For example, each show may include many episodes, but each episode belongs to exactly one show. A name can be given to the entire association and to both ends of it. In the example, there is a single unique association between each pair of classes, so the names have been omitted. Various other restrictions can be attached to the association ends.

Generalization is the relationship between a general class and a more specific variety of it. The general class (superclass) can be specialized to yield one of several specific classes (subclasses). This has several benefits. The data structure and behavior of the superclass can be inherited by each subclass without the need for repetition. Most importantly, an operation in a superclass can be defined without an implementation, with the understanding that an implementation will be eventually supplied by every subclass. This is called polymorphism, because the abstract operation takes on a different concrete form in each subclass. A class lacking implementations for one or more operations is an abstract class; it cannot be instantiated directly, but it can serve as a data type declaration for variables that can be bound to objects of any subclass. This means that operations can be written in terms of abstract classes, with the actual implementations dependent on the actual object types at run time. It also permits collections of objects of mixed subclasses to be freely manipulated; the operation implementation appropriate to each object in the collection is determined dynamically at run time.

Generalization is shown as a line between classes with a large triangle attached to the more general class. A group of common subclasses can be shown as a tree sharing a single triangle to the superclass. In the example, RecordRequest is a superclass of "RecordEpisode" and "RecordSeries." This indicates that a request for a recording may either indicate a single episode or all the episodes of a given series. Both RecordEpisode and RecordSeries inherit the attribute ("Priority") and operations ("Add" and "Cancel") from the superclass RecordRequest. The main advantage of inheritance is that subsequent changes to the superclass, such as the addition of new attributes or operations, are automatically incorporated in the subclasses.

From the example, we see that a series recording request can be indirectly related to a set of episodes. This might suggest that the implementation of the internal recording mechanism needs only to deal with episodes. The relationship between episodes and shows can be maintained within the request queue. Class diagrams permit this kind of analysis to be made quickly.

Several other kinds of relationships such as various kinds of dependencies and the derivation of certain attributes from others are less frequently used.

Use Case Diagram

A use case diagram shows the externally visible behavior of a system as it might be perceived by a user of the system. Such users can include humans, physical systems, and software systems. The diagram is used to factor and organizes the high-level behavior of a system into meaningful units during early design, but detailed implementation is usually referred to other, more specific models, such as activity diagrams or collaboration diagrams; Figure C.2 shows a use case diagram for a DVR (again, much simplified).

An actor represents a set of external users playing a similar role in interacting with the system being modeled. For example, actors that interact with the DVR are the viewer and the network. A use case represents a meaningful unit of behavior performed by the modeled system for the benefit of or at the request of a particular actor. For example, the viewer can request a recording, play an existing recording, and buy a pay-per-view show. (A real DVR would have a number of additional use cases that would make the example too large.) The network is involved in the "BuyShow" use case initiated by the user and also participates in the "DownloadGuide" use case, which represents downloading the program guide periodically in the background without the involvement of the viewer. The granularity of a use case is somewhat subjective, but it should be large enough to be meaningful to the actor, rather than trying to capture the implementation of the system.

For example, it is not appropriate to include minor user interface actions such as paging through the program guide, selecting a show, and specifying the recording parameters as separate use cases; these are best modeled as details of the user interface that implement the use case. The purpose of enumerating use cases is to capture the externally visible behavior of a system at a high level.

The target system or class is shown as a rectangle containing a set of use cases, each shown as an ellipse containing the name of the use case. In the example, the main focus is the DVR, its

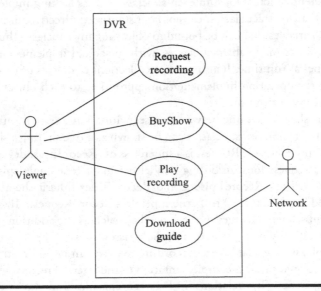

Figure C.2   Use case diagram.

use cases, and the objects that communicate with it directly. An actor is shown as a stick person outside the bounds of the system rectangle. In the example, the viewer and network are actors with respect to the DVR. A line connecting an actor to a use case shows the association between the actor and the use case in which it participates. Each actor can participate in multiple use cases, and each use case can interact with multiple actors. Details of the actors themselves are not included in the use case diagram; the focus of the use case diagram is the behavior of the target class, in this case the DVR. If some of the actors are also systems, they might have their own use case diagrams.

Various relationships among use cases can be shown, such as generalization and the inclusion of one use case as a part of another.

The use case diagram does not contain a great deal of detailed information. It is primarily of use in the early stages of design to organize the system requirements, particularly from a user viewpoint. The behavior of a use case can be documented using sequence diagrams, activity diagrams, or, in most cases, simple text descriptions.

Sequence Diagram

A sequence diagram shows an interaction among two or more participants as a set of time-sequenced messages among the participants. The main focus is the relative sequence of messages, rather than the exact timing. It is used to identify and describe typical as well as exceptional interactions to ensure that system behavior is accurate and complete. Sequence diagrams are particularly useful for interactions among more than two participants, as two-party conversations are easily shown as simple text dialogs alternating between the two parties. Figure C.3 shows an example of a sequence diagram for viewing a live show using the DVR.

A sequence diagram involves two or more roles, that is, objects that participate in the interaction. Their names (and optionally types) are shown in small boxes at the top of the diagram, with vertical lines (lifelines) showing their lifetime in the system. The example shows three roles: the viewer, the DVR, and the satellite network. In this example, all three objects exist for the duration of the interaction; therefore, their lifelines traverse the entire diagram.

A message is a single one-way communication from one object to another. A message is shown as a horizontal line between lifelines with an arrowhead attached to the receiving lifeline. The name of the message is placed near the line. In a more detailing diagram, the parameters of the messages can be shown. The example begins when the viewer requests to see the program guide, which the DVR then displays. The viewer may ask to scroll the guide, which the DVR redisplays. This subsequence is placed inside a "loop" box to show that it may be repeated. The viewer then requests to view a show. In this example, the show happens to be a pay-per-view show, so the DVR asks the viewer whether the viewer wants to buy the show. The user does buy the show, so the DVR submits the request to the network. The example shows two possible outcomes in the two sections of the alt (for "alternative") box: If the purchase is successful, the network sends an authorization and the DVR displays the shows; if the purchase fails (maybe the viewer's credit limit has been exceeded), an error message is displayed. Although a sequence diagram can display closely related alternatives, it is not intended to show all possible interactions. It is meant to show typical interaction sequences to help users understand what a system will do and to help developers think about possible interactions among features. Instead of trying to show everything on a single sequence diagram, a model should include multiple sequence diagrams to represent variant interactions. It is particularly important to include at least one sequence diagram for every possible error condition or exception that can occur.

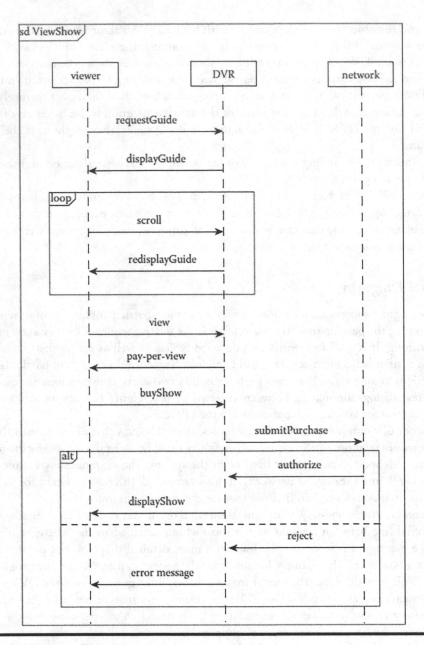

Figure C.3   Sequence diagram.

State Machine Diagram

A state machine diagram shows the states that an object (often a system) can assume and the transitions that move the object among states in response to various events that the object may perceive. It is used to identify and structure the control aspect of system behavior. Simple state machines have been important in electrical engineering and computing for many decades. UML uses an enhanced version of state machines based on the work of David Harel, which allows grouping of states into higher-level states. This is a kind of generalization of states that avoids the repetition of the same transition in many low-level states. State machine diagrams are useful for traditional control applications as well as for understanding the life history of objects.

Figure C.4 shows an example of a state machine diagram showing the life history of a recording request in the DVR. The same state machine is applicable to every recording request. Many recording requests can exist simultaneously, and each has its own distinct state.

A state is shown as a rounded rectangle containing the name of the state. For example, the "Queued" state means that a recording request has been created, but the start time of the show has not yet arrived. A transition is shown as an arrow from a source state to a target state. The transition is labeled with the name of an event that may occur during the source state and, optionally, with a slash (/) character followed by an action. If an event occurs when an object is in a state that is the source state for a transition for that event, the action on the transition is performed and the object assumes the target state specified by the transition. For example, if the start time of a show arrives while a recording request is in the Queued state, the DVR starts recording, and the recording request assumes the "Recording" state. That state in turn lasts until the stop time of the show arrives, after which the recording ceases and the recording request assumes the "Saved" state.

The creation of an object is shown by a small black disk. The disk is connected by a transition to the initial state of the object; the transition may also specify an initial action. For example, when a recording request is created, a list item is allocated for it and it assumes the Queued state.

The destruction of an object is shown by a small bull's eye. If a transition causes the object to reach this state, any action on the transition is performed and the object is destroyed. For example, if the "Delete" event occurs while a recording request is in the Recording state, the memory for the recording is deallocated and the request is destroyed; if it occurs in the "Queued" state, no recording has yet occurred, so the request is just destroyed.

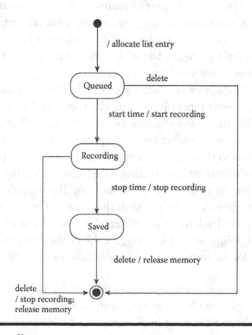

Figure C.4   State machine diagram.

The full version of state diagrams supports concurrent states. The state of an object can have two or more separate parts, each of which progresses independently. For example, a more elaborate version of the DVR state machine diagram would show the ability both to record a show and to play an earlier portion of the recording simultaneously. Although the concurrent states are mostly independent, they can have some dependencies. For example, if the viewer fast-forwards the playback of the recording so that it catches up with the live recording, the viewer cannot jump forward any more.

Substates are also supported by nesting substates within superstates. For example, the Delete event could be attached to a single transition on a high-level superstate containing multiple detailed substates. If the event occurs during any of the detailed substates, the high-level transition would take precedence. The ability to apply a single transition to many nested substates is particularly useful for exception conditions and forced exits, because it becomes unnecessary to attach the error handling to each nested substate explicitly.

Activity Diagram

An activity diagram shows the sequencing of actions within an overall process. In general, it includes concurrent processes as well as sequential processes, so it has the form of a directed graph rather than a simple chain. This diagram is used to understand the control and data dependencies among the various pieces of detailed behavior that compose a high-level system behavior, to ensure that the overall behavior is adequately covered and that detailed behaviors are performed in the correct orders.

The sample activity diagram in Figure C.5 shows the process of changing a channel on the DVR. An activity diagram can be divided into multiple columns, each showing the actions performed by one of the participants. The participants in this example are the viewer, the DVR program guide, and the satellite receiver subsystem. An action is shown as a rounded rectangle. This is the same symbol as a state in a state diagram because an action is a state of a process. A decision is shown as a diamond with two or more outgoing arrows, each labeled by a condition. An arrow shows a temporal dependency between two actions; the first must be completed to enable the second. Synchronization is shown by a heavy bar with two or more inputs or outputs; a synchronization bar initiates or terminates concurrent actions.

In the example, initially the viewer selects a channel. This triggers the program guide subsystem to determine if the viewer subscribes to the channel and if it is currently broadcasting. If the channel is not available, the viewer receives an error message. If the channel is available, the receiver must perform three concurrent actions to display the channel, as shown by the synchronization bar followed by arrows to three concurrent actions. The decision performed by the program guide enables the three concurrent actions: tuning the receiver, selecting the correct satellite input, and setting the television format to low or high definition. The initiation of concurrent activity is called a fork. Once concurrent actions are enabled, they proceed independently until a subsequent synchronization bar joins them back together. When all three actions are complete, the synchronization bar on the bottom is triggered, which triggers the guide unit to start recording the current channel. The synchronization of concurrent activity is called a join. Finally, the user is allowed to view the channel.

In the complete version of activity diagrams, conditional and iterative behavior may be represented, as well as more complicated kinds of synchronization.

Activity diagrams can be used to plan implementations of operations. They are also useful for modeling business processes.

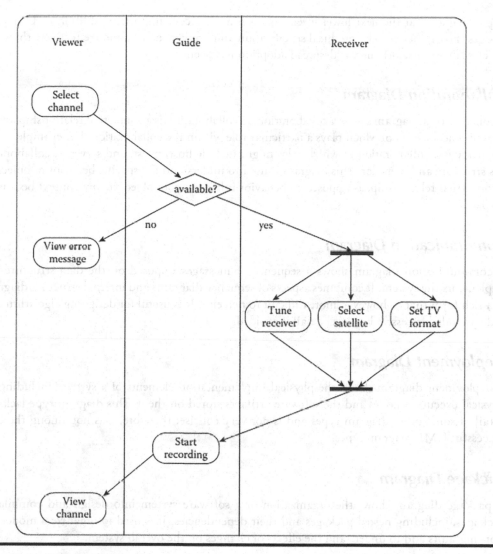

Figure C.5 Activity diagram.

Other Diagram Types

UML has several other diagram types that are not illustrated in this entry. See the references for more details.

Some of the most important types are briefly described later in the entry.

Internal Structure Diagram

An internal structure diagram is a variant on the class diagram that shows the internal structure of an object in terms of its constituent parts and their relationships to each other and to ports connecting the object to the external world. This diagram type is used for hierarchical design, in which an external view of a class at one level is expanded into a particular realization or internal implementation at the next lower level. This is a more recent important addition to UML that bridges the gap between hierarchical specification and implementation more effectively than simple class diagrams, although widespread adoption has been slow.

Collaboration Diagram

A collaboration diagram shows a collaboration. A collaboration is a contextual relationship among a set of objects, each of which plays a particular role within the collaboration. For example, a dinner party is a collaboration in which roles might include host, guest, and server. A collaboration has structure and behavior. This diagram is used to understand the specific behavior of objects in a contextual relationship, as opposed to behavior inherent to an object in any context because of its class.

Communication Diagram

A communication diagram shows a sequence of messages imposed on the data structure that implements the system. It combines aspects of sequence diagrams and internal structure diagrams in a rich form that is, however, more difficult to perceive. It is useful for designing algorithms and code provided excessive low-level detail is avoided.

Deployment Diagram

A deployment diagram shows the physical implementation elements of a system, including the physical execution nodes and the software artifacts stored on them. This diagram type lacks the detail of some other diagram types and is not very concise; therefore, it is not among the most successful UML diagram types.

Package Diagram

A package diagram shows the organization of a software system into design and compilation packages, including nested packages and their dependencies. It is used to organize a model into working units and to understand the effects of changes on the overall system.

Profiles

UML is a large general-purpose modeling language, but frequently the full power of UML is not needed in a particular usage. At the same time, sometimes a model builder wants to adopt more stringent assumptions than those made by the full UML, which must accommodate a wide range of uses. A UML profile is intended for this kind of usage. For example, UML in general assumes full concurrency, but a profile for a sequential application domain might remove this assumption to simplify modeling within that domain.

A profile is a named, specified subset of UML together with a limited set of additions and constraints that fit the profile for a particular area of usage. Profiles are often constructed and supported by standardization workgroups responsible for a particular application domain. They provide a way to adapt UML to a particular application area rather than constructing a new complete modeling language.

Because profiles are layered on standard UML using a limited set of extension mechanisms, they can be constructed and maintained by general-purpose UML modeling tools.

A lot of modeling energy has gone into the construction of profiles. These include profiles for various programming languages, database schemas, system engineering, real-time systems, and a wide variety of application domains.

Usage

UML can be used in many different ways for different purposes. One advantage of using UML rather than specialized languages for various domains is that most real-world applications include subsystems of many different kinds, and most specialized languages are somewhat restrictive in their scope, leading to problems in combining models built using different languages. A number of commercial and open-source tools support editing of UML models; the tools vary widely in completeness and other features, such as support for code generation.

Programming Design

UML can be used to perform the high-level design of software systems by first capturing requirements using class diagrams, use case diagrams, and sequence diagrams. The class diagrams can be elaborated to design classes that will be implemented in programming languages. Package diagrams can help to organize the structure of the program units.

Programming Detail

Detailed class diagrams can be used to generate class declarations in languages such as C++ and Java. A number of commercial software design tools can perform this kind of code generation. Specification of algorithms can be done using activity diagrams and state machine diagrams, although these can sometimes be as expansive as the final code. Code generation using patterns can produce a significant amplification of detail and may be the most productive use of modeling to generate code.

Database Design

Class diagrams can model the information to be represented within a database. A number of UML extensions have added database features (such as keys and indexes) to basic UML. The database community has been somewhat wary of UML, however, often preferring to use older modeling languages explicitly designed for databases, but often lacking in expressiveness. One advantage of using UML is that both database schemas and programming language class declarations can be generated from the same information, reducing the danger of inconsistency.

Business Process Design

Activity diagrams are good for modeling business processes. Class diagrams can model the data implicit in business processes, usually with more power than traditional business modeling languages.

Embedded Control

State machine diagrams can represent the control aspects of a real-time system, but class diagrams and internal structure diagrams can also represent the data structures that the system uses.

Application Requirements

Use case diagrams can capture the high-level functionality of a system. Class diagrams can capture the information in a particular application domain. These can then be elaborated to produce a more detailed design model.

UML History

Modeling languages for various purposes dates to at least the early 1970s. Although some were widely used in specific application domains, such as telephone switching systems, none achieved widespread general usage. In the early 1990s, a handful of object-oriented modeling languages developed for general use gained some popularity with no single language being dominant. These languages incorporated ideas from various earlier modeling languages but added an object-oriented spin. The first version of UML was created in 1995 by James Rumbaugh and Grady Booch by uniting their object modeling technique (OMT) and Booch methods. The bulk of that union still forms the core of UML, including the class diagram, the state machine diagram (including Harel's concepts), the sequence diagram (subsequently much extended with ideas from message sequence charts), and the use case diagram of Jacobson. Subsequent contributions by Ivar Jacobson and a team of submitters from a number of companies resulted in the standardization of UML 1.0 by the Object Management Group (OMG) in 1997. The rights and trademarks to UML were then transferred to the OMG. OMG committees generated several minor updates to UML and a major update to UML 2.0 in 2004 that added profiles and several new diagram types, including internal structure diagrams and collaboration diagrams. OMG continues to maintain the standard. The complete standard and information on standardization activities can be found on their Web site at http:// www.omg.org by following links to UML.

Assessment

UML has achieved its goal of uniting and replacing the multitude of object-oriented modeling languages that preceded it. It is frequently used in the literature to document software designs. It has been less successful in penetrating the practices of average programmers. Partly this is due to the misperception that UML is primarily just a notation, whereas its main purpose is to supply a common set of abstract concepts useful for modeling most kinds of discrete systems. Another factor, however, is that the various parts of UML are widely uneven in quality, approachability, and detail. Some parts, such as class diagrams, are widely used, whereas others, such as deployment diagrams, fail to live up to their promise. One problem is that the committee process of an organization such as the OMG encourages messy compromises, redundancy, and uneven work at the expense of cohesion. There is also an unresolved tension in what the modeling language should be, from an elegant and minimal theoretical basis at one extreme to a "programming through pictures" tool at the other extreme, with the more useful abstract modeling language somewhere in the middle. The tension about what UML should be has worked itself out differently in different submodel types and even in various parts of the same submodel type, which can lead to confusion and uneven coverage of modeling. To its credit, UML is a universal language with features to support a wide range of systems. Such languages, including practical programming languages, must be large and messy to some extent if they are to be used by a wide range of users on a wide range of problems. UML would undoubtedly benefit from a serious trimming of redundant and less successful concepts, but whether that can occur within the limits of a community process is doubtful. UML will probably remain a useful but far-from-perfect tool, but the window for a more coherent modeling language may well be closed.

Further Reading

See the OMG Web site for a detailed specification of UML (UML Specification), including the latest additions to the standard as well as ongoing standardization activity. The UML Reference Manual (Rumbaugh et al. 2005) provides the most coherent detailed reference to UML, more highly organized than the highly detailed standard documents. The UML User Guide (Booch et al. 2005) provides an introduction to UML organized by diagram type. UML Distilled (Fowler 2004) is a brief and highly selective introduction to the parts of UML that its author feels are most important, including some overall suggestions on a recommended design approach. Many other books describe UML either in whole or focused on particular usages.

References

Booch, G., Rumbaugh, J., & Jacobson, I. (2005). *The Unified Modeling Language User Guide*, 2nd edition. Addison-Wesley, Boston, MA.

Fowler, M. (2004). *UML Distilled: A Brief Guide to the Standard Object Modeling Language*, 3rd edition. Addison-Wesley, Boston, MA.

Rumbaugh, J., Jacobson, I., & Booch, G. (2005). *The Unified Modeling Language Reference Manual*, 2nd edition. Addison-Wesley, Boston, MA.

UML Specification. http://www.uml.org (accessed January 2022)

Appendix D: User Stories

Introduction

Figure D.1 shows user stories as the initiation point for a software development project whether based on a traditional development model—represented by a V-shaped software development life cycle (SDLC)—or an Agile development model—represented by a generic Agile SDLC. Although first defined within the eXtreme Programming (XP) development model as user story cards, user stories are used as the initiation and first requirements capture process for all SDLCs, whether traditional waterfall or an Agile. User stories serve as the mechanism for identifying the project's overall requirements and those features that will be incorporated in each individual software application iteration. User stories were first associated with the XP flavor of Agile development but were rapidly adapted to all types of Agile development models: Scrum, Agile Unified Process, dynamic systems development method, Essential Unified Process, feature-driven development, Open Unified Process, and lean software development processes.

As with all journeys, software development begins with the first small step of writing the first set of user stories. This can be initiated, with the customer and the development team, at a whiteboard, with a stack of 3×5 index cards, or at a laptop with a capture program as simple as a text editor. Initially, it is easiest to begin the process with the stack of index cards. The 3×5 card provides a physical representation of what is being done with the captured user information. For the first set of requirements gathering meetings, these cards can be put on a wall or whiteboard, moved around into logical groupings, and easily marked up. For building complex user applications, a tool with the attributes described here is necessary to build complete iteration requirements and test cases.

The "3×5 card" captures the basic user story information of the following:

1. User story title—Figure D.2 "Name" Field
2. User story—Figure D.2 "Description" Field
3. Test case—Figure D.3

These three pieces of information are all that are needed in the first pass of the user story collection with the customer. For visual information processing, it is best to tape the cards to a large wall and then go through the process, with the customer, of prioritizing the timing of completing each user story. This priority will determine in which iteration the customer receives the capability told in the story. This initial "requirements gathering" of the user stories can take several days. The end result is a starting point for the system and a definition of the first release of the application. Another outcome of this first meeting is the definition for this project of what time increment constitutes an iteration as shown in the Agile life cycle. As a rule of thumb, a release should be no

Generic agile life cycle

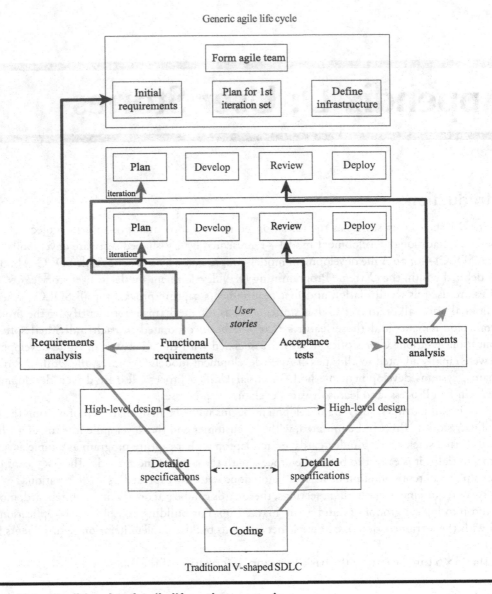

Traditional V-shaped SDLC

Figure D.1   Traditional and Agile life cycle user stories.

General

ID:	US142
Name:	Create SQL script(s) that configure ADA 1.3.0
Tags:	
Description:	As a Drilling Engineer, I can create a file containing an SQL script to update ADA on a rig with (1) the wells and formation tops for the wells on the current pad, and (2) limits for drilling parameters for a well
Attachments:	
Owner:	BillO Package: SDM

Figure D.2   General user story information.

TC21	If well exists do not change state
TC47	SQL must not change if there is a current wel (active well) for the pad
TC48	If new pad delete exisiting wells
TC49	Purge following tables
TC50	If same pad
TC51	Repopulate following tables
TC52	Make script transactional
TC53	For now, Layer and LayerMinMax should one FK for each row
TC54	Set Auto-Increment to 1 for any table that is purged

Figure D.3   Test cases.

more frequent than weekly or later than monthly. After a month of waiting, the customer might as well be subject to a waterfall life cycle model.

After this initial requirements gathering session, the development team completes the following information on each user story:

1. User story owner—Figure D.2 "Owner" field
2. Application capability set—Figure D.2 "Package" field
3. Story points—Figure D.3 "Plan Est." field
4. Task estimate in hours—Figure D.3 "Task Est." field
5. Small release assigned story for delivery—Figure D.4 "Iteration" field

"Story points" is a user story estimating concept. These are relative, not absolute measures that take into account not just the customers' concept of priority but also technology and tools being used, problem domain, development team size, development team composition and expertise, working environment, and the expertise and skill of the estimators. They measure the effort required to implement a story as one number that tells the team the story's difficulty. A story point range is 1, 2, 4, 8, or 16. Story points do give some indication of how much time was spent in an iteration. For example, at the end of a 10-day iteration, a five-person team completes 50 story points. To equate that to hours you are measuring, 10 days at 8 hours is 80 working hours times five developers, or 400 person hours. Dividing the 400 by the 50 story points completed gives the development team an 8 hour per story point velocity of development.

The development team will use these relative points to group together the stories that will make them into an "iteration." This is a learning process for the team in understanding how many points they can develop and deliver in the release time allotted. Keeping development metrics and calculating the "velocity of development" per iteration will refine the team's estimating abilities. The time estimate is not done until the first estimate of the points going into the release is determined.

Notice that the test cases in Figure D.3 have nonsequential numbering; they jump from test case 21 to test case 47. Even though a user story stands alone, the test cases gathered at this point in the life cycle need to be saved and reused for future user stories that have a relationship to previous ones, and to form the basis for a regression test suite for the entire application. Once the application is completed and goes into maintenance—as all successful software development projects do—the most valuable artifact of development next to the code is the suite of tests for each component and release.

Schedule				
State:	Accepted		Blocked:	false
Release:	SDM Config 1.1		Iteration:	Sprint 5
Plan Est:	8.0 Points		Task Est:	15.0 Hours
Actual:	0.0 Hours		To Do:	0.0 Hours

Figure D.4 User story scheduling.

Custom

Theme: | SQL Generation

Notes

Notes: | Assumptions

- ADA will contain a maximum of 1 limits configuration at a time
- The script will export the limits for the well that is currently being configured
- Well definitions for all the wells on the pad will be included (including formation tops)
- If pad or any of the wells are not defined, the script will create them.
- If pad, well or formation top information is already on the ADA system, this data will be updated

Figure D.5 User story customization.

Figure D.5 shows further information that can be included within the user story. The idea of a theme is useful when discussing the feature set being delivered in a release. Themes are larger than a release but smaller than the application. It is another way for the user to understand the magnitude of the capabilities being released and the percentage completed of the entire project. User stories and iterations look at the development process from the developers' perspective of satisfying the user story. The user sees the delivery of the iterations combined into releases but at times loses sight of the overall project. Themes are a midway point in helping the user and development team ensures that groups of iterations become meaningful releases and that the releases have a reference point within the overall project.

Implementing User Stories in a Development Organization

Based on the information contained in a single user story as shown by the above figures, you will soon run out of space on your index card. Unless you are building a simple "Hello world" software application, you'll need some automated tools. You can roll your own using open-source software like MySQL or any of the commercial off-the-shelf products Microsoft provides. You need to decide based on the characteristics of your development organization and your customer, factoring in the time frame dictated by your release schedule. A model to look at for Web-based user story tools is provided by Rally Software (https://wiki.rallydev.com/display/ rlyhlp/Rally+ Community+ Overview) Community Edition for managing user stories. The user story figures used in this entry were taken from a multiyear, real-time process optimization software development project using Agile development.

At the highest level of managing an Agile project, the project manager needs visibility into the backlog of user stories, the release underway, and the testing needed for each release. The project manager has the responsibility of rationalizing the user stories into iterations to generate releases that satisfy the customer's goals for the entire development effort. Figure D.6 shows the planned releases and the themes addressed by the user stories.

When the project manager needs to drill down to a release, for example, SDM 1.3 (Smarter Drilling Manager 1.3), the open user stories for that release can be easily generated and shown in Figure D.7. Referring back to Figure D.1, "Iterations" are tracked in keeping with the Agile life cycle flow.

Both Figures D.6 and D.7 show the user story plan estimates in story points and the task estimates in person hours. It becomes quite easy to gauge relative release and iteration sizes based on user story points.

Implementing user stories within a development organization requires embracing Agile methods. User stories will provide requirements gathering, acceptance testing, and a relative, quantifiable method for breaking the software development process into user-acceptable releases.

User Stories Drive Agile Projects

Adoption of user stories as your organization's Agile requirements gathering technique, backed up with an adequate automated system, is the driver for successful Agile software development projects. User stories must be:

1. Understandable to the customer and the developers
2. Implementable within one iteration of a multi-iteration release
3. Estimateable by both relative story points and person hours of effort
4. Testable to ensure that they were successfully implemented

Name	Theme	Start Date	Release Date ▼	Resources	Plan Est	Task Est	To Do	Actuals	State
									Planning ✓
SDM 1.3	BP branding, rules changes and config tool plot shading	2009-08-27	2009-08-27	3.0 Points	0.0	0.0	0.0		Planning
SDM 1.2	Config Tool security achedule A	2009-08-10	2009-08-16	29.0 Points	0.0	0.0	0.0		Planning
Rule Response Simulator 1.0		2009-04-13	2009-05-04	53.0 Points	494.0	176.0	0.0		Planning
ADA 1.3.1		2009-03-17	2009-04-30	31.0 Points	112.0	0.0	0.0		Planning
SDM Config 1.1	.NET version of configuration tool that allows historical data to be imported, advise to be graphically edited, and a configuration exported. Standalone version running against a local SQL Server database	2009-01-01	2009-03-31	171.5 Points	191.0	20.0	0.0		Planning
ADA 1.3		2009-01-01	2009-03-05	50.0 Points	119.0	0.0	16.0		Planning

Figure D.6 Product releases.

ID #	Name	Release	Iteration	State	Plan Est ▼	Task Est	To Do
		SDM Config 1.1 ✓	All ✓	All ✓	80.0	118.0	0.0
US133	When configuring, view plots of drilling data from offset wells	SDM Config 1.1	Sprint 4	D P C A	40.0	2.0	0.0
US18	Graphically edit the limits (max and min) for drilling parameters for a well	SDM Config 1.1	Sprint 4	D P C A	13.0	14.0	0.0
US142	Create SQL script(s) that configure ADA 1.3.0	SDM Config 1.1	Sprint 5	D P C A	8.0	15.0	0.0
US175	Edit WSL Boundaries for a rig	SDM Config 1.1	Sprint 5	D P C A	5.0	33.0	0.0
US177	Display WSL Boundaries on configuration plot	SDM Config 1.1	Sprint 5	D P C A	5.0	17.0	0.0
US17	Select offset wells and benchmark well for configuration	SDM Config 1.1	Sprint 4	D P C A	3.0	1.0	0.0
US174	Define New Rigs	SDM Config 1.1	Sprint 5	D P C A	2.0	10.0	0.0
US159	When configuring, restrict configuration limits to be within Boundaries from WSL	SDM Config 1.1	Sprint 5	D P C A	1.0	9.0	0.0
US120	Able to create wells and pads from spreadsheet	SDM Config 1.1	Sprint 4	D P C A	1.0	6.0	0.0
US19	Save the configured limts for a well	SDM Config 1.1	Sprint 4	D P C A	0.5	1.0	0.0
US181	Color-coding option for plots based on MSE quartile	SDM Config 1.1	Sprint 5	D P C A	0.5	1.5	0.0
US176	Associate Rig with a well	SDM Config 1.1	Sprint 5	D P C A	0.5	5.0	0.0
US182	Record well location	SDM Config 1.1	Sprint 5	D P C A	0.5	3.5	0.0

Figure D.7 Release user stories.

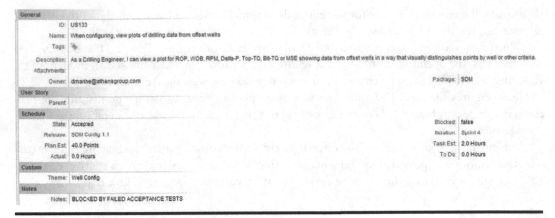

Figure D.8   User story #133 details.

ID ▲	Name	Work Product	Type	Priority	Owner	Package	Method
			All ▾	All ▾	All ▾	All ▾	All ▾
TC8	auto-zoom	US133: When configuring, view plots of drilling data from offset wells	Acceptance		MarkP	SDM	Manual
TC12	'ROP' not 'Quality' in listbox	US133: When configuring, view plots of drilling data from offset wells	Acceptance		MarkP	SDM	Manual
TC13	Visually verify color-code by quality color distribution	US133: When configuring, view plots of drilling data from offset wells	Acceptance		MarkP	SDM	Manual

Figure D.9   User story #133 test cases.

That's all that has to be included. You can add more features to your process, but the above four points will drive the implementation of user stories. Figures D.8 and D.9 show these basics implemented along with extensive information found useful to improving the user story process.

Example User Story Descriptions

The following are real user stories that were successfully developed, tested, and implemented within a single iteration:

1. As a drilling engineer, I need to be able to graphically edit limits (maximum and minimum) for rate of penetration (ROP), weight on bit (WOB), revolutions per minute (RPM), Delta-P, top torque, bit torque, and mechanical specific energy (MSE).
2. As a drilling engineer, I can create a file containing an SQL script to update ADA on a rig with (i) the wells and formation tops for the wells on the current pad and (ii) limits for drilling parameters for a well.
3. As a drilling engineer, I can define the well site leader (WSL) maximum/minimum boundaries for each drilling parameter for each formation for the rig.
4. As a drilling engineer, I can see the WSL boundaries on the plot while I'm setting limits.
5. As a drilling engineer, for each well, I need to (i) select offset (comparable) wells whose data will be used as a guide for setting expected ranges and (ii) select a benchmark (reference) well whose ranges will be used as the starting ranges.
6. As a drilling engineer, I can define new rigs for a field.
7. As a drilling engineer, when I'm editing limits, the limits will be constrained to lie between the boundaries defined for the rig.

8. As a drilling engineer, I create wells and pads whenever I import ADA data using a spreadsheet. I can also change the data for an ADA well by reimporting it.
9. As a drilling engineer, I need to define new wells in an integrated ADA database and to load and save my work on a well.
10. As a drilling engineer, when I look at a drilling parameter plot, I'd like to know which points are associated with MSE quartiles.

Keys to Implementing User Stories

1. Define your organization's Agile life cycle.
2. Make a commitment to user stories for requirements gathering.
3. Make a commitment to user stories for acceptance test identification.
4. Refine your Agile process with user story-based metrics.

Summary

User stories were first described as the starting point for XP. They are used for all SDLCs—whether or not based on Agile techniques. User stories provide two of the most critical inputs to the SDLC: requirements and test scenarios. What does a user story look like? "As a drilling engineer, I can create a file containing an SQL script to update ADA on a rig with (i) the wells and formation tops for the wells on the current pad and (ii) limits for drilling parameters for a well." This user story is from a user (drilling engineer) with a requirement to be able to create, as the user, SQL scripts within an application called ADA.

A user story is a promise made to the customer that the software development team will have a future conversation about the identified functionality.

User stories and iterations look at the development process from the developers' perspective of satisfying the user. They are the basis for time estimates for the system releases under construction. User stories eliminate the need for a large requirements document. User stories drive the creation of the acceptance tests. In order to verify the user story, one or more acceptance tests are defined. This is a powerful technique for the user to understand not only the "what" of their requirements but to also define, at the onset of the iteration, the "how" of ensuring its inclusion in the system.

Further Readings

Extreme Programming—A Gentile Introduction. http://www.extremeprogramming.org/ (accessed January 2022).

Introduction to User Stories. http://www.agilemodeling.com/artifacts/userStory.htm (accessed January 2022).

Object Management Group (OMG)—OMG™ is an International, Open Membership, not-for-Profit Computer Industry Consortium. http://www.omg.org (accessed January 2022).

Rally Software Development. http://www.rallydev.com/ (accessed January 2022).

User Story. http://en.wikipedia.org/wiki/User_story (accessed January 2022).

Appendix E: Use Cases

Introduction

Conceptually, requirements are requested system's features, often described as properties that help a system fulfill its purpose. Requirements are the product of the requirements elicitation process, a central part of requirements engineering (cf. Requirements Engineering: Principles and Practice, p. 949) intended to determine the needs of the system stakeholders. In that context, functional requirements capture the intended behavior of the system in question. This behavior may be expressed as services, tasks, or functions a system is required to perform. Here, one of the most important goals of requirements engineering is to help system stakeholders understand the behavior of the system in question. Among the different techniques for capturing system behavior, use cases are of major importance because their approach ensures expressing requirements in a way understandable to both users and developers.

This entry discusses use cases as a requirements engineering approach to capturing functional requirements. The entry elaborates use case concepts, best practices, and traps to avoid. The following subsection gives an overview of use cases and their implication in software development. The Use Case Concepts section emphasizes the use case concepts in terms of actors, use case description, scenarios, pre- and postconditions, diagrams, and best practices. The ATM Use Case Model: A Case Study section presents a case study example and finally the Summary concludes the entry.

Understanding Use Cases

Functional requirements capture the intended behavior of the system. This behavior may be expressed as services, tasks, or functions the system is required to perform (IEEE Std 1998). Use cases have been recognized as a valuable approach to discovering and presenting functional requirements. Craig Larman calls use cases "stories of using the system," providing an excellent technique to both understanding and describing requirements (Larman 2004). The advantage is that system users may see just what a new system is going to do. This is possible because a use case is expressed with a flow of events presenting scenarios of system use understandable to both users and developers. Moreover, the use case approach is considered as a practical technique for documenting requirements where use cases emphasize system users' goals and demonstrate how the system works with users to achieve these goals (Bittner et al. 2003; Larman 2004). In that context, a use case can be defined as a software engineering technique that defines a goal-oriented set of interactions between system users called actors and the system under consideration.

> Example: A user (actor) logs on to (goal) the automated teller machine (ATM) (system) using an ATM card and entering personal identification number (user–system interactions).

In general, use cases are text documents. However, the UML (Booch et al. 2005) provides a special use case diagram notation to illustrate the use case model of a system. Hence, a use case model consists of use case diagrams depicted in UML and use case description providing detailed information about the requirements. Here, the description is considered as the most important part because it documents the flow of events between the actors and the system in question. Use case diagrams provide a sort of road map of relationships between the use cases in a model and help system stakeholders better understand the user–system interaction process. It is important to mention though that a use case model does not impose the structure of the system in question; that is, use cases do not represent objects or classes of the resultant system design.

Use-Case-Driven Development Process

Use cases were first proposed by Jacobson in 1987 and have since become one of the key techniques for modeling requirements. Nowadays, use cases impose use-case-driven development as a vital part of the so-called unified process (UP) (Jacobson et al. 1999). The latter has emerged as a popular software development process targeting object-oriented systems. The following elements present some of the impacts the use cases have on the software development process:

- Because use cases impose an efficient approach to recording software requirements, they have become the primary (preferred by software development teams) requirements technique. Although popular in the past, other techniques such as functions list have become secondary.
- Use cases emphasize activities from three stakeholder perspectives—problem, solution, and project (Spence and Bittner 2004).
- Use cases promote the so-called iterative planning in the rational UP (Royce 1998; Spence and Bittner 2004), which is a framework for iterative software development. In this approach, the software development process is divided into iterations to gain better control over the system development and to mitigate risk (Larman 2004; Spence and Bittner 2004).
- Use cases are not object oriented; that is, the use case approach does not require object-oriented analysis. Hence, use cases may be applied to nonobject-oriented projects employing procedural languages, such as C, or functional languages, such as Scheme or Lisp. This increases the applicability of use cases as a software requirements technique (Larman 2004).
- Use case realizations drive the software design.
- Conceptually, use cases describe features of the system, and although the use case approach is not object oriented, it helps developers determine the design classes.
- Use cases may be used by developers to derive test cases and test case scenarios. In fact, use cases and test cases may be much coupled, thus forcing the development team to think about testing from the very beginning. In such a case, requirements engineering and test development are synchronized processes.

■ Clear and effective use cases promote better communication between the system's stakeholders; for example, the use cases present the system's features in a format understandable to all the stakeholders. In addition, the so-called business use cases (Larman 2004) help developers understand and business managers express potential business situations that the system may face. This helps to understand the context of a new system in the business environment.

To summarize, by targeting the system behavior the use cases focus on capturing the functions of a system, thus eventually driving the process of software development and providing traceability from requirements to design via test cases. Therefore, creating a robust, effective, and comprehensive set of use cases covering all functional requirements might be the most important task in requirements engineering.

Use Case Concepts

The general objective of use cases is to capture all functional requirements of a software system. Conceptually, a use case represents an activity that actors (e.g., users) may perform with a system. Such an activity is presented via possible scenarios identifying different flows of events driving the activity into different paths. A use case typically includes name and description, captured functional requirements, constraints, activity scenarios, use case diagrams, scenario diagrams, and additional information. A use case model of a particular system is a set of related use cases and actors.

Use Case Actors

The concept of actor is central to use case modeling. An actor represents a class of system users playing a specific role in use cases. Note that several users may play the same role in a system and an actor represents a single role, not a single user. Here, all the actors in a use case model from a list of all possible roles one may play when interacting with the system. Moreover, a single user may take one or more roles. Both actors and use cases determine the scope of the system in question. Actors are external to the system and use cases happen within the system.

The roles presented by actors may be taken not only by humans but also by organizations, machines, or machine components, and other software systems or software components. Note that actors may have a collaborative or participatory role in a use case. Here, actors may have goals in the system in question and participate in use cases to fulfill those goals, or they may provide support to the system. From this perspective, we distinguish three kinds of actors:

■ Primary actors have goals that can be met by using the services (or functions) of the system. Note that identifying all the primary actors is of major importance because by knowing these actors we may identify their goals driving the use cases.
■ Example: A client (primary actor) deposits money (goal) in the bank (system).
■ Supporting actors provide supporting services to the system. Such an actor is often a computerized system or machine, but it can also be another system, organization, or person. It is important to identify the supporting actors because they help the design team discover external interfaces and protocols needed by the system in question.
■ Example: Any money transfer operation requires data exchange with the ATM server (supporting actor).

■ Offstage actors are not directly involved in any use cases but have an interest in some aspects of the system behavior, that is, in some of the use cases. Although not of major importance, the identification of these actors helps requirements engineering ensure that all stakeholder interests are identified and satisfied. Often, offstage actors are not presented in the use case model unless they are explicitly named by other system stakeholders.
■ Example: When pressed, the emergency button turns on the hidden camera and calls the police department (offstage actor).

Use Cases

In the use case model approach, primary actors use the system via use cases. Here, the latter describe what the system does for the primary actors, what services the system requires from the supporting actors, and eventually what the offstage actors' interest in the system is. Use cases can be written in different formats. There are three possible degrees of detail in use case description (Larman 2004):

■ **Brief**: the use case in question is summarized in one paragraph by emphasizing only the basic scenario (cf. the Basic Scenario section).
■ **Casual**: informal text usually expressed in multiple paragraphs that cover various use case scenarios (cf. the Use Case Scenarios section).
  – **Fully dressed**: all steps of all the possible scenarios are explained in detail; supporting sections are used to add more information, such as actors, related use cases, preconditions, postconditions, and failure conditions (cf. Table E.1).

There is no single-format template for writing use cases. Table E.1 shows an example of a fully dressed use case description borrowed from Vassev (2009). As shown, the use case description is structured in a tabular format and presents detailed information helping the readers understand actors' goals, tasks, and requirements. Note that use cases may be related through special relationships. As described, the Get Demand use case is included in a use case and itself includes another one (cf. Related Use Cases in Table E.1). The relationships among the use cases in a single use case model may be graphically presented by a special use case diagram (cf. the Black-Box and White-Box Use Cases section).

Use Case Scenarios

Conceptually, a single use case may result in multiple possible scenarios each forming a single use case instance; that is, one path through the use case (Larman 2004). Often a use case defines a collection of possible scenarios each presenting a distinct flow of events that may occur during the performance of a use case by an actor. Thus, a use case scenario is a sequence of actions and interactions between actors and the system in question. A use case scenario is usually described in text and depicted visually with a sequence diagram (cf. the Alternative Scenario section). The description of a use case may encompass three different types of scenarios (paths)—basic scenario, alternative scenario, and failure scenario. Here, a use case may be defined as a collection of related scenarios that describe how actors use a system to achieve a goal.

To fully comprehend the diversity of use case scenarios, it is important to remember that a use case scenario is a single path through a use case. In that context, Figure E.1 depicts a generic view of the possible paths through a use case. As shown, the straight arrow represents the basic scenario,

Table E.1   Fully Dressed Use Case Description Example

Name	Get Demand
Actor	Worker
Description	A worker seeks and gets any pending demand from the demand space (DS).
Related use cases	This use case is included in the use case "Dispatch demand," and it includes the use case "Migrate demand."
Preconditions	A pending demand is stored in the DS.
Basic scenario	1. The worker completes its work and makes its associated transport agent (TA) listen to the DS for any pending demand. 2. The TA discovers a pending demand through its associated dispatcher proxy (DP). 3. The DP changes the status of that demand from pending to in process. 4. The DP takes a copy of that demand and passes it to the TA. 5. The TA migrates the demand copy to the worker (refers to the use case Migrate demand).
Alternative	1.1. The dispatch demand (DD) is local to the worker, and the worker makes its associated DP listen to the DS for any pending demand. 1.1.1. The DP discovers a pending demand. 1.1.2. The DP changes the flag of the demand from pending to in process. 1.1.3. The DP takes a copy of that demand and passes it to the worker.
Failure	1.1. The worker cannot find its associated TA.
Failure conditions	The TA is no longer available due to a network failure or due to a TA failure.
Postconditions	1. The copy of the demand is delivered to the worker. 2. The state of the original demand is changed to "in process."

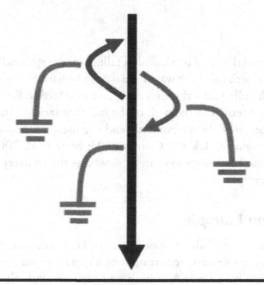

Figure E.1   Use case scenarios—paths through a use case.

and the curves represent the alternative paths some of which could result in failure scenarios. Here, some paths return to the basic scenario, and some end the use case.

Basic Scenario

A basic scenario (also called a basic flow of events [Bittner et al. 2003]) describes the main sequence of actions and interactions where a use case is completed correctly and is straightforward. Note that every use case comprises one sole basic scenario that shows the most realistic path of succeeding the use case objective. In such a scenario, no interruption in the basic course of action occurs nor are alternative paths taken (cf. the basic scenario in Tables E.1 and E.2). Due to these Description characteristics, Cockburn (2001) and Larman (2004) refer to the basic scenario as the main success scenario.

Alternative Scenario

Besides the basic scenario, a use case may have one or more alternative (optional) scenarios (cf. alternative scenarios in Table E.2). Larman (2004) calls the alternative scenarios' use case extensions and describes them as branches from the main success scenario. According to Cockburn (2001), an alternative scenario is an alternative course that has an individual goal that leads to success and a minimal guarantee. Here, if the basic scenario represents the main route to successful use case goal accomplishment, the alternative scenarios may be considered as detours (Bittner et al. 2003). Thus, the alternative scenarios complement the basic one. Although some may be essential to the success of the system, often the alternative scenarios are extraneous. To define such scenarios, we may simply define where those branch from the basic scenario, the alternative sequence of steps, and where the original scenario is resumed. Sometimes, the alternative scenarios may explicitly end the use case without returning to their original scenario (e.g., the basic scenario). It is also possible to have alternative scenarios branching from other alternative scenarios (cf. Figure E.1). In such a case, the original scenario is no longer the basic scenario but the first alternative scenario.

Failure Scenario

Some of the alternative scenarios may be viewed as failure (or exceptional) scenarios. Such scenarios describe errors that can occur and the way those should be handled by the system (Bittner et al. 2003). Failure scenarios handle paths through a use case that lead to the failure of the overall use case goal. Such paths never return to the path of the basic scenario or another alternative success scenario. Thus, failure scenarios always explicitly end the use case. Some authors termed failure scenario as use case exception (Kulak and Guiney 2003; Metz et al. 2003; Simons 1999). Here, the interaction path of a use case exception can be viewed as the recovery statements of a handled programming language exception.

Scenario Description Example

Table E.2 presents another use case description example. Here, the so-called Place Order use case comprises, in addition to the sole basic scenario, three alternative and one failure scenarios.

Because neither Alternative #1 nor Alternative #2 scenario ends the use case (both join the basic scenario), and because these two scenarios are not mutually exclusive, their combination

Table E.2   Use Case "Place Order"

Name	Place Order
Actor	Customer
Description	The customer provides address information and a list of product codes. The system confirms the order.
Preconditions	The customer must be logged on to the system.
Basic scenario	1. Customer enters name and address. 2. Customer enters product code for items they wish to order. 3. The system supplies a product description and price for each item. 4. The system keeps running the total of ordered items as they are entered. 5. The customer enters credit card information. 6. The system validates the credit card information. 7. The system issues a receipt to the customer.

Alternative #1	3.1. The product is out of stock:
	   3.1.1. The system informs the customer that the product cannot be ordered.
	   3.1.2. Continue with 2.

Alternative #2	6.1. The system rejects the credit card.
	   6.1.1. The system informs the customer that the credit card information is not valid.
	   6.1.2. Continue with 5.

Alternative #3	Alternative #1 and Alternative #2
Failure	6.1.2.1. The customer cancels the order.
Failure Conditions	The credit card is not valid.
Postconditions	An order has been submitted.

results in Alternative #3 scenario. Note that both alternative and failure scenarios are not presented with their full list of steps, but only with steps that branch from the basic scenario (e.g., Alternative #1 to #3 scenarios) or from another alternative/failure scenario (e.g., failure scenario). Moreover, as described, all the alternative scenarios follow paths that join the basic scenario at a particular step. For example, Alternative #1 scenario joins the basic scenario at step 3.1.2 (cf. Table E.2). Figure E.2 depicts the paths through the Place Order use case as described in Table E.2. Following these paths, we can easily determine all the possible scenarios and eventually discover new ones. Note that only the paths covering the basic scenario and the failure scenario end the use case.

Here, the sequence of steps for the different scenarios are as follows:

■ Basic: 1, 2, 3, 4, 5, 6, and 7.
■ Alternative #1: 1, 2, 3.1, 3.1.1, 3.1.2, 2, 3, 4, 5, 6, and 7.
■ Alternative #2: 1, 2, 3, 4, 5, 6.1, 6.1.1, 6.1.2, 5, 6, and 7.

- Alternative #3: 1, 2, 3.1, 3.1.1, 3.1.2, 2, 3, 4, 5, 6.1, 6.1.1, 6.1.2, 5, 6, and 7.
- Failure: 1, 2, 3, 4, 5, 6.1, 6.1.1, and 6.1.2.1.

Note that more scenarios are possible when the use case iterates multiple times over the alternative paths. For example, two iterations over Alternative #1 path will result in the following sequence.

1, 2, 3.1, 3.1.1, 3.1.2, 2, 3.1, 3.1.1, 3.1.2, 2, 3, 4, 5, 6, 7

Use case scenarios may be presented graphically in the UML sequence diagrams (cf. the System and Business Use Cases section).

Use Case Preconditions and Postconditions

A use case may be defined with the required states of the actors and the system itself at the time a use case is to be started. This specific definition is known as use case preconditions. Thus, with the preconditions, we define conditions that must be met in order to perform a use case. Note that the preconditions are not a description of the events that start the use case in question, but a description of the specific conditions under which the use case is applicable (Bittner et al. 2003). For example, the Get Demand use case (cf. Table E.1) can be performed only if a pending demand is stored in the DS, or the customer must be logged on to the system in order to perform the Place Order use case (cf. Table E.2). Here, the preconditions are necessary conditions but not sufficient to perform a use case; that is, only an actor may perform a use case.

In addition to the preconditions, a use case may also be defined with the so-called postconditions used to define the required states of the system at the end of a use case. There could be different postconditions describing the states in which the system can be when different use case

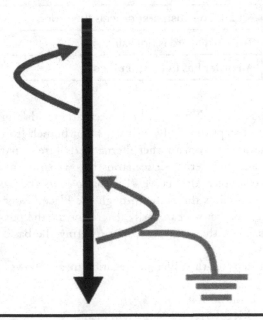

Figure E.2   Paths through the Place Order use case.

scenarios are performed. Larman calls the postconditions success guarantee (Larman 2004) to state what must be true on successful completion of a use case. Cockburn (2001) considers two types of postconditions—minimal guarantee and success guarantee—to address the fact that the use case goal may be fully delivered, partially delivered, or abandoned. The minimal guarantee corresponds to the failure scenarios' result and specifies the least promise of a use case. The success guarantee extends the minimal guarantee when the use case goal is fully achieved. Here, the minimal guarantee is implicitly included in the success guarantee (Metz et al. 2003). In the examples presented in this entry, the term postconditions is used to denote the success guarantee, and the term failure postconditions is used to represent the minimal guarantee. For example, the postconditions (success guarantee) defined in the Place Order use case are "an order has been submitted" (cf. Table E.2), and the Get Demand use case defines the postconditions "the copy of the demand is delivered to the worker" and "the state of the original demand is changed to in process" (cf. Table E.1).

Use Case Types

In addition to the three degrees of detail in use case description (cf. the Use Cases section), use cases can be divided by their system visibility into black-box and white-box use cases. Moreover, although the use case model approach targets software systems, sometimes use cases may be used to describe business processes.

Black-Box and White-Box Use Cases

In practice, the black-box use cases are the most common type because they emphasize the functionality of the system in question without revealing any internal details of the system. Note that use cases are a requirements engineering approach (see Requirements Engineering: Principles and Practice, p. 949), and thus, they should not deal with the system's internal workings, which should be targeted by the system design. The black-box use cases do not reveal the internal mechanisms of the system in question. Instead, the use cases are presented in the form of "responsibilities" (Larman 2004) of the system, presenting what the system must do but not how it will do it.

On the contrary, the so-called white-box use cases look inside the system. Such use cases involve internal system components and their behavior. Moreover, when dealing with white-box use cases, we must consider internal actors as well as external ones (Cockburn 2001).

System and Business Use Cases

A use case model targets the functional requirements of a system. Here, often the use cases forming a use case model are called system use cases. Because such use cases are usually intended to capture the users' goals in the system in question, they are also called user-goal use cases. In terms of system visibility, the system use cases are of black-box or white-box format. Note that it is a good practice, though, to write the system use cases as fully dressed black-box use cases (Cockburn 2001). For example, both Get Demand (cf. Table E.1) and Place Order (cf. Table E.2) use cases are such system use cases.

Usually, the use case model of a system tackles particular business requirements imposed by the business that uses the system. Although the use case model of a system does not capture the business itself, the use case approach can be used to define business processes. In this case, the use cases are called business use cases (Bittner et al. 2003; Larman 2004). Thus, with such use cases, we describe a business not a system. For example, in a coffee store, a possible business use case could be "Make Coffee." The actors of such use cases are actors in the business environment (customers, suppliers, shareholders, etc.) (Larman 2004). Business use cases might be very useful if there is a need to understand the business context where a system is going to be used.

Use-Case-Associated Diagrams

UML (Larman 2004) provides notations for drawing use case diagrams, where we distinguish actors performing use cases. In addition, UML sequence diagrams may be used to depict use case scenarios.

Use Case Diagram

A use case diagram visualizes and conveys the structure of the use case model of a system. Here, a use case diagram shows interactions between a system and the use case actors (cf. the Use Case Actors section). An actor represents a specific role in one or more interactions with the system expressed through use cases. Conceptually, a use case diagram is a graph where the nodes represent both actors and use cases, and the lines represent relationships among use cases and actors. Naturally, the use case diagrams have low information content, and they are complementary documents meant to assist the use case textual descriptions (cf. Tables E.1 and E.2).

UML (Booch et al. 2005) provides appropriate notations for depicting actors, use cases, and relationships. Here, a use case is represented by an ellipse that can be labeled. An actor is usually drawn as a named stick figure or, alternatively, as a class rectangle with the «actor» keyword. As for the relationships, UML introduces notations for four different types of relationships found in use case models:

- **Association (UML notation: solid line).** An association determines the communication path between an actor and the associated use case. In general, an association exists whenever an actor is involved in an interaction described by a use case; that is, an actor is involved in a use case. Thus, associations connect use cases with primary, supporting, and offstage actors. The UML notation for associations is a solid line with an optional arrowhead at one end of the line. The arrowhead indicates the direction of the initial invocation of the actor–use case interaction.
- **Generalization (UML notation: solid line with closed arrowhead).** The generalization relationship denotes a relationship between a more generic use case (or actor) and a more specific use case (or actor), where the latter inherits features from the former. Here, use case features are scenarios and actor features are behavior. A generalization is drawn as a solid line with a closed arrowhead pointing toward the generalized use case (or actor), as shown in the example below. From the generalization relationship perspective, a use case diagram can be viewed as a special UML class diagram with two special kinds of classes, where classes of the same kind can be connected by a generalization arrow.
- **Extend (UML notation: dashed line with open arrowhead and the «extend» keyword).** The extend relationship denotes the insertion of additional behavior (use case scenarios) into a base use case where the latter does not know about it. In general, the incorporation of the extension use case depends on what happens (and under what conditions) when the base use case executes. Note that the extension use case owns the extend relationship; that is,

the arrow points toward the extended use case. Also, it is possible to specify several extend relationships for a single base use case.

■ **Include (UML notation: dashed line with open arrowhead and the «include» keyword).** The include relationship denotes the additional behavior included in a base use case that explicitly defines the inclusion. Thus, the base use case owns the Include relationship; that is, the arrow points toward the included use case. It is a good practice to use the include relationship when multiple use cases have a common function that can be shared by all.

A use case diagram may also include an optional system boundary box, which is denoted as a rectangle around all the use cases. A system boundary box indicates the scope of the system in question. Anything within the box represents functionality that is in the scope of the system and anything outside the box is not. Although system boundary boxes are rarely used, they can be used to identify functionality of different releases of a system, that is, which functional requirements (denoted by use cases) will be fulfilled by which major releases of a system.

Figure E.3 is an example of a use case diagram covering most of the UML notations for use cases. This is a simple solution to the famous elevator use case model problem. As shown, this use case model consists of three actors and five use cases connected through different relationship links. Here, the security personnel is the only actor performing the "go to basement" use case, but this actor could be a regular visitor as well (denoted by the generalization relationship). In addition, the "go to floor" use case is extended by the "request help" use case, and "service elevator" is a complex use case including two other use cases. As you can see, the diagram itself does not say much about the use cases. In fact, the diagram helps us perceive the overall system functionality in terms of actors' goals depicted by the use cases. However, we need the textual use case description (use case scenarios) to understand how the actors interact with the system in order to achieve their goals.

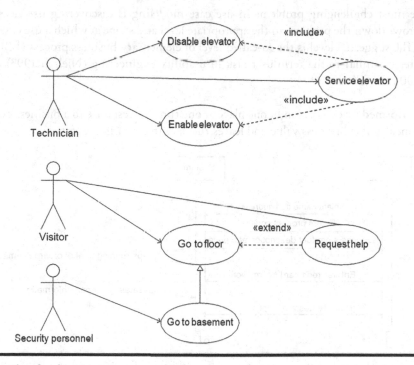

Figure E.3   Simple elevator use case model.

Use Case Scenario Diagrams

In addition to the UML use case diagrams, UML sequence diagrams (Booch et al. 2005) may be used to graphically depict the use case scenarios. Although UML sequence diagrams are typically used at the software design stage to describe object- oriented software systems, they may be successfully used as process flow diagrams to depict the sequence of actions and interactions forming the steps of a use case scenario. The following elements describe some of the remarkable UML notations for sequence diagrams (SDs) (Booch et al. 2005):

- The SDs are two-dimensional. Here, the horizontal dimension is used to depict objects participating in interactions (e.g., actors and the system in question), and the vertical dimension is used to depict the time line.
- There is no significance to the horizontal ordering of the interacting objects.
- Each object is shown in a separate column.
- Each object has a lifeline denoted by a dash or a double solid line. The dash line means the object is created but not active. The double solid line means the object is active.
- The objects interact via messages.
- Each message is shown as an arrow from the lifeline of the object that sends the message to the lifeline of the object that receives the message.

Figure E.4 depicts the graphical representation of the basic scenario of the Place Order use case (cf. Table E.2). Note that the system and the DP actor are depicted in the diagram as single objects.

Best Practices

One of the most challenging problems in use case modeling is discovering use cases. Larman (2004) narrows down the problem to the appropriate level and scope at which a use case should be expressed. The suggested level is the so-called level of elementary business process (EBP).

The latter is similar to the term user task in usability engineering (Nielsen 1993) defined by Larman (2004) as follows:

A task performed by one person in one place at one time, in response to a business event, which adds measurable business value and leaves the data in a consistent state.

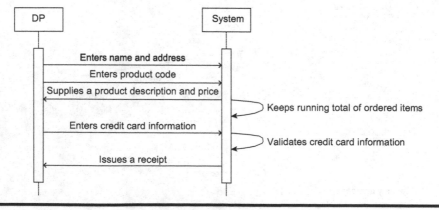

Figure E.4   UML sequence diagram: "place order" basic scenario.

For example, Get Demand (cf. Table E.1) and Place Order (cf. Table E.2) are EBP-level use cases. Unable to define the right level of EBP for the system in question, developers may define many use cases at a very low level; for example, a single step of action. Here, a wrong use case will be "push button."

Larman (2004) calls the EBP-level use cases user goal-level use cases. Such use cases are used by the primary actors (cf. the Use Case Actors section) to achieve their goals in the system. This approach conceives the following recommended procedure for discovering user goal-level use cases (Larman 2004):

1. Find the user goals of every primary actor.
2. Define a distinct use case for each user goal.

Cockburn (2001) calls use cases actor's contracts with the system under design. Here, in order to easily discover use cases, we should think about a use case as "the primary actor's goal, calling upon the responsibility of the system under design." Here, identifying the user goals is of major importance for developing a successful use case model.

In addition, we often have to consider the so-called subgoals that support a user goal. Larman (2004) calls such subgoals subfunction goals. Use cases capturing subgoals are usually written as supplementary use cases included (cf. the Use Case Diagram section) in the user goal-level use cases. For example, both Disable Elevator and Enable Elevator use cases are subgoal use cases included in the Service Elevator use case (cf. Figure E.3).

Algorithm for Determining Use Cases

Considering the fact that use cases are functional requirements defined to help system users achieve their goals in the system under consideration, Larman (2004) proposes an algorithm for defining use cases. The following is an extended version of this algorithm:

1. Determine the system boundary. The latter clearly separates the actors' intentions from the system's responsibilities.
2. Identify the primary actors. Remember that the primary actors have goals that should be achieved by using the services of the system.
3. For each primary actor identify the user goals. The primary actor goals should be defined at the level of EBP.
4. Define use cases that will help primary actors achieve their user goals. It is a good practice to name use cases according to their goal. These are the user goal-level use cases.
5. Identify the subgoal use cases. Usually, these are low-level use cases included in the user goal-level use cases.
6. Identify the supporting actors providing supporting services to the system. The supporting services should be required by the use cases.
7. Identify the offstage actors. Because these actors are not directly involved in any use case, they should be explicitly named by the system stakeholders or logically deducted.

Use Case Benefits and Traps to Avoid

Today, use cases are commonly used as a primary and effective technique in requirements engineering for capturing and communicating functional requirements for software development.

Much of the use case popularity is probably due to the many benefits that result because of their use. Some of the use case benefits are given as follows:

- Emphasizes user goals and objectives.
- Users and other stakeholders become more involved in the process of requirements engineering.
- Provides common notation understandable to both professionals and nonprofessionals.
- Use cases are action-oriented and focused on system functionality—system functionality is decomposed into a set of discrete tasks.
- Helps developers derive test cases and some design insights.
- It may be reused for user documentation (both text and diagrams).
- It may be used as a basis for planning iterative work.
- It may be used as the so-called þ1architecture view of the 4þ1 design view model (Kruchten 1995), where a small set of useful and important scenarios are used to demonstrate how a system is going to work.

However, in order to benefit from use cases, there are a few well-known traps that we must avoid when applying the use case approach:

- Too many use cases. This problem arises when there are use cases not defined at the right EBP level. Moreover, often a use case represents a single requirement. Note that it is a good practice to define use cases representing a common group of requirements.
- Scenario duplication across use cases. Scenario duplication means redundancy. To avoid duplication, define supplementary use cases and use the include relationship (cf. the Use Case Diagram section).
- Use case name duplication. Use cases must be named with unique names across a use case model.
- User interface design included in use cases. A use case should not emphasize the system user interface, but the system functionality.
- Data definitions included in use cases. Note that data requirements are nonfunctional and should be tackled by other requirements engineering approaches (cf. Requirements Engineering: Principles and Practice, p. 949), such as the domain model approach (Larman 2004).

ATM Use Case Model: A Case Study

To demonstrate the use case concepts revealed in the Use Case Concepts section, this section presents a case study on how to develop a use case model for an ATM.

Problem Description

Users log on to the system using their ATM card and personal identification number (PIN). After logging in, public clients can deposit, transfer, or withdraw money, or pay their public services bills. Commercial clients can deposit money or print out their monthly transaction report. The system technician can upload a statistics report on the ATM server (and optionally print it on the ATM machine's printer), add some money in the machine, or shut it down. Any operation requiring a database access requires a data exchange with the ATM server. Any user can press an emergency button that will turn on the hidden camera and call the police department.

Discovering Actors and Use Cases

To describe the ATM use case model, we need to determine both actors and use cases. Next, we need to describe the use cases and relate them with the appropriate relationships. Here, we applied the algorithm described in the Algorithm for Determining Use Cases section:

1. Determine the system boundary. The system is ATM.
2. Identify the primary actors. All the actors should be nouns, entities with behavior. All the actors interact with the system, but only the primary actors have goals in the system. Here, the discovered primary actors are as follows:
   a. Public client
   b. Commercial client
   c. System technician
3. For each primary actor identify the user goals.
   Recall that the primary actor goals should be defined at the level of EBP. The user goals of the different primary actors are as follows:
   a. Public client: deposit money, transfer money, withdraw money, pay bills, emergency (the act of emergency call).
   b. Commercial client: deposit money, print transaction report, emergency.
   c. System technician: add money, shut down ATM, upload statistics report, print statistics report, emergency.
4. Define use cases that will help primary actors achieve their user goals. Here, the identified user goal-level use cases have the same names as the user goals identified at 3.
5. Identify the subgoal use cases. Recall that these are low-level use cases usually included in the user goal-level use cases. Here, the identified subgoal use cases are as follows:
   a. Log on—a user must be logged on to the system in order to perform operations.
   b. Data exchange—any operation requiring a database access requires a data exchange with the ATM server.
6. Identify the supporting actors providing supporting services to the system. Recall that any supporting service needs to be required by the use cases. Following are the identified supporting actors:
   a. ATM server—maintains data needed by the ATM.
   b. Printer—prints reports.
7. Identify the offstage actors. Both the police department and the camera may be considered as offstage actors.
   a. Police Department—eventually, secures the ATM.
   b. Camera—eventually, observes the ATM.

Writing Use Cases

Due to space limitations, only the description of one use case is presented here. Table E.3 presents a fully dressed description (cf. the Use Cases section) of the "log on" use case.

Table E.3   Use Case "Log On."

Name	Log on
Actor	Public client, commercial client, system technician
Description	Users log on to the system by using their ATM card and PIN.
Related use cases	This use case is included in the use cases "withdraw money," "deposit money," "transfer money," "pay bills," "add money," "shut down ATM," "print transaction report," "upload statistics report," and "print statistics report." In addition, this use case includes the use case "data exchange."
Preconditions	The user has an ATM card and a PIN number.
Basic Scenario	1. A user puts the card in the card reader slot. 2. The ATM validates the card information. 3. The ATM requires the user's PIN. 4. The user enters the PIN. 5. The ATM validates the PIN. 6. The ATM grants the user withaccess to specific ATM operations.

Alternative	5.1. The PIN number is not valid.
	    5.1.1. ATM informs the user that the PIN is invalid.
	    5.1.2. Continue with 4.

Failure #1	2.1. The ATM rejects the card.
	    2.1.1. ATM informs the user that his/her card is invalid.
	    2.1.2. ATM returns the card.

Failure #2	5.1.2.1. The user cancels the operation. 5.1.2.1.1. ATM returns the card.
Failure conditions #1	The card is not valid.
Failure conditions #2	The PIN is not valid.
Postconditions	The user is logged on to the system.

Drawing Use-Case-Associated Diagrams

Based on the discovered use cases and actors (cf. the Discovering Actors and Use Cases section) and on the description of the use cases (cf. the Writing Use Cases section; note that in reality all the use cases should be described), we draw the use case diagram representing the use case model of the ATM system (cf. Figure E.5).

In addition, UML SDs may be used to graphically represent the use case scenarios. These diagrams will have similar format and features as the one presented in the Use Case Scenario Diagrams section.

Summary

This entry has presented the concepts and techniques behind use cases, which are widely recognized as a valuable approach to discovering and documenting functional requirements.

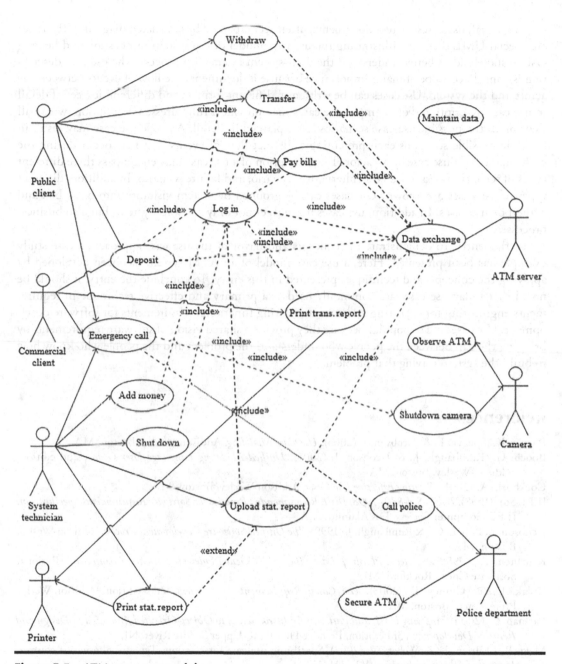

Figure E.5   ATM use case model.

Conceptually, use cases are special stories of using a system expressed with a flow of events defining scenarios of system use. The emphasis is on the user goals and on how the system works with the users to achieve their goals. Here, the system users are presented as actors playing a specific role in use cases. The advantage is that use cases are understandable to both users and developers, and users and other stakeholders become more involved in the process of requirements engineering.

In general, use cases are text documents, often accompanied by associated diagrams. The latter are special UML diagrams illustrating the use case model of the system in question and helping system stakeholders better understand the user–system interaction process. The use case description is considered to be of major importance because it documents the flow of events between the actors and the system. Use cases can be written in different formats and different degrees of detail in use case description. For example, use cases may provide a fully dressed description, where all steps of all the possible use case scenarios are explained in detail. A single use case may result in multiple possible scenarios each presenting a distinct flow of events that may occur during the performance of a use case by an actor. The description of a use case may encompass three different types of scenarios—basic scenario, alternative scenario, and failure scenario. In addition, different types of use cases are known. Use cases can be grouped by system visibility into black-box and white-box use cases. In addition, use cases may target not only software systems but also business processes.

In this entry, to demonstrate the concepts and flavor of the use case approach, a case study example has been presented. Here, a use case model of an ATM system has been developed by applying the concepts and techniques presented in this entry. To conclude the entry, it should be noted that today use cases are commonly used as a primary and effective technique in requirements engineering for capturing and communicating functional requirements for software development. Use cases are so popular because they provide a shared vision of software requirements by bridging the gap between the people who understand the problem and the people who know how to build the system solving that problem.

References

Bittner, K., Spence, I., & Jacobson, I. (2003). *Use Case Modeling*. Addison-Wesley, Boston, MA.

Booch, G., Rumbaugh, J., & Jacobson, I. (2005). *Unified Modeling Language User Guide*, 2nd edition. Addison-Wesley, Boston, MA.

Cockburn, A. (2001). *Writing Effective Use Cases*. Addison-Wesley, Boston, MA.

IEEE Std. (1998). *IEEE-Std-830–1998: IEEE Recommended Practice for Software Requirements Specification*. IEEE Computer Society, Los Alamitos, CA.

Jacobson, I., Booch, G., & Rumbaugh, J. (1999). *The Unified Software Development Process*. Addison-Wesley, Boston, MA.

Kruchten, P. (1995). *Architectural Blueprints—The "4p1" View Model of Software Architecture*. Rational Software Corp., Rockford, MI.

Kulak, D., & Guiney, E. (2003). *Use Cases: Requirements in Context*, 2nd edition. Addison-Wesley Professional, Boston, MA.

Larman, C. (2004). *Applying UML and Patterns: An Introduction to Object-Oriented Analysis and Design and Iterative Development*, 3rd edition. Prentice Hall PTR, Upper Saddle River, NJ.

Metz, P., O'Brien, J., & Weber, W. (2003). Specifying use case inter-action: Types of alternative courses. *Journal of Object Technology*, 2(2): 111–131.

Nielsen, J. (1993). *Usability Engineering (Interactive Technologies)*. Morgan Kaufmann, San Francisco, CA.

Royce, W. (1998). *Software Project Management: A Unified Framework*. Addison-Wesley, Boston, MA.

Simons, A. (1999). Use cases considered harmful. In R. Mitchell, A. C. Wills, J. Bosch & B. Meyer (Eds.), *Proceedings of the Technology of Object-Oriented Languages and Systems (TOOLS—29 Europe)* (pp. 194–203). IEEE Computer Society, Los Alamitos, CA.

Spence, I., & Bittner, K. (2004). *Managing Iterative Software Development with Use Cases*. White Paper. IBM Corp. http://www.ibm.com/developerworks/rational/library/5093.html (accessed November 2009).

Vassev, E. (2009). *General Architecture for Demand Migration in Distributed Systems*. LAP Lambert Academic Publishing, KÖln, Germany.

Appendix F: IBM DOORS Requirements Management Tool

Gavin Arthurs

History of DOORS

Dynamic **O**bject-**O**riented **R**equirements **S**ystem (DOORS) is a requirements management tool. Requirements are managed as separate artifacts consisting primarily of text and a configurable number of attributes. In addition to requirements, DOORS provides a mechanism to configure, create, and analyze traceability links between requirements. This dual capacity to behave like a document-style word processor and provide querying and viewing like a database provides DOORS with its unique character that differentiated it from other solutions in the market of the time. These features, along with a powerful built-in customization facility (via DOORS scripting language, DXL), have made the DOORS product a huge market success that continues today.

DOORS was invented by Dr. Richard Stevens, in the early 1990s. The first commercial version of DOORS was brought to the market by QSS Ltd and Zycad Inc. in April 1993. In 1996, Zycad and QSS Ltd formed the QSS Inc., headquartered in New Jersey. For the next few years, DOORS' market share grew rapidly, and in 2000, Telelogic Inc. acquired QSS Inc. and DOORS became part of a suite of Systems Engineering and Software development solutions. In 2008, IBM acquired Telelogic where it remains today.

By any measure, DOORS has been fantastically successful in the market for the last 25 years, a remarkable accomplishment for any software tool. DOORS is used in virtually every industry worldwide from aerospace to automotive to energy and utilities and has a user base of over 1 million.

At the time of the Telelogic acquisition, IBM had an existing requirements management tool called RequisitePro as well as a few other tools that overlapped with Telelogic's tools. To consolidate and address the demands of modern development, IBM developed a strategy for federating development artifacts and tools based on the concepts of linked data. This has manifested in the Jazz platform and associated tools. The requirements management capability is provided by a tool called DOORS Next Generation (DNG). Today, DOORS and DNG are developed, sold, and marketed side by side, but DNG represents a superset of capabilities by incorporating the best aspects of DOORS while adding new capabilities needed for success in today's market.

Overview of Modern Requirements Management

Challenges confronting developers of modern complex products have placed new demands on how effective requirements management is implemented. So much of today's system functionality is implemented in software and indeed many cases software is the only option to achieve much of the required functionality. Efficient software development requires agility and an ability to accept and manage rapid change. In fact, accepting (even embracing and leveraging) change is now seen as a competitive advantage and must be "baked in" to the capabilities of an enterprise's development process.

Other Development Artifacts

In addition to change, other approaches developed to manage complexity in system development have blurred the perception what a *requirement* is. The development of modeling languages such as Unified Modeling Language (UML) and Systems Modeling Language (SysML) with their graphics-based notations and strong semantic has become de facto standards in model-based development approaches for systems engineering and software development. For instance, in most model-based systems engineering processes, the SysML model is used to develop and define the major components of the system, interfaces, and behavior, as well as to define the way how components interact. Essentially, the model is a specification of the *requirements* of the "to be" built system. A modern requirements management tool must explicitly accommodate and integrate with these model artifacts as well as textual ones.

Need for Multiple Contexts

Requirements management tooling (like DOORS) initially focused on the creation and management of large numbers of textual requirements, providing capabilities to create, manage, and analyze traceability between multiple levels and ultimately produce requirements documents and specifications. Configuration management features were added to formalize changes and create a versioning capability within the requirements space. Fundamentally, the requirements engineer viewed requirements primarily in the context of a textual document and the tooling was built to support that view. Even today, this is the approach used by many companies. This, however, is changing as the pressures of shorter schedules, demands for lean and agile behavior, and the acknowledgment that requirements are more than the text statements.

A requirements tool today must view requirements management as just one capability in an ecosystem of capabilities needed to develop a system. A traditional requirements document or specification is just one important context to view and reason about; however, there are other contexts that are equally useful, for instance, requirements in the context of related tests and test results or requirements in the context of supporting design or modeling artifacts. Here is a list of artifacts potentially used in the development process:

- Tests
- Test results
- Use cases
- Model artifacts
- Scenarios
- Behavior descriptions

- Interfaces
- Hazards
- Safety goals
- Schedule tasks
- Schedules
- Change requests
- Iterations

This list is not exhaustive; it should be clear that many development artifact types are possible. It follows that to enable this kind of context development, a tooling environment is required to support a mechanism that can facilitate the creation, management, and analysis of traceability between *all* such artifact types. The tools associated with each domain (i.e. requirements, test management) need to support accepting "links from" and creating "links to" the other artifact types and their associated tooling (Figure F.1).

Other Considerations

There are a few other features needed for tooling to support today's development realities. These features are not about requirements management specifically but are related to deployment and use of the tool. A full discussion of these items is beyond the scope of this book, but these are mentioned here for consideration.

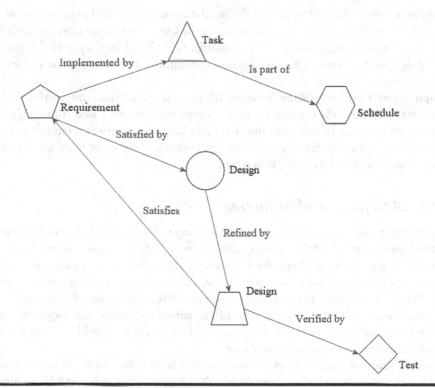

Figure F.1   Example of how a requirement could be viewed in context with other development artifacts.

- **Deployment Options**: This covers how the tool capabilities are delivered to the end user—thin or thick client architecture, cloud or on-premises, traditional licensed product or software as a service (SaaS) model. These features can have significant impact on the successful deployment of a tool within an organization.
- **User Control and Access**: This includes user identity, read/write access to projects, groups of requirements, or individual requirements. User administration; creation, deletion, assignment of licenses, adding to projects, etc.

IBM Doors Next Generation

DNG is IBM's flagship requirements management tool. While it shares the "DOORS" name, it in no way leverages or shares any of the original DOORS tool's architecture, code base, UI, etc. From an architecture, user experience, and administration point of view, it is a completely different tool. DNG is built on the IBM jazz platform (jazz.net). DNG incorporates not only many of the features that made DOORS a successful tool but also features that support the concepts outlined previously.

The IBM Jazz Platform

The jazz platform architecture is built on two foundational concepts:

Linked Data: Jazz applies the concepts of linked data to definition and management of development artifacts. These concepts are embodied in the Open Service for Lifecycle Collaboration (OSLC—http://www.oasis-oslc.org/) standards. OSLC and jazz separate the implementation of development tools (i.e. DNG) from the definition and access to the artifacts (i.e. the requirements).

Common Set of Tool Integration Services: All the jazz tools share cross-cutting features and functions. These include the user UI which is typically browser based (although other technologies are supported), user administration and licensing, query and reporting, and even a common configuration management strategy. Following is a list of highlights of DNG; the tool is best understood by using it in person.

Requirement Capture and Authoring

At the beginning, projects usually have some form of input documentation (requirements). These can be formal (such as standards, existing specifications, and drawings) or informal lists or examples of similar existing products. Regardless of the format, it is important to capture this information as part of the process for understanding the needs of the project. DNG has several approaches to do this. Electronic documents can be imported into DNG and managed as a single artifact or certain formats (such as MS Word document) can be parsed and individual requirements created representing specific statements in the document. This is useful to provide the appropriate level tracing granularity to the source documentation.

Of course, authoring requirements directly in DNG is possible where the tool is most suited, and DNG has powerful features to support this. DNG supports full rich-text editing, use of standard templates, and project-wide glossaries. Users can work in a document-centered view (i.e. like

DOORS) with the use of modules. Modules are specification structures that support concepts of hierarchy together with headings and formatting to present and work with requirements in a traditional document format.

If a user's preference is more loosely structured with focus on working with and defining requirements in a more database style, DNG supports the use of collections. Collections allow the user to easily organize requirements into views based on attributes, filters, types, or any arbitrary selection. For many agility-based teams, this is the preferred way of working.

In addition to textual requirements definition and capture, DNG provides an integrated graphical capability. It supports many popular notations including BPMN, UML/SysML, and Flow charting. While these are just drawings (as opposed to models), DNG supports traceability between individual elements on the diagram and requirements. For systems with user interfaces, DNG provides a robust set of tools supporting UI mock-ups and storyboards.

Requirement Tracing

Requirements traceability is the primary feature of a robust requirements management tool after capturing or authoring. Requirements tracing is an important way in which requirements are given context for understanding and the mechanism by which many types of analysis are performed. The ability to easily create, manage, and analyze traceability is critical in today's complex projects where there are hundreds or thousands of requirements, and as described previously, we need to remain efficient and agile in the face of constant requirements change.

DNG supports traceability through the OSLC standard defining the linking mechanisms and the common link types. Links can be created by the user in several ways (via "drag-and-drop" and context menus) and can be created between individual requirements as well as groups of requirements. Users can define custom link type to be used at the project level giving the user the ability to control the semantic of the linking structure for a project. DNG supports an arbitrary number of links per artifact to support the largest and most demanding projects.

Reporting and Analysis

Capturing and linking requirements is of limited value unless this information can be viewed, analyzed, and shared. DNG provides several features to accomplish this. In DNG, a "view" is a primary mechanism that allows users to create, view, and share useful presentations of DNG data. Views are named artifacts that define what and how this information is displayed. Views act like filters or queries of the information in the DNG project repository. They can be applied or removed anytime, and once created they can be saved and reused. Views can take the form of tables or lists or can include hierarchy based on linking and can be displayed in a browser-like structure (Figure F.2).

The DNG links explorer is another feature to show requirements in context. This is an automatically generated diagram depicting how requirements are related to each other based on the links between them. The diagram is dynamic and users can choose to expand the linking depth to the extent possible and to filter on link type or to focus on a particular requirements relationship (Figure F.3).

Both *views* and *requirements explorer* are important DNG features to support requirements analysis activities. They can be configured to support impact analysis, gap analysis, completeness assessment, validation and verification status, and many more.

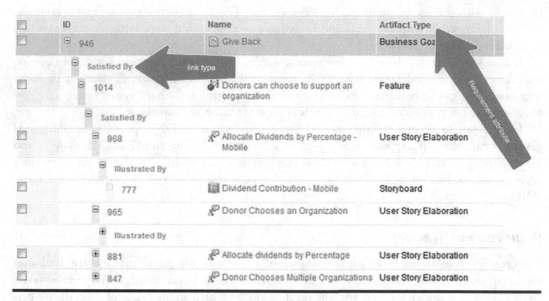

Figure F.2   Example of a DNG view showing hierarchy.

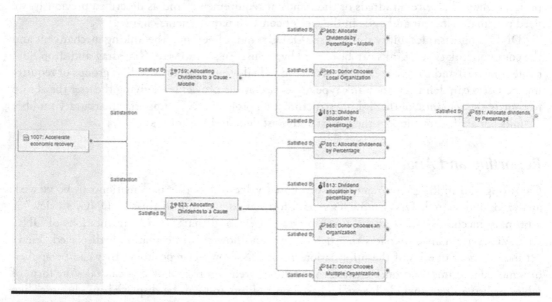

Figure F.3   Example of a DNG links explorer diagram.

The other important capability provided by a requirements management tool is information used to assess project status and health, which is generally known as "reporting." DNG provides several powerful features for this activity. Reporting is a certain style of document used for reports that are run periodically. As DNG is a Jazz-based tool, the reporting functions are primarily provided in real-time formats such as dashboards. Dashboards are "landing" or "home" pages that can be customized and scoped. Personal dashboards are for individuals; others can be scoped for the project team or for management. In either case, dashboard content is fully customizable. Examples of content are assigned work, recent requirement changes, test status, and traceability status. Much of this information is available "out-of-the-box," but custom information and displays are possible. A DNG dashboard can display project metrics and can be "live" information or historical information. There are requirement management-specific displays for:

- Count
- Coverage
- Volatility
- Verification status
- Traceability gaps

DNG also provides *document* reporting in a variety of formats through the jazz reporting service.

Configuration Management

As discussed earlier, efficiently managing requirements change is critical for the success of today's complex system development. DNG provides an implementation of a full suite of configuration management concepts to support this task (Figure F.4).

DNG requirements can be "baselined." Baseline is read-only snapshot of a set of requirements that is named and stored. There is also the concept of a *stream* that is the current set of read/write requirements. Over time as changes are made to requirements, baselines are generated to create a version of the requirements that can always be returned to or reused. Changes can be managed informally by controlling access to the requirements via user privileges. Changes can also be managed by using a *change-set*. A change-set allows a user to make edits to requirements set in isolation and later deliver that set back to the stream after it has been reviewed and approved. Users can also branch from an existing baseline. This creates a parallel stream that can exist and be managed

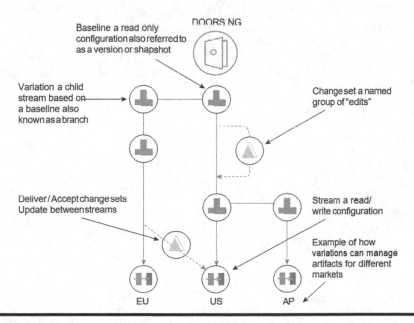

Figure F.4   Configuration management in DOORS next generation.

separately from the parent stream indefinitely. This is a key capability to exploit requirement reuse, as in many circumstances users must manage several versions of a system that are mostly similar but have a small set of features that differentiate them. In these cases, most of the requirements are the same and so the ability to easily reuse them is key. DNG also supports the movement of change-sets between branches when a set of changes is applicable to more than one version of the system.

Trying IBM DOORS Next Generation

DNG is available free for trial like the cloud. Users can find instructions for setting up the trail at IBM's jazz.net website: https://jazz.net/products/rational-doors-next-generation/.

There are also several third-party resources and services both free and chargeable to get up and running with DNG.

Glossary

abuse case: a form of behavioral description using use cases that describes unwanted systems behaviors due to malicious intent.

adaptability: the ease with which conformance to standards can be checked.

affective computing: an interdisciplinary field spanning computer science, psychology, and cognitive science, and focuses on the study and development of systems that can recognize, interpret, process, and simulate human affects.

affordability engineering: activities related to managing expectations and estimating and managing costs of a project.

agency problem: a conflict of interest that can affect a stakeholder's objectivity.

agile development: class of "lightweight" software development methodologies that emphasize people over processes and feature a series of short development increments.

Analytical Hierarchy Process (AHP): a quantitative technique for decision-making that is effective in identifying inconsistencies.

antipattern: a situation involving an identifiable and nonunique pattern of negative behavior. The solution to an antipattern is called refactoring.

artificial intelligence (AI): the science and engineering of developing computer programs which when fed into a machine make the machine exhibit the intelligence of humans.

assertion: a relation that is true at a certain instant in the execution of a program.

availability: quality that refers to the frequency of system outages that lead to unavailability of the system usage by the users.

blockchain: a type of distributed ledger technology that is characterized by the CoDIFy-Pro characteristics: Consensus, Decentralization, Immutability, Finality, and Provenance.

brainstorming: requirements elicitation technique that uses informal sessions with customers and other stakeholders to generate overarching goals for the systems.

Capability Maturity Model Integration (CMMI): a systems and software quality model consisting of five levels.

capability, responsibility, collaboration (CRC) card: a brainstorming tool used for object-oriented software development to determine the class hierarchy and functional relationships between classes.

card sorting: requirements elicitation technique that has stakeholders' complete set of cards that include key information about functionality. The requirements engineer then groups the information logically or functionally.

client: see *customer*.

cloud computing: a model for enabling ubiquitous, convenient, on-demand network access to a shared pool of configurable computing resources (e.g., networks, servers, storage, applications, and services) that can be rapidly provisioned and released with minimal management effort or service provider interaction.

CMMI: see *Capability Maturity Model Integration*.

COCOMO: a tool that estimates software project effort and duration based on a set of parameters that characterizes the project. COCOMO is an acronym for "COnstructive COst Model."

cone of uncertainty: a graphic depiction of the increasing accuracy that is possible for estimates as the details of a project become more known over time.

confidentiality: the quality that refers to the access to the data. Only authorized persons can get an access to the data in a system.

configuration management: the identification, tracking, and control of important artifacts of the system.

Conops: common abbreviation for "Concept of Operations," a very short document or statement describing the overall intent of and scope for a system. The Conops is generally viewed as an elaboration of the mission statement.

context diagram: any group of informal or formal diagrams showing a system and the systems it interacts with for the purposes of defining operational boundaries.

COSMIC: a method to measure a standard functional size of a piece of software. COSMIC is an acronym of "COmmon Software Measurement International Consortium."

COSYSMO: a tool derived from COCOMO that estimates effort and duration in hardware/software systems. COSYSMO is an acronym for "COnstructive SYStem Model."

crowdsourcing: a collection of information, opinions, or work from a group of people, usually sourced via the Internet.

customer: the class (consisting of one or more persons) who is commissioning the construction of the system. Also called *client* or *sponsor*.

desiderata: that which is wanted (from the Latin for "desired things").

design constraint requirement: a requirement related to standards compliance and hardware limitations.

design traceability: a form of traceability that links from the requirements to the design.

designer as an apprentice: a requirements discovery technique in which the requirements engineer "looks over the shoulder" of the customers to enable him/her to learn enough about the customers' work to understand their needs.

DevOps: a set of practices intended to reduce the time between committing a change to a system and the change being placed into normal production, while ensuring high quality

discounted payback period: the payback period determined on discounted cash flows rather than undiscounted cash flows.

divergent goals: an antipattern in which two or more major players in a situation work against each other.

domain analysis: any general approach to assessing the "landscape" of related and competing applications to the system being designed.

domain requirement: a requirement that is derived from the application domain.

dubious requirement: a requirement whose origin and/or purpose is unknown.

earned value analysis (EVA): a family of techniques for determining project progress based on metrics.

ethnographic observation: any technique in which observation of indirect and direct factors informs the work of the requirements engineer.

Extreme Programming (XP): an agile software development methodology.

feature points: an extension of function points that is more suitable for embedded and real-time systems.

formal method: any technique that has a rigorous, mathematical basis.

function points: a duration and effort estimation technique based on a set of project parameters that can be obtained from the requirements specification.

goal: a high-level objective of a business, organization, or system.

goal-based approaches: any elicitation techniques in which requirements are recognized to emanate from the mission statement, through a set of goals that leads to requirements.

goal-question-metric (GQM): a technique used in the creation of metrics that can be used to test requirements satisfaction.

gold plating: specifying features that are unlikely to be used in the delivered system.

group work: a general term for any kind of group meetings that are used during the requirements discovery, analysis, and follow-up processes. JAD is a form of group work.

hazard: system safety concerns that must be reflected in the requirements specification as "shall not" behaviors.

Hofstede's cultural dimensions: a set of metrics that characterizes an individual's values based on cultural influences. The dimensions are power distance, individualism vs. collectivism, masculinity vs. femininity, uncertainty avoidance, and term orientation.

informal method: any technique that cannot be completely transliterated into a rigorous mathematical notation.

inspection: an organized examination or formal evaluation exercise. It involves the measurements, tests, and gauges applied to certain characteristics in regard to an object or activity.

integrity: the ability of a system to withstand attacks to its security.

internal rate of return (IRR): a way to calculate an artificial "interest rate" for some investment for the purposes of comparing the investment with alternatives.

Internet of Things (IoT): uniquely identifiable objects/things and their virtual representations in an internet-like structure.

interoperability: the ability of two or more systems or components to exchange information and to use the information that has been exchanged.

introspection: developing requirements based on what the requirements engineer "thinks" the customer wants.

Joint Application Design (JAD): elicitation technique that involves highly structured group meetings or mini-retreats with system users, system owners, and analysts in a single venue for an extended period of time.

Kanpan: a lean method to manage and improve work across human systems by aiming to manage work by balancing demands with available capacity, and by improving the handling of system-level bottlenecks.

laddering: a technique where a requirements engineer asks the customer short prompting questions ("probes") to elicit requirements.

lean project development: any form of systems engineering that uses lightweight processes.

lightweight process: any form of systems engineering process that minimizes the amount of documentation artifacts, mandatory meetings, and emphasizes collaboration and early delivery.

metrics abuse: an antipattern in which metrics are misused through misunderstanding, incompetence, or malicious intent.

misuse case: a form of behavioral description using use cases that describe unwanted systems due to incompetence or accident.

model checker: a software tool that can automatically verify that certain properties are theorems of the system specification.

negative stakeholders: stakeholders who will be adversely affected by the system.

net present value (NPV): a calculation to determine the value, in today's dollars, of some asset obtained in the future.

NFRs framework: is a process-oriented and goal-oriented approach that is aimed at making NFRs explicit and putting them at the forefront in the stakeholder's mind.

nonfunctional requirement (NFR): requirements that are imposed by the environment in which the system is to operate.

obsolete requirement: a requirement that has changed significantly or no longer applies to the new system under consideration.

payback period: the time it takes to get the initial investment back out of the project.

performance requirement: a static or dynamic requirement placed on the software or on human interaction with the software as a whole.

Planguage: a language with a rich set of keywords that permits precise specifications of requirements including qualities.

POS: point of sale system, an example used throughout this book.

process clash: an antipattern in which established processes are incompatible with the goals of an enterprise.

product line: a set of related products that have significant shared functionality.

profitability index (PI): the net present value of an investment divided by the amount invested. Used for decision-making among alternative investments and activities.

proto typing: involves the construction of models of the system in order to discover new features.

protocol analysis: a process where customers and requirements engineers walk through the procedures that they are going to automate.

Quality Attribute Scenario: description of a quality attribute response given a stimulus.

Quality Function Deployment (QFD): a technique for discovering customer requirements and defining major quality assurance points to be used throughout the production phase.

quality requirement: the totality of characteristics of an entity that bear on its ability to satisfy stated and implied needs.

refactoring: a solution strategy to an antipattern.

reliability: the ability of a system or component to perform its required functions under stated conditions for a specified period of time.

repertory grids: elicitation technique that incorporates a structured ranking system for various features of the different entities in the system.

requirement: a statement that specifies (in part) how a goal should be accomplished by a proposed system.

requirements agreement: the process of reconciling differences in the same requirement derived from different sources.

requirements analysis: is the activity of analyzing requirements for problems.

requirements churn: refers to changes in requirements that occur after the requirements are agreed upon.

requirements engineering: the branch of engineering concerned with the real-world goals for, functions of, and constraints on systems.

requirements traceability: a form of traceability that links between dependent requirements.

return on investment (ROI): the value of an activity after all benefits have been realized.

rich picture: a holistic approach to system boundary and stakeholder identification focusing on essential stakeholder wants and needs.

ridiculous requirement: a requirement that would be difficult to deliver given the state of technology at the time it is proposed. A requirement that, though possible to realize, is unlikely to be used.

Scaled Agile Framework (SAFe): a system for implementing Agile, Lean, and DevOps practices at scale. The Scaled Agile Framework is the most popular framework for leading enterprises because it works: it's trusted, customizable, and sustainable.

scenarios: informal descriptions of the system in use that provide a high-level description of system operation, classes of users, and exceptional situations.

scope creep: unchecked growth of functional requirements beyond what was originally intended, often exceeding the purposes and intent of the system.

scrum: an agile software methodology involving short development increments called sprints and a living backlog of requirements.

security: a measure of the system's ability to resist unauthorized attempts at usage and denial of service while still providing its services to legitimate users.

semi-formal technique: a technique that though appears informal has at least a partial formal basis.

"shall not" requirement: a requirement that describes forbidden behavior of the system.

singularity of a requirement: a quality indicating that a requirement is specifying a single behavior without having conjunctions.

small businesses: enterprises that employ fewer than 50 persons and whose annual turnover or annual balance sheet total does not exceed EUR 10 million.

soft systems methodology(SSM): a holistic approach to system engineering that focuses on accurate identification of all system entities, stakeholders, goals, and interactions early in the process. See also *rich picture*.

source traceability: a form of traceability that links requirements to stakeholders who proposed these requirements.

sponsor: see *customer*.

stakeholder: a broad class of individuals who have some interest (a stake) in the success (or failure) of the system in question.

Systems Modeling Language: a collection of modeling tools for systems. SySML and UML are closely related.

task analysis: a functional decomposition of tasks to be performed by the system.

technical debt: a metaphor referring to the eventual consequences of poor system architecture and system development within a codebase.

traceability: a property of a systems requirements document pertaining to the visible or traceable linkages between related elements.

Unified Modeling Language (UML): a collection of modeling tools for object-oriented representation of software and other enterprises. UML and SysML are related.

usability: the ease with which a user can learn to operate, prepare inputs for, and interpret outputs of a system or component.

use case: a depiction of interactions between the system and the environment around the system, in particular, human users and other systems.

use case diagram: a graphical depiction of a use case.

use case points: an effort estimation technique based on characteristics of the project's use cases.

user: a class (consisting of one or more persons) who will use the system.

user stories: short conversational text used for initial requirements discovery and project planning.

validation: the process to establish the correspondence between a system and users' expectations.

value engineering: activities related to managing expectations and estimating and managing costs during all aspects of systems engineering, including requirements engineering.

verification: the process that establishes the correspondence of an implementation phase of the software development process with its specification.

viewpoints: a way to organize information from (the point of view of) different constituencies.

virtual reality: a computer-generated simulation where the user can interact with a virtual environment.

virtual world environment: complex simulation of a real-world environment using enhanced graphics and various specialized input devices such as tactile pressures.

Wideband Delphi: a consensus-building technique based on multiple iterations of alternative ranking and discussion.

Wiki: a collaborative technology in which users can format and post text and images to a website.

workshop: a formal or informal gathering of stakeholders to determine requirements issues.

Index

Note: **Bold** page numbers refer to tables; *italic* page numbers refer to figures and page numbers followed by "n" denote endnotes.

Printed in the United States
by Baker & Taylor Publisher Services